WindEurope Summit 2016

Journal of Physics: Conference Series Volume 749

Hamburg, Germany
27 – 29 September 2016

Editors:

Michael Muskulus
Sandrine Aubrun

ISBN: 978-1-5108-2996-1
ISSN: 1742-6588

Printed from e-media with permission by:

Curran Associates, Inc.
57 Morehouse Lane
Red Hook, NY 12571

proceedings.com

Some format issues inherent in the e-media version may also appear in this print version.

Printed by Curran Associates, Inc. (2016)

For permission requests, please contact the Institute of Physics at the address below.

Institute of Physics
Dirac House, Temple Back
Bristol BS1 6BE UK

Phone: 44 1 17 929 7481
Fax: 44 1 17 920 0979

techtracking@iop.org

Additional copies of this publication are available from:

Curran Associates, Inc.
57 Morehouse Lane
Red Hook, NY 12571 USA
Phone: 845-758-0400
Fax: 845-758-2633
Email: curran@proceedings.com
Web: www.proceedings.com

Table of contents

Volume 749

WindEurope Summit 2016
27–29 September 2016, Hamburg, Germany

Accepted papers received: 1 September 2016
Published online: 14 September 2016

Preface

Papers
Resource assessment

Turbine technology

Supply chain, O&M and logistics

Other topics

Preface

On behalf of the European Academy of Wind Energy (EAWE[1]) and WindEurope (formerly the European Wind Energy Association, EWEA, until April 2016), we have the pleasure of presenting the Scientific Proceedings of the WindEurope Summit 2016.

WindEurope Summit 2016 took place in Hamburg from 27-29 September 2016 and was one of the largest and most important meeting places of the wind energy industry, featuring 50 thematic sessions with more than 300 oral and 200 poster presentations. Known previously as *EWEA's Annual Conference* and as the *European Wind Energy Conference* before that, it has included a Science & Research component since 2007. The event is a well-known meeting place that allows engineers and scientists to present their latest findings to a mixed audience consisting of both researchers and industry experts. The industry involvement helps to focus the research, while the scientists often demonstrate relevant, novel ideas and solutions to problems of current industrial interest. This often leads to interesting comments and discussions that have an impact well beyond the conference boundaries.

In the course of continuously improving the conference experience, in 2015 the organisers decided to combine scientific and general content in all conference sessions, rather than reserving the presentation of the Science & Research contributions in a separate track as was done at previous editions. Sessions at EWEA Annual Event 2015 were therefore combinations of general and of scientific presentations, which allowed for a more flexible choice of topics. In particular small research topics could also be accommodated in the context of a wider session. This successful setup of the conference has been carried over to this year's event in Hamburg.

The improvements continue this year, with an agreement between WindEurope and IOP Publishing, which allows for the scientific contributions to be published in the Journal of Physics: Conference Series, an Open Access journal managed by the Institute of Physics (IOP). The European Academy of Wind Energy has for some time recommended such an arrangement, and we are very happy to see this happen. The most important advantage is that conference papers will be much more visible, being fully indexed and online available free of charge to anyone, while the authors retain their copyright under the permissive IOP Proceedings License. The IOP also provides long-term archival and access.

All papers published in this volume originate from scientific contributions presented in the oral sessions of WindEurope Summit 2016, and have been peer-reviewed in order to ensure strict, international research standards. Selected papers, suitably extended and rewritten, will additionally be invited and considered for the new *Wind Energy Science* journal of the EAWE.

The proceedings in this volume consist of 21 papers covering a wide spectrum of wind energy research. The topics include aerodynamic investigations and blade design, controller design for load mitigation, forecasting of power production, investigations related to wind turbine wakes, atmospheric stability and monitoring of turbines, as well as contributions concerned with floating support

[1] http://www.eawe.eu/

structures, optimal strategies for bidding on the energy market and mathematical failure modeling for offshore wind turbines, among others.

As with previous events, the European Academy of Wind Energy is responsible for organizing the review process for scientific contributions, has contributed to develop the sessions, and provides scientific chairs for all sessions. All papers were peer-reviewed by a Scientific Committee, consisting of scientists from EAWE member institutes and their associates. We thank all authors for their willingness to take part in this procedure, including the strict deadlines imposed by the tight publication schedule, and the reviewers for their hard work alongside their daily business. WindEurope is responsible for the organisation and logistics of the conference, and we thank their highly professional staff and their associates for the excellent collaboration.

Michael Muskulus
Editor Scientific Proceedings WindEurope Summit 2016
Scientific Programme Coordinator WindEurope Summit 2016
Norwegian University of Science and Technology NTNU

Sandrine Aubrun
Editor Scientific Proceedings WindEurope Summit 2016
President European Academy of Wind Energy EAWE
Université d'Orléans

Peer review statement

All papers published in this volume of *Journal of Physics: Conference Series* have been peer reviewed through processes administered by the proceedings Editors. Reviews were conducted by expert referees to the professional and scientific standards expected of a proceedings journal published by IOP Publishing.

Generation and Validation of Spatial Distribution of Hourly Wind Speed Time-Series using Machine Learning

F Veronesi & S Grassi

Institute of Cartography and Geoinformation – ETH Zurich
Stefano-Franscini-Platz 5, 8093 Zurich

fveronesi@ethz.ch

Abstract. Wind resource assessment is a key aspect of wind farm planning since it allows to estimate the long term electricity production. Moreover, wind speed time-series at high resolution are helpful to estimate the temporal changes of the electricity generation and indispensable to design stand-alone systems, which are affected by the mismatch of supply and demand. In this work, we present a new generalized statistical methodology to generate the spatial distribution of wind speed time-series, using Switzerland as a case study. This research is based upon a machine learning model and demonstrates that statistical wind resource assessment can successfully be used for estimating wind speed time-series. In fact, this method is able to obtain reliable wind speed estimates and propagate all the sources of uncertainty (from the measurements to the mapping process) in an efficient way, i.e. minimizing computational time and load. This allows not only an accurate estimation, but the creation of precise confidence intervals to map the stochasticity of the wind resource for a particular site. The validation shows that machine learning can minimize the bias of the wind speed hourly estimates. Moreover, for each mapped location this method delivers not only the mean wind speed, but also its confidence interval, which are crucial data for planners.

Keywords: wind resource assessment, time-series, confidence intervals.

1. Introduction

Wind resource assessment is a key component of the planning phase for every wind farm project. In particular, wind speed time-series at fine resolution (i.e. hourly or sub-hourly) are helpful to estimate the temporal changes of the electricity generation and indispensable to design stand-alone systems that are affected by the mismatch of supply and demand. In addition, the variability of wind speed is important to evaluate the impact of the electricity injection into the power grid. Wind speed and direction data are measured worldwide by weather stations at various time-intervals. However, their coverage is inadequate to identify suitable locations to build new wind farms. In many cases, after a potential site has been identified, the planning phase continues with at least a year of continue wind measurements using a meteorological mast [1], which is a device that measures wind at various heights. This step highly increases the total cost of the project and therefore has a non-negligible impact on the rate of return of the initial investment. One way to decrease the planning costs is by employing techniques to estimate the temporal variability of the wind resource on the potential site, i.e. wind resource assessment.

Several techniques have been developed over the years to estimate the spatio-temporal pattern of the wind resource. The majority of these techniques are part of a set referred to as numerical wind flow models, which estimate the wind resource solving the physical equations that govern the motion of air in the atmosphere [2]. These methods have varying level of complexity, derived by the type and amount of equations they include. The simplest ones are the mass-consistent models [3], first developed in the 1970s, which only solve the equation of conservation of mass. On the other end of the complexity spectrum are numerical weather prediction models (NWP; [2]), which solve all the computational fluid-dynamics equations plus others that govern the energy exchanges between soil and atmosphere. These methods are able to estimate both the long term wind resource and its time variability, even though they tend to be time-consuming and computationally expensive [4].

Another branch of research has been dedicated to the development of techniques for wind resource assessment based purely on statistical algorithms. Statistical models correlate wind speed data from weather stations, with remotely sensed physical parameters, to infer the wind spatio-temporal pattern. As demonstrated in previous work [4], statistical methods are accurate, computationally efficient, and less time-consuming than physical models. These methods have been tested in the literature for estimating both the long term pattern of the wind resource (e.g. [4]–[8]) and for time-series estimations with models such as ARMA (Auto Regressive Moving-Average; [9], [10]), Markov chain [11] and autoregressive models [12]. However, the spatio-temporal prediction, i.e. the estimation of the hourly wind speed pattern in areas where no direct observations are available, of wind speed time-series using machine learning techniques is a recent research topic [13]. The major problem in wind resource assessment is the large amount of uncertainty involved, which ranges from malfunctions of the weather stations to the extrapolation of the wind speed profile in complex terrains. Assessing this uncertainty is difficult with numerical wind flow models, but straightforward with statistical wind resource assessment, which can precisely account for all these sources of uncertainties [8].

In this work, we present a new generalized statistical methodology, based on machine learning, to generate the spatial distribution of wind speed time-series, using Switzerland as a case study. This research is a continuation of the work we presented at EWEA 2015 [14], and demonstrates that statistical wind resource assessment can be successfully used for estimating wind speed time-series. In fact, this method is able to efficiently obtain reliable wind speed estimates and to propagate all the sources of uncertainty, from the measurements to the mapping process, so that the final confidence intervals allow a reliable estimation of the stochasticity of the wind resource for a particular site.

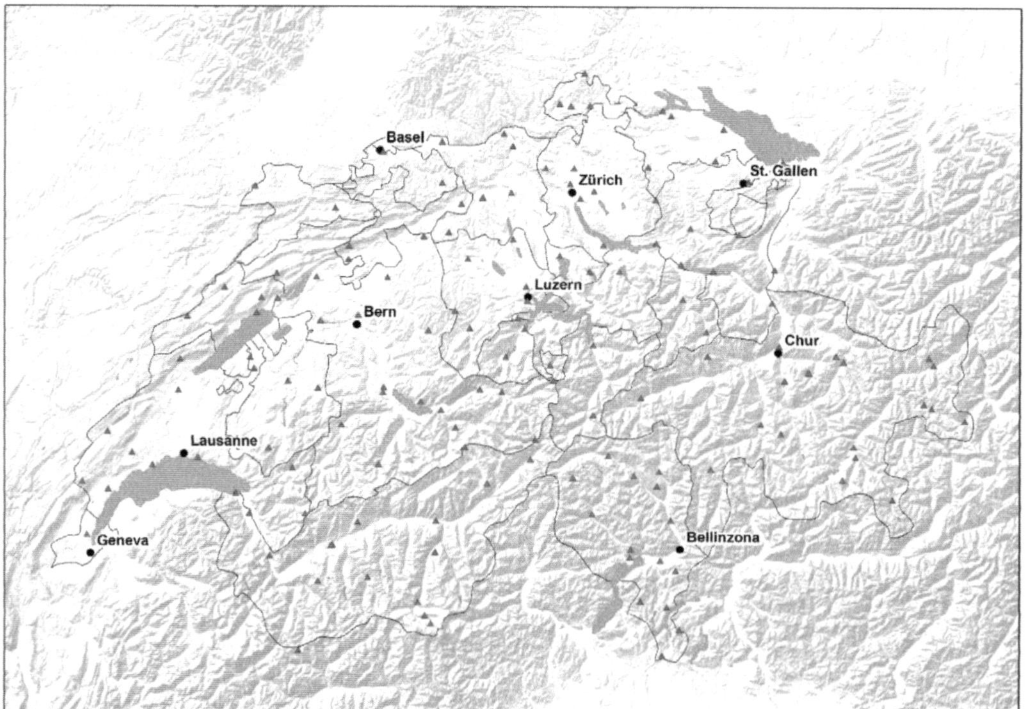

Fig. 1: Location of the automatic weather stations (red triangle) belonging to the MeteoSwiss network.

2. Materials and Methods

2.1. Dataset

We collected 10 min average wind speed data over 5 years from 161 stations of the MeteoSwiss (Swiss Federal Office of Meteorology and Climatology) network. The stations are sparsely located across Switzerland (the exact location is depicted in Fig.1) and measure wind at a height of 10 m AGL These data provide the variable for the machine learning model to estimate it in locations where no weather observations are available.

The wind speed measurements were correlated with around 8'000 predictors, consisting of environmental and climate data with different time and spatial resolutions covering Switzerland. A detailed list of all the data employed for this study, which were collected from various open data repositories, is presented in Table 1.

2.2. Machine Learning Approach

This research employed the wind resource assessment method, based on machine learning, developed by Veronesi et al. [4] to estimate the long term wind speed and direction distributions over large regions. Machine learning algorithms are very flexible since they are data driven, meaning that the method works the same way with wind long term averages or hourly observations. In this research we employed the method developed in [4] to estimate wind hourly speeds, thus the machine learning algorithm was run for a total of 8'790 times, one for each hour of the year, to provide detailed wind speed time-series.

This approach was developed to statistically estimate the wind field not only in space but also in time. In the paper we presented last year at the EWEA annual event [14], we used this approach to estimate the long term wind speed distribution in Switzerland over a 1 Km grid. With this research we went one step further, adding a temporal component to the analysis. In this work the machine learning algorithm was employed to estimate a long term hourly time-series for each grid cell. For long term time series, we mean that the 10 min data were aggregated over the entire sampling period to produce average time-series, with uncertainty, which include the range of variation of wind speed for every hour of the year. In other words, the results we present can be used to assess the average hourly range of variation of the wind resource.

One of the main objective of this study was to provide planners with more informative wind speed time-series, compared with other wind resource assessment methods. The crucial difference between statistical and physical wind resource assessments is in the way they handle uncertainty. Machine learning is perfectly able to assess its own accuracy, thus providing planners with high-resolution error information that physical methods cannot provide. This is true for long term planning, as demonstrated by Veronesi et al. [4], but it is also important in time-series analysis. Having access to a detailed uncertainty estimation may in fact allow for a more precise estimation of the risk involved in developing a site, with a finer spatio-temporal granularity compared to long term averages.

To reach this objective we devised the method in such a way as to include every possible source of uncertainty, from the variability present in the data to the error that affects the machine learning estimations. The latter is relatively easy to include, since contrary to numerical wind flow models, statistical methods are by default able to assess their uncertainty. On the contrary, the variability in the weather observations is more difficult to incorporate in the model. For example, in the 5 years of data we gathered for this experiment it is safe to assume that in certain hours the anemometers would record higher than average wind speed, maybe caused by storms. Therefore, when the 10 min data are aggregated some of them will present high values, and other low values. Using only the average speed for each hour means losing the information in regards to the variability in the data. It is important for planners to know that certain hours present historically more variable speeds, while others generally present calmer winds. For this reason, and to account for the data variability, we used a resampling approach called Bootstrapping [15] to fit multiple statistical models to the 30 hourly observations (i.e. 6 samples/hour over 5 years). This way we were able to propagate the variability from the anemometer data to the wind speed predictions, while also accounting for all other sources of uncertainty. This allowed us to provide a full uncertainty estimation, in terms of confidence intervals, for each location and each time frame we estimated.

Table 1: List of global data used as predictors for the machine learning algorithm to estimate hourly wind speed in Switzerland

Data Provider	Variables	Temporal Resolution	Spatial Resolution
NOAA [16]	Normalized Difference Vegetation Index Brightness Temperature Smoothed Brightness Temperature Vegetation Condition Index Temperature Condition Index Vegetation Health Index Precipitation	Daily	4 x 4 Km
NASA [17]	Global Horizontal Radiation Minimum Global Horizontal Radiation Maximum Global Horizontal Radiation Diffuse Radiation Minimum Diffuse Radiation Maximum Diffuse Radiation Direct Normal Radiation Minimum Direct Radiation Maximum Direct Radiation Latitude Tilt Radiation Minimum Latitude Tilt Radiation Maximum Latitude Tilt Radiation Air temperature Daily Minimum Air temperature Daily Maximum Air temperature Relative humidity Atmospheric pressure Earth temperature	Monthly	1 Degree
CM-SAF [18]	Evaporation Precipitation Evaporation Vertically Integrated Water Vapour Latent Heat Flux Vertically Integrated Liquid Water Near Surface Specific Humidity Precipitation Surface Albedo Surface Downward Longwave Radiation Surface Incoming Shortwave Radiation Surface net longwave radiation Surface net shortwave radiation Surface outgoing longwave radiation Surface radiation budget Near Surface Wind Speed	Monthly	0.5 x 0.5 Degrees
USGS [19]	Digital Elevation Model – GMTED2010		7.5 ArcSeconds

2.3. Validation

The validation of the estimates was carried out with a five-fold spatio-temporal cross-validation. We created a cross-validation loop consisting of 1'000 iterations. For each iteration, 20% of the weather stations were randomly excluded from the training process of the machine learning algorithm. The algorithm was then trained using data from one particular hour of the year, chosen randomly. Then the testing was done comparing the estimates from the machine learning algorithm with the wind observation from the same hour in the stations excluded from training.

The bias, i.e. the difference between wind observations and machine learning estimates, was calculated after each iteration of the validation loop. The results were then aggregated by month, to show the monthly accuracy of the spatio-temporal wind speed model. Instead of just presenting the average bias, which is not much informative, we included its full distribution as monthly box-plots.

2.4. Extrapolation to Hub Height

For extrapolating the wind speed from 10 m, i.e. the height of the weather stations, to hub height, we employed the process presented in Grassi et al. [20], briefly described here.

This process exploits the logarithmic relation between wind speed and terrain roughness indicated by Equation 1 from [21]:

$$\bar{v}_H = \bar{v}_{ref} \cdot \frac{ln(H/z_o)}{ln\left(H_{ref}/z_0\right)} \qquad (1)$$

Where $\bar{v}_H$ is the wind speed at hub height (m/s), $\bar{v}_{ref}$ is the wind speed at reference height (m/s), in this case 10 m, H is the hub height (m), H_{ref} is the reference height (m), i.e. 10 m. The important parameter of this equation is z_0, the roughness height. This is the parameter that controls the increase in wind speed with altitude and depends on terrain and land-use. Essentially, this parameter described the changes in the vertical wind shear caused by the friction with the earth surface, which depends directly upon the type of land cover. On areas of bare land or small vegetation, this parameter is close to zero, whereas in areas of taller vegetation or covered by buildings this parameter is closer to 1.

Equation 1 represents a standard method to extrapolate wind speed with elevation, but its accuracy is strongly dependent upon the precision of the roughness parameter. This is generally taken from literature and assigned based on the land-use and vegetation types of the location of the planned wind farm. However, this approach has some limitations, since it considers the wind shear as something that depends only on local terrain conditions and land cover. On the contrary, the downwind terrain conditions have also a large impact on the wind shear in the area considered for the turbine placement. For example, imagine we want to build a wind farm on arable land, which on its South-East side is bordered by a forest. The wind coming from this direction would have a much lower speed compared to the one coming from other directions. If we only consider the land cover of the wind farm area, i.e. arable, we will overlook this aspect, thus potentially overestimating the electricity production of the wind farm.

In Grassi et al. [20] the authors present an approach carried out on GIS applications that is perfectly able to take the directionality of the wind into account for computing the roughness height. This approach not only considers the terrain roughness of the local area under investigation, but also the downwind areas and their terrain conditions in respect to 12 wind directions. This approach allows a more accurate computation of the z_0 factor and thus a better understanding of the potential of a site to become suitable for wind farm planning.

3. Results and Discussions

This research presents the results of a case study carried out in Switzerland to test a method, based on statistical wind resource assessment, to map the spatio-temporal pattern of wind speed across the country. The final output of the model is a wind speed map of Switzerland at a spatial resolution of 1 km, and a temporal resolution of 1 h. Statistical wind resource assessment is an efficient method of estimation, meaning that the computational time and load are minimal, compared to numerical wind flow models. This allows its integration with simulation techniques capable of estimating the error propagation every step of the process, aiming not only to estimate the average wind speed time-series, but also to provide planners with detailed confidence intervals. In fact, we carried out simulations using the 5 years of data we collected at 10 min of temporal resolution, to specifically propagate the variance of the measurements into the statistical model. The model is then perfectly able to assess its own accuracy, which means that the confidence intervals we present are the sum of all the variance and uncertainty sources. This is a crucial information that allows the precise calculation of the risk involved in developing a site where the uncertainty is high, since the amount of electricity generated can potentially fluctuate substantially in time.

Figure 2 shows the results of the validation process: the monthly bias (i.e. difference between observed and estimated values) for the percentiles 5th and 95th. We present only these values because they provide practitioners with a way to calculate 90% of the hourly variability of the wind resource. On average, this bias resulted -0.08 m/s for the 5th percentile and -0.11 m/s for the 95th percentile. By looking at the plot in Figure 2 we can clearly observe that the variation of the bias of the 5th percentile is relatively small, while it is more acute for the 95th, where the bias is above absolute 1 m/s. In essence, on average the bias of both percentiles is very low, and this indicates that we are able to provide planners with a reasonable estimate of the full range of variability of the wind speed (i.e. 90% of the time the wind speed is within this range).

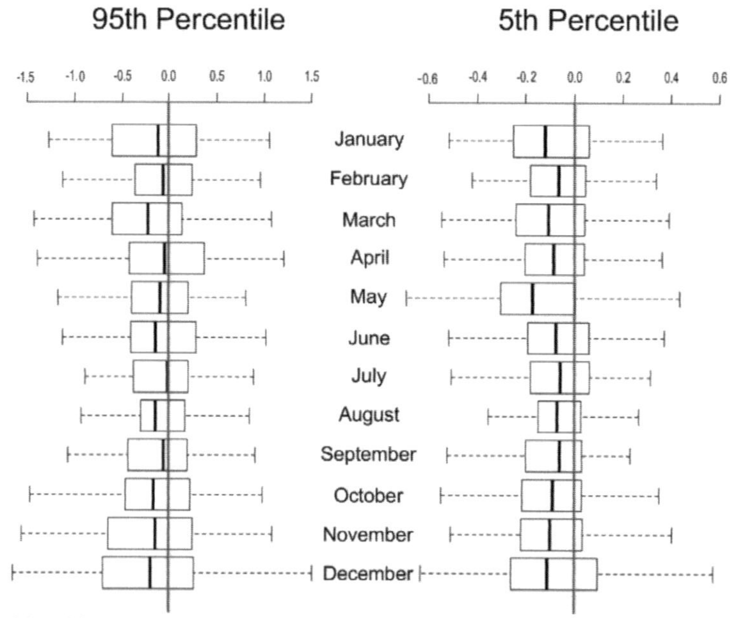

Fig. 2: Monthly distribution of the bias of the estimated 5th and 95th percentiles.

Figure 3 shows an excerpt of the generated time series with the mean wind speed (black line) and the percentile 25th, 50th, 75th and 90th (dotted lines). In order to show the impact of the wind variability of the electricity generation, we calculated the energy output of a single Enercom turbine with a nominal power of 2.3 MW and 90 m hub height. The purpose of this test is only to show that considering the uncertainty around the mean wind speed is crucial for planners to identify correctly the interval of probability of the power output. For this reason, we considered a single turbine, which is often not a realistic scenario for wind farms, where multiple turbines are located relatively close to each other and factors such as the wake effect need to be considered to calculate the total power output. However, since we are only interested in showing the effect of considering uncertainty when computing the potential power output, considering a single turbine is sufficient. The results are shown in Figure 4.
In this plot we not only included the mean wind speed, but also all the percentiles we calculated in our study. Considering only the mean wind speed, certain time intervals seem productive. However, what we can tell by adding a measure of the local variability of the wind resource is that even in these intervals the productivity of the wind farm may be well below average, for certain years. This can be clearly identified using the confidence intervals we provide. This is a crucial information since it may allow planners to not only calculate the average power output, but also its upper and lower boundary and assign precise probability of occurrence to each of them. This certainly allows a very precise assessment of the risk involved in developing a site.

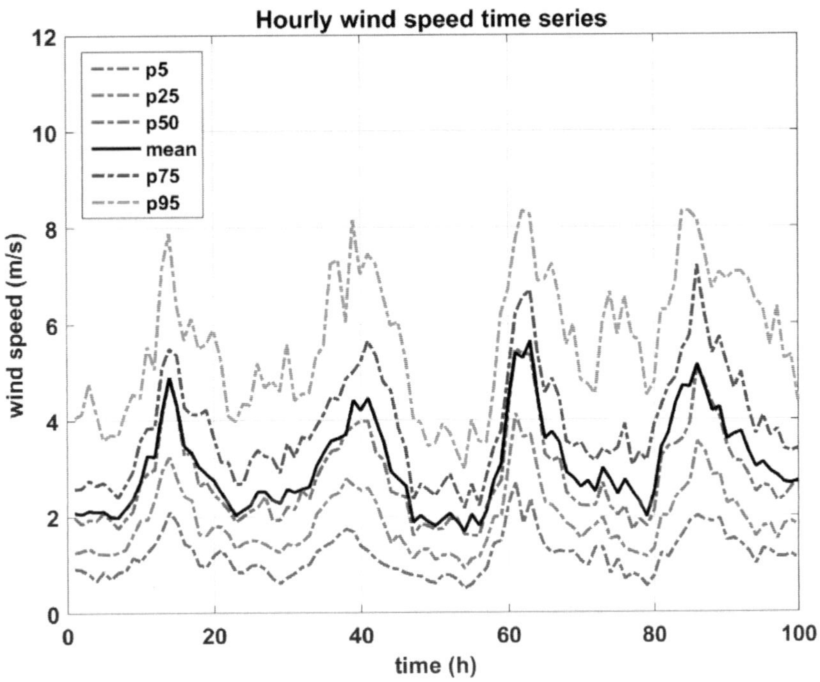

Fig. 3: Excerpt of the generated mean wind speed time-series (at 90 m AGL) and the corresponding percentiles.

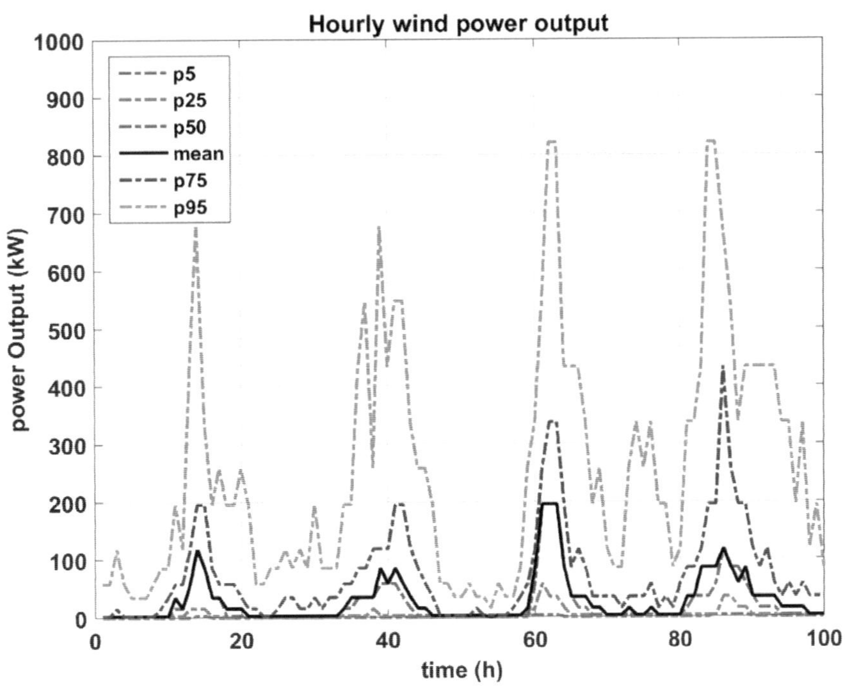

Fig. 4: Excerpt of the generated mean power output time-series and the corresponding percentiles.

4. Conclusions

This research is an example of the use of statistical wind resource assessment for estimating hourly time-series, including a precise range of uncertainty. Estimating the wind resource at high spatio-temporal resolution using statistical methods is certainly much faster than with traditional methods based on computational fluid dynamics, as demonstrated in [4], and this makes this method useful for the industry. In fact, being able to provide practitioners with hourly confidence intervals at high spatial resolution can be advantageous not only for planning but also for wind farms operations.

The method presented here is based on machine learning and estimates the spatio-temporal pattern of the wind speed including a detailed uncertainty estimation, in the form of confidence intervals. With such a method practitioners are able to determine *a priori* the range of variability of the wind resource along the year, and in every mapped location. This may be useful to estimate the power production in specific weeks or months, but it cannot be used for forecasting. In fact, the results we generate are only accurate in providing a range of variation, not in providing the exact wind speed in future time intervals.

Moreover, machine learning contrary to numerical weather prediction, is a data driven approach. This means that in areas with very low coverage of weather stations, for example in the Swiss Alps, its estimates are affected by a higher uncertainty. This could only be decreased by either adding weather observations or potentially using numerical wind flow models to increase the climatic information for these areas. The important thing to remember about statistical wind resource assessment, which makes it an effective method, is that it is able to estimate its own accuracy, thus allowing practitioners to spot problematic locations. This is very difficult to achieve with numerical weather prediction models.

5. Acknowledgment

The authors acknowledge the CTI - Commission for Technology and Innovation (CH), and the SCCER-FURIES - Swiss Competence Center for Energy Research - Future Swiss Electrical Infrastructure, for their financial and technical support to the research activity presented in this paper.

References

[1] P. Argyle and S. J. Watson, "Assessing the dependence of surface layer atmospheric stability on measurement height at offshore locations," *Journal of Wind Engineering and Industrial Aerodynamics*, vol. 131, pp. 88–99, 2014.
[2] M. Brower, *Wind resource assessment: a practical guide to developing a wind project*. John Wiley & Sons, 2012.
[3] G. T. Phillips, "Preliminary user's guide for the NOABL objective analysis code. Special report, 15 June 1977-15 June 1978," Science Applications, Inc., La Jolla, CA (USA), 1979.
[4] F. Veronesi, S. Grassi, and M. Raubal, "Statistical learning approach for wind resource assessment," *Renewable and Sustainable Energy Reviews*, vol. 56, pp. 836–850, 2016.
[5] H. Aksoy, Z. Fuat Toprak, A. Aytek, and N. Erdem Ünal, "Stochastic generation of hourly mean wind speed data," *Renewable Energy*, vol. 29, no. 14, pp. 2111–2131, Nov. 2004.
[6] W. Luo, M. C. Taylor, and S. R. Parker, "A comparison of spatial interpolation methods to estimate continuous wind speed surfaces using irregularly distributed data from England and Wales," *International Journal of Climatology*, vol. 28, no. 7, pp. 947–959, 2008.
[7] L. Foresti, D. Tuia, M. Kanevski, and A. Pozdnoukhov, "Learning wind fields with multiple kernels," *Stochastic Environmental Research and Risk Assessment*, vol. 25, no. 1, pp. 51–66, 2011.
[8] M. Cellura, G. Cirrincione, A. Marvuglia, and A. Miraoui, "Wind speed spatial estimation for energy planning in Sicily: A neural kriging application," *Renewable Energy*, vol. 33, no. 6, pp. 1251–1266, 2008.
[9] Castellanos and V. I. Ramesar, "Characterization and estimation of wind energy resources using autoregressive modelling and probability density functions," *Wind Engineering*, vol. 30, no. 1, pp. 1–14, 2006.
[10] K. Philippopoulos and D. Deligiorgi, "Statistical simulation of wind speed in Athens, Greece based on Weibull and ARMA models," *Int. J. Energy Environ*, vol. 3, no. 4, pp. 151–158, 2009.
[11] A. Shamshad, M. A. Bawadi, W. W. Hussin, T. A. Majid, and S. A. M. Sanusi, "First and second order Markov chain models for synthetic generation of wind speed time series," *Energy*, vol. 30, no. 5, pp. 693–708, 2005.
[12] P. Poggi, M. Muselli, G. Notton, C. Cristofari, and A. Louche, "Forecasting and simulating wind speed in Corsica by using an autoregressive model," *Energy conversion and management*, vol. 44, no. 20, pp. 3177–3196, 2003.

[13] O. Ohashi and L. Torgo, “Wind speed forecasting using spatio-temporal indicators.,” in *ECAI*, 2012, pp. 975–980.
[14] F. Veronesi, S. Grassi, and M. Raubal, “Satellite data to improve the accuracy of statistical models for wind resource assessment,” presented at the EWEA 2015 Annual Event, Paris.
[15] B. Efron and R. J. Tibshirani, *An introduction to the bootstrap*. CRC press, 1994.
[16] National Oceanic and Atmospheric Administration (NOAA), “Global Surface Summary of the Day - GSOD.” 2016.
[17] P. W. Stackhouse, “Surface meteorology and solar energy,” 2011.
[18] J. Schulz, P. Albert, H.-D. Behr, D. Caprion, H. Deneke, S. Dewitte, B. Dürr, P. Fuchs, A. Gratzki, and P. Hechler, “Operational climate monitoring from space: the EUMETSAT Satellite Application Facility on Climate Monitoring (CM-SAF).,” *Atmospheric Chemistry & Physics*, vol. 9, no. 5, 2009.
[19] J. J. Danielson and D. B. Gesch, “Global multi-resolution terrain elevation data 2010 (GMTED2010),” US Geological Survey, 2011.
[20] S. Grassi, S. Junghans, and M. Raubal, “Assessment of the wake effect on the energy production of onshore wind farms using GIS,” *Applied Energy*, vol. 136, pp. 827–837, 2014.
[21] H. Erich, *Wind turbines: fundamentals, technologies, application, economics*. New York: Springer, 2000.

The consideration of atmospheric stability within wind farm AEP calculations

Jonas Schmidt[1], Chi-Yao Chang[1], Martin Dörenkämper[1], Milad Salimi[2], Tim Teichmann[3], Bernhard Stoevesandt[1]

[1] Fraunhofer IWES, Küpkersweg 70, 26129 Oldenburg, Germany
[2] Technische Universität Cottbus, Platz der Deutschen Einheit 1, 03046 Cottbus, Germany
[3] Nordex Energy GmbH, Langenhorner Chaussee 600, 22419 Hamburg, Germany

E-mail: jonas.schmidt@iwes.fraunhofer.de

Abstract. The annual energy production of an existing wind farm including thermal stratification is calculated with two different methods and compared to the average of three years of SCADA data. The first method is based on steady state computational fluid dynamics simulations and the assumption of Reynolds-similarity at hub height. The second method is a wake modelling calculation, where a new stratification transformation model was imposed on the Jensen an Ainslie wake models. The inflow states for both approaches were obtained from one year WRF simulation data of the site. Although all models underestimate the mean wind speed and wake effects, the results from the phenomenological wake transformation are compatible with high-fidelity simulation results.

1. Introduction

The atmospheric stratification largely affects the flow conditions in the atmospheric boundary layer (ABL) relevant for both off-shore [1] and on-shore [2] wind energy sites. The atmospheric stability is driven by temperature gradients and buoyant forces and has strong impact on the generation of turbulence and thus the wind turbine wake recovery. It also determines the amount of mixing between different layers of air, and hence it is crucial for the transport of energy into the wind farm from above (and below).

Stable conditions often imply longer and more pronounced wind turbine wakes, due to the above mentioned reasons, unstable conditions have the reverse characteristics. In general, wakes affect both the energy yield and the loads of the wind turbines, hence also their lifetime and the operational costs of the wind farm. Thermal stratification is therefore linked to wind farm planning, and, more generally, to the calculation of the annual energy production (AEP).

However, stratification effects are often ignored for such calculations, and sometimes even the average of different stability situations is considered, due to the lack of better methods. This is the point that we aim to address in this work, we demonstrate that the effects can be represented in computational fluid dynamics (CFD) calculations as well as in the engineering model wake superposition approach.

In the following, we include the effect of atmospheric stability into the AEP calculation of an existing onshore wind farm in quasi-flat terrain for the year 2015. The time series of inflow wind and stability conditions are obtained from a mesoscale simulation with the Weather Research and Forecasting (WRF) model [3]. Afterwards, a classification based on wind speed, direction and the Obhukov length (atmospheric stability) is carried out.

2. The site

The on-shore site under investigation is located in the USA and consists of 17 Nordex wind turbines with hub height $H = 100$ m, rotor diameter $D = 100$ m and nominal power $P = 2.5$ MW. The layout is shown in Fig. 1. The terrain is quasi-flat and the site is dominated by agricultural land, which is modelled by a uniform roughness length of $z_0 = 0.15$ m.

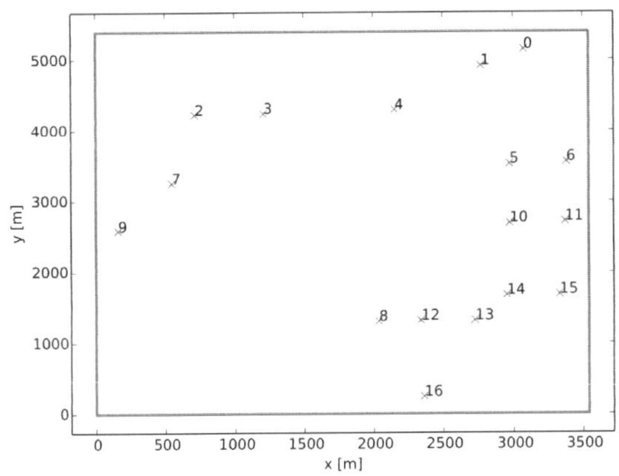

Figure 1. The wind farm layout.

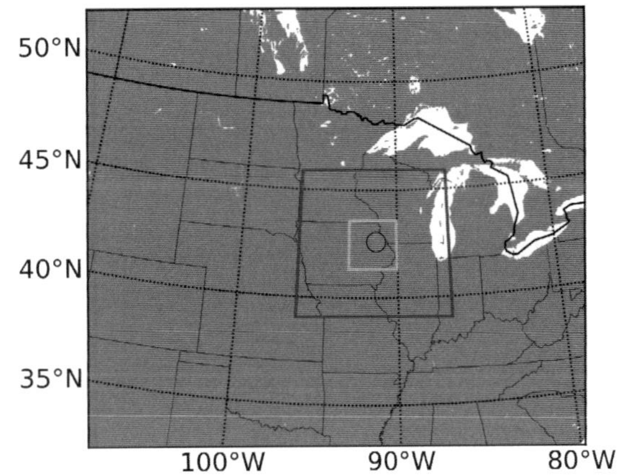

Figure 2. Distribution of the domains D1 (size of the plot), D2 (blue), D3 (orange). The location of the wind farm is marked by the red dot.

3. Method

In this paper the results from one-year of mesoscale model simulations are used as boundary condtions for subsequent microscale and engineering model simulations to calculate the AEP for an onshore site. Therefore, this section subdivides into the description of the mesoscale, the microscale and the engineering model setup.

3.1. Mesoscale model simulations

The mesoscale model simulations presented in this publication were conducted with the WRF model [3]. The model is well established in research and industrial wind energy applications (e.g. [4], [5]). We used a setup similar to the one in [1] which was specifically chosen for wind energy applications and thus vertically well resolves the heights most relevant for wind turbines.

The setup consisted of three domains (D1-D3) nested within each other. Fig. 2 shows the distribution of the domains centered around the wind farm location. The location of interest is located in eastern Iowa in mildly complex terrain. Tab. 1 gives an overview of the WRF version as well as the parametrisations and boundary conditions used. An entire year (2015) was simulated separated into four simulations each with a duration of three months using spectral 3D analysis nudging [3]. The first two days (48 hours) of each quarterly simulation were disregarded in the subsequent analysis to allow for enough spin-up time for the model.

Table 1. WRF setup used within this study. For a more detailed description of model and its options see [3].

Option	Value
WRF-version	3.6.1
Domains & Resolution	D1 ($\Delta_{x,y}$ = 18.9 km), D2 (6.3 km), D3 (2.1 km)
Vertical Resolution	62 Eta-Level (out of which 20 $<$ 1 km)
Initialisation	2-way nesting
Shortwave radiation	Dudhia scheme
Longwave radiation	Rapid Radiative Transfer Model (RRTM)
Turbulence scheme	MYNN 2.5 level scheme
Nudging	MERRA-Reanalysis [6]

The results from the model calculations were put out every 10 min. Afterwards, the original wind data (on eta-levels) were interpolated to hub height and together with the dimensionless Obhukov length $L_{90} = H_{hub}/MOL$ extracted from the models result files. Based on these three parameters a binning into bins of $1.0\,ms^{-1}$ in wind speed, 10° in wind direction and five stability classes in ΔL_{90} (see Tab. 2) was performed.

Table 2. Binning of the stability parameter L_{90}

Range of L_{90}	Classification
L_{90} < -0.45	very unstable
-0.45 < L_{90} < -0.05	unstable
-0.05 < L_{90} < 0.05	neutral
0.05 < L_{90} < 0.45	stable
L_{90} > 0.45	very stable

Fig. 3 and 4 show the resulting wind and stability roses (distributions by wind directions). The roses generally demonstrated prevailing south south-westerly winds with a secondary maximum for north-westerly winds. Generally all wind direction sectors are of importance as none of the twelve sectors has a probability of occurrence of less than 5 %. Stably stratified conditions are dominating at the wind farm location.

The second postprocessing step for the WRF simulation data was a selection of the relevant states based on the distributions of these three parameters. These states then served as boundary conditions for the subsequent set of microscale simulations to determine the AEP at the wind farm location.

3.2. Microscale model simulations

All CFD results for this work were generated with OpenFOAM [7] (version 2.3.1) and in-house extensions by Fraunhofer IWES (e.g. [8–10]). Thirty-six wind directon sectors were simulated for one characteristic wind speed, one neutral condition and four dimensionless Obukhov lengths (see Tab. 2). The weights for adding simulation results are derived from the mesoscale

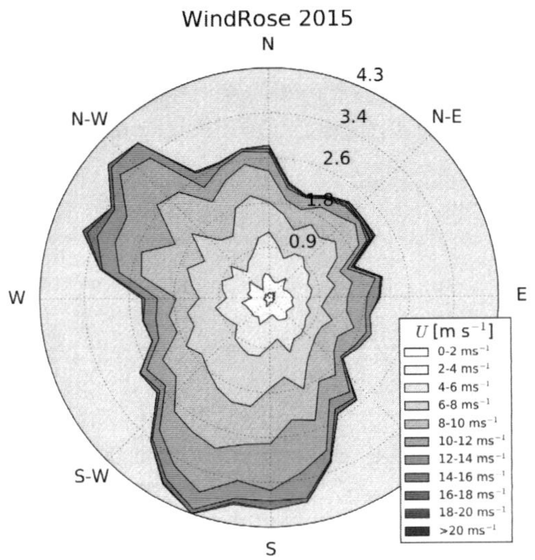

Figure 3. Figure caption for first of two sided figures.

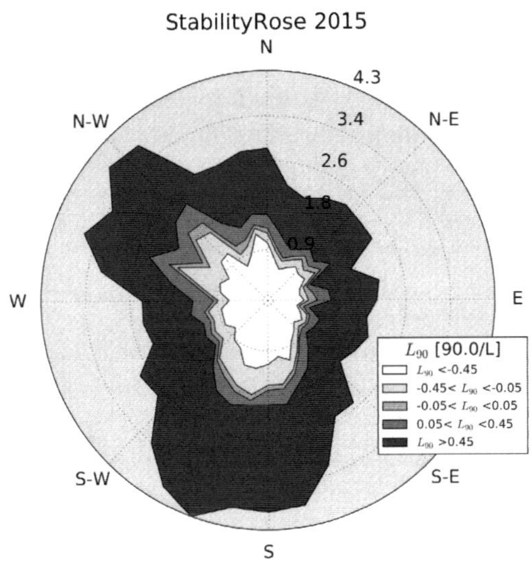

Figure 4. Distribution of the atmospheric stability L_{90} over the wind direction

results. The mesh has cylindrical topology with 2.3 mio. cells and the lowest cell height is 1 m. The terrain is quasi-flat and was generated with the in-house *terrainMesher*.

The applied turbulence model is a buoyant adaption of the k-ϵ-f_P turbulence model [11] with parameters as suggested in the reference. The inflow boundary profiles for wind velocity U, turbulent kinetic energy k, turbulent dissipation ϵ and potential temperature Θ were consistently generated by a one dimensional cyclic precursor run. The roughness length is uniformly set to $z_0 = 0.15$ m, forest canopy can be ignored at the site. The solver is an in-house development for steady state stratified atmospheric boundary layer flow, based on the SIMPLE algorithm [12]. The wind turbines were modelled by uniformly loaded actuator disks with about 500 cells each, switched on after running the background wind field simulation first.

For the AEP calculation the hypothesis of Reynolds-similarity is applied. This means that from a single simulation in a certain wind direction sector with a characteristic wind speed at hub height of 10 ms^{-1}, the wind speeds in each cell for other inflow wind speeds are obtained by local re-scaling. The error that is made by this simplifying assumption can be expected to be smaller than the error that corresponds to a direct evaluation of the sector based on a single simulation. This assumption reduces the number of simulations for a wind rose with 12 sectors to $36 \times 5 \times 1 = 180$ simulations.

3.3. Engineering model simulations

All engineering model results were generated with the wind farm modelling software flapFOAM [9, 10, 13, 14]. flapFOAM is based on the principle of wake superposition, and some of its modelling components are briefly described below. For more details, the reader is referred to the references.

3.3.1. Wake models All calculations in this paper are performed with wake models that are based on the models by Jensen [15] and Ainslie [16]. For reasons that will become apparent later, they are filtered by flapFOAM's wake transformation option 'Gaussianize'. This transformation is based on the assumptions of axial symmetry and neglects radial velocity deficit components. It can be applied to arbitrary wake models and translates the deficit into three functions $R(x)$, $A(x)$ and $r_0(x)$ on a one dimensional grid, for which we choose a spacing of $\Delta x = 10$ m. The function $R(x)$ represents the wake radius in m at a distance x behind the disk, determined by the condition that the deficit has dropped below 1% of the central value. The other two functions represent a Gaussian function with respect to the radial coordinate r,

$$f(x,r) = A(x)\exp\left[-\left(\frac{r}{r_0(x)}\right)^2\right] \tag{1}$$

and they are interpolated at fourth order accuracy as functions of the continuous variable x. The amplitude $A(x)$ is fixed by the central value ($r = 0$) of the underlying wake model, and the width of the Gaussian function by the requirement that the total mass flux of the Gaussian function should be identical to the calculated flux $F(x)$ of the original wake model within a disk of radius $R(x)$, yielding $r_0(x) = \sqrt{2F(x)/A(x)}$.

The increase of turbulence intensity in a single wake is modelled according to Frandsen [17]. For this work we chose quadratic summation of wind velocity and turbulence intensity deficits in multiple wake situations.

3.3.2. Inflow states and AEP calculation The meteorological variables that are particularly of interest for the wake flow are the wind direction ϕ, the wind speed v and inverse dimensionless Obukhov length L_{90} at hub height. With binning $\Delta\phi = 10°$, $\Delta v = 1.0$ ms^{-1} and L_{90} as in Tab. 2 we obtain 2106 states with non-vanishing probability from the mesoscale results. For each of these states one calculation is performed with flapFOAM, and the AEP is then obtained by taking the weighted sum of the predicted wind farm power. For each state and each turbine the latter is obtained from the power curve of the wind turbine model, evaluated at the effective local wind speed as measured by the turbine. For simplicity we consider the wind speed at the disk centre only and ignore partial wake and wind shear effects.

The ambient wind field is modeled by standard log profiles with uniform roughness length $z_0 = 0.15$ m and thermal stratification corrections.The ambient turbulence intensity I is obtained from the Richards-Hoxey solution [18] of the k-ϵ turbulence model

$$I^2(z) = \frac{2\hat{u}_*^2}{3\sqrt{C_\mu}}\,, \quad \hat{u}_* = \frac{\kappa}{\ln(z/z_0)} \tag{2}$$

where $C_\mu = 0.033$ and $\kappa = 0.4$ are constants. Notice that stability corrections of turbulence intensity are ignored in the engineering modelling approach throughout this work.

3.3.3. Thermal stratification effects on wakes A new phenomenological model for the effect of thermal stratification on wakes has been proposed in [19] and implemented in flapFOAM. The model has been obtained by fitting parametrised curves to central wind speed deficit results of CFD RANS actuator disk simulations. For eight different values of the inverse Obukhov length parameter ξ a single uniformly loaded actuator disk was simulated with methods from Section 3.2. The deviation of the central wake deficit in the stratified situation compared to the neutral case for identical inflow wind speed U_0 at hub height is

$$\Delta\text{Def}_\xi(x) = \left(1 - \frac{U_\xi(x)}{U_0}\right) - \left(1 - \frac{U(x)}{U_0}\right) = \frac{U(x) - U_\xi(x)}{U_0} \tag{3}$$

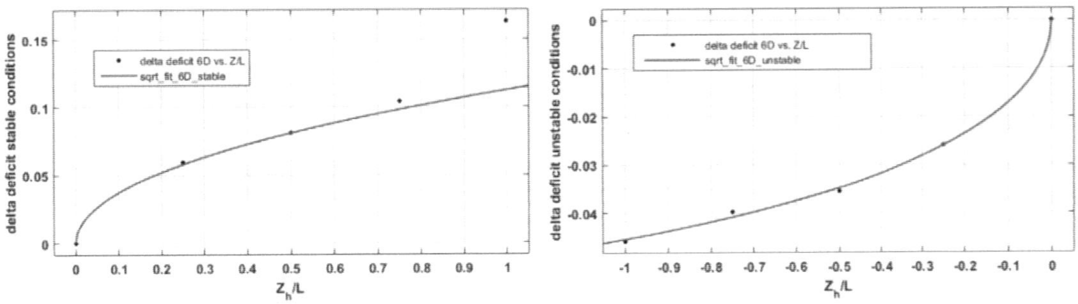

Figure 5. Change of wake deficit compared to the neutral case as a function of hub height z_h over MOL L, for stable (left) and unstable (right) stratification. The blue curve is the parametrisation from Eq. (4). The very stable result $z_h/L = 1$ was excluded for the model fit.

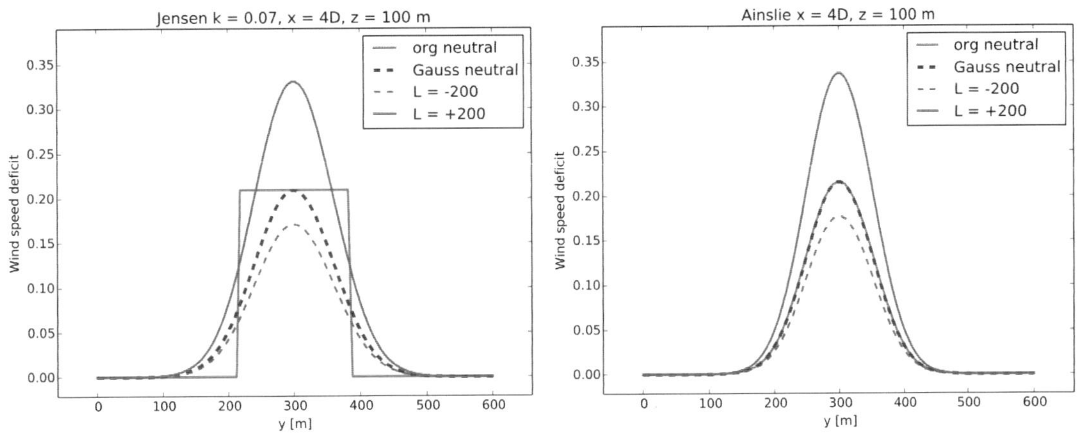

Figure 6. Exemplary wind speed deficits at 4D behind the rotor for the Jensen (left) and Ainslie (right) wake models, and their Gaussian and thermal stability transformed equivalents.

where U denotes the horizontal wind velocity component and the subscript ξ denotes stratified flow conditions. As shown in Fig. 5 this dimensionless quantity can be modelled by the function

$$f(s,\xi) = a(s)\sqrt{|\xi|} + b\xi \ , \quad \text{with} \left\{ \begin{array}{lll} \text{stable:} & a(s) = c_1 \exp\left[-\left((s-c_2)/c_3\right)^2\right] \ , & b = c_4 \\ \text{unstable:} & a(s) = c_5 s - c_6 \ , & b = c_7 \end{array} \right. \tag{4}$$

and constants $c_1 = 0.1695$, $c_2 = 2.156$, $c_3 = 6.73$, $c_4 = 0.033$, $c_5 = 0.0026$, $c_6 = 0.074$, $c_7 = -0.011$. Here $s = x/D$ is the distance from the disk in units of the rotor diameter D. For the stable case the model peaks at $x = c_2 D \approx 2.2D$, for larger distances the stability effect falls off following a Gaussian. For unstable stratification the sign of $f(s,\xi)$ is determined by the sign of $a(s)$. Since simulations show a reduction of the wake deficit for unstable conditions this sign must be negative, and the model is only valid for $x \leq c_6/c_5 \ D \approx 28D$.

Furthermore the actuator disk simulations did not show a strong variation of the width of the wake [19]. In flapFOAM the model is realized by two steps. First, the chosen base wake model is transformed into a Gaussian function, as described in Section 3.3.1. Then, the value $U_0 f(x/D,\xi)$ is added to the corresponding amplitude and the scale $r_0(x)$ is not modified.

Examples for the Jensen and the Ainslie wake models and an Obukhov length $L = \pm 200$ m are shown in Fig. 6.

4. Results and discussion

The to the simulations corresponding year (2015) of SCADA data are available for the site from Section 2. The data was filtered according to a status flags of the turbines. The resulting average wind speed and the AEP of the wind turbines is compared to simulated results in Figs. 7-8 and 9, respectively. In figures 7 and 9, the results have been normalized to the average of each simulation, while in 8 the normalisation was done with respect to the average of the SCADA data to allow for an estimation of the bias between the simulation methods. For flapFOAM both neutral and stratified results are shown, for Jensen and Ainslie wake models.

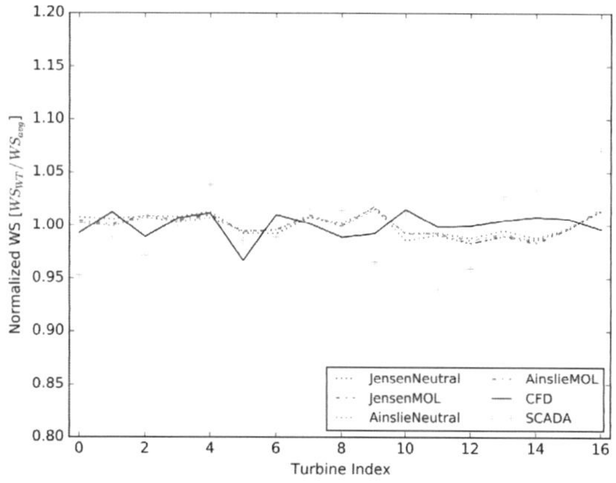

Figure 7. Wind speed calculation results normalized by the average of each data-source.

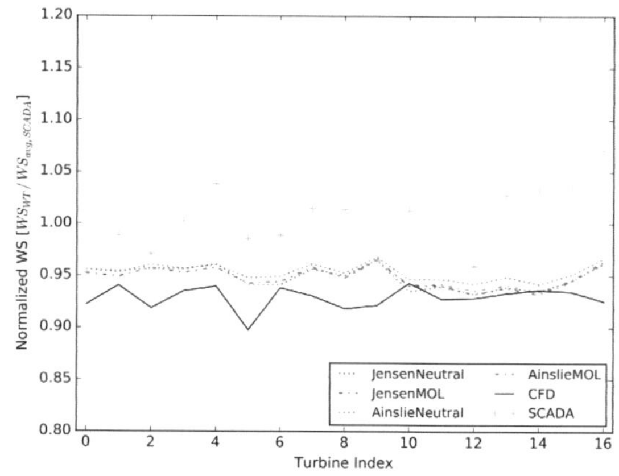

Figure 8. Wind speed calculation results normalized by the average of the SCADA-data.

Generally, the variations in the wind speed at the turbines are much lower in case of the flapFOAM results compared to the SCADA data. One reason for this is that the variation in the wind speed at the measurement location (downstream of the rotor disk) is generally very strong. CFD and flapFOAM seems to underestimate the wind speeds at the locations of the turbines, which is also visible in flapFOAM results (Fig. 8). However, another possibility is that the SCADA wind speed is biased towards high values.

The lowest mean wind speed and AEP results according to the SCADA data are found for turbines 2, 5, 6 and 12. From the wind farm layout in Fig. 1 and the wind rose from Fig. 3 it is apparent that these turbines are subject to wakes from southerly winds, which are well represented in the wind rose. The wind turbines 4, 15, 16 have the highest average wind speeds and AEP, they are neither hit by southern nor by north-easterly winds or (in case of turbine 4) the next upstream turbine is several tenth of rotor diameters upstream. flapFOAM is generally capable in simulating this variation between the turbines. Here, turbines 9 and 16, which are both located at the edges of the wind farm show the largest wind speeds and AEP.

None of the models, CFD or engineering, seems to able to convincingly capture the complexity of the wake situation at the site. This is likely due to the relatively coarse wind direction resolution of 10°. Since some wake situations are missed by 36 wind directions, the non-wake situations statistically dominate the results for most of the turbines and no clear picture emerges.

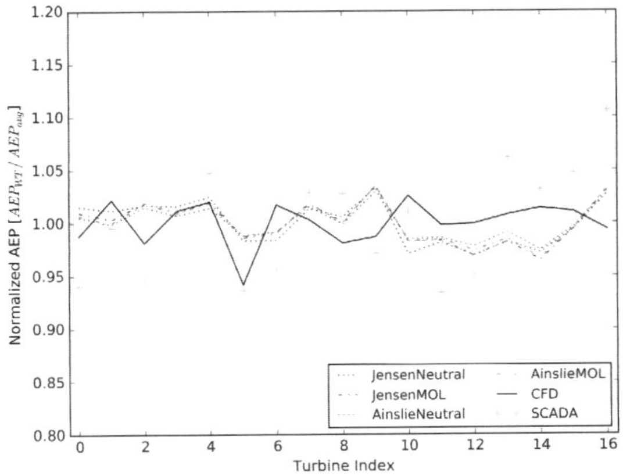

Figure 9. AEP calculation results normalized by the average of each data-source.

The difference between the original Jensen and Ainslie wake models and their thermally transformed counter parts is around 5% or less. However, the stability corrected wake models yield better results compared to measurements. The small impact of the stability transformation of the wake models on the resulting AEP is due to the small wake effect that was observed in all simulations. Since for each state the wind speed at hub height was fixed to the WRF result, and since no shear effects have been considered for the calculation of the effective wind speed at the rotors, the wake effect exclusively enters the flapFOAM calculation through the imposed wake deficit transformation in the chosen setup.

5. Conclusions

In this work we studied an existing wind farm in North America and compared two simulation approaches for the AEP calculation with averaged one year SCADA data. The two methods are steady-state CFD and the engineering model software flapFOAM. The input for both was based on the same results of a one-year mesoscale simulation of the site with WRF. By imposing a binning in the three variables wind direction, wind speed and dimensionless inverse Obukhov length, and by sorting and filtering according to probability, the number of states was reduced to 2106 states.

For CFD this number is still too high, it was reduced to 180 by assuming Reynolds-similarity for the local wind speed at hub height. However, the results show an underestimation of the wind speed and thus the AEP. Also, variation of the SCADA results due to the wake effect are not captured well. Since this is also observed for the engineering model results we conclude that the sectors of the wind rose were most likely not sufficient for the observation of significant wake effects.

For the very same reason we do not observe a large impact of the presented engineering model for modelling thermal effects on wake deficit models. However, the transformed Jensen and Ainslie wake models have better agreement with the SCADA AEP results than CFD, which underpredicts the wind speed. The stratification transformation of arbitrary wake effects was obtained by fitting the variation of the central wake deficit with Obukhov length to corresponding CFD actuator disk results. This process yielded a phenomenological model that can now be used for fast AEP calculations that include thermal stratification. This model was first applied to an existing wind farm in this work and compared to measurements. We conclude that further validation and more detailed simulations are necessary, also the dependency on roughness length and the performance of the model in complex terrain remain to be investigated.

References

[1] M. Dörenkämper, M. Optis, A. Monahan, and G. Steinfeld. On the offshore advection of boundary-layer structures and the influence on offshore wind conditions. *Boundary Layer Meteorol.*, 3(155):459–82, 2015.

[2] S. Wharton and J. K. Lundquist. Assessing atmospheric stability and its impacts on rotor-disk wind characteristics at an onshore wind farm. *Wind Energy*, 4(15):525–46, 2012.

[3] W. Skamarock, J. Klemp, J. Dudhia, D. Gill, D. Barker, M. Duda, X. Huang, W. Wang, and J. Powers. A description of the advanced research wrf version 3. Technical Report NCAR/TNâĂŞ475+STR, NCAR - National Center for Atmospheric Research, Boulder, Colorado, USA, 2008.

[4] A. Peña, A. Hahmann, C. Hasager, F. Bingöl, I. Karagali, J. Badger, M. Badger, and N. Clausen. South Baltic wind atlas. Technical Report Ris-R-1775(EN), Technical University of Denmark, 2011.

[5] T. Lorenz and I. Barstad. A dynamical downscaling of ERA-Interim in the North Sea using WRF with a 3 km grid - for wind ressource applications. *Wind Energy, early online release*, 2016.

[6] M. M. Rienecker, M. J. Suarez, R. Gelaro, R. Todling, J. Bacmeister, E. Liu, M. G. Bosilovich, S. D. Schubert, L. Takacs, G.-K. Kim, and et al. MERRA: NASAs Modern-Era Retrospective analysis for Research and Applications. *J. Climate*, 14(24):3624–48, 2011.

[7] OpenFOAM. http://www.openfoam.org, 2016. [Online; accessed 04-August-2016].

[8] C Peralta, H Nugusse, SP Kokilavani, J Schmidt, and B Stoevesandt. Validation of the simplefoam (RANS) solver for the atmospheric boundary layer in complex terrain. In *ITM Web of Conferences*, volume 2. EDP Sciences, 2014.

[9] J. Schmidt and B. Stoevesandt. Wind farm layout optimisation using wakes from computational fluid dynamics simulations. In *EWEA conference proceedings*, Barcelona, Spain, 10–13 March 2014.

[10] J. Schmidt and B. Stoevesandt. Wind farm layout optimisation in complex terrain with cfd wakes. In *EWEA conference proceedings*, Paris, France, 17–20 November 2015.

[11] M. Paul van der Laan, Niels N. Sorensen, Pierre-Elouan Rethore, Jakob Mann, Mark C. Kelly, Niels Troldborg, J. Gerard Schepers, and Ewan Machefaux. An improved k-epsilon model applied to a wind turbine wake in atmospheric turbulence. *Wind Energy*, 18(5):889–907, 2015.

[12] Joel H Ferziger and Milovan Peric. *Computational methods for fluid dynamics*. Springer Science & Business Media, 2012.

[13] Jonas Schmidt and Bernhard Stoevesandt. Modelling complex terrain effects for wind farm layout optimization. *J. Phys.: Conf. Ser.*, 524(1):012136, 2014. The Science of Making Torque from Wind, Copenhagen, Denmark.

[14] Jonas Schmidt and Bernhard Stoevesandt. The impact of wake models on wind farm layout optimization. *J. Phys.: Conf. Ser.*, 625(1):012040, 2015. Wake Conference, Visby, Sweden.

[15] N. O. Jensen. A note on wind generator interaction. Technical Report Risø-M-2411, Risø National Laboratory, 1983.

[16] J. F. Ainslie. Development of an eddy viscosity model for wind turbine wakes. In *BWEA Wind Energy Conference*, Oxford, UK, 1983.

[17] S. Frandsen, R. Barthelmie, S. Pryor, O. Rathmann, S. Larsen, J. Højstrup, and M. Thøgersen. Analytical modelling of wind speed deficit in large offshore wind farms. *Wind Energy*, 9:39–53, 2006.

[18] P. J Richards and R. P. Hoxey. Appropriate boundary conditions for computational wind engineering models using the $k - \epsilon$ turbulence model. *J. Wind Eng. Ind. Aerod.*, 46-47(0):145–153, 1993. Proceedings of the 1st International on Computational Wind Engineering.

[19] M. Salimi. Modelling atmospheric stability effects on wakes by actuator disk simulations. Master Thesis, TU Cottbus, Germany. In preparation., 2016.

Journal of Physics: Conference Series **749** (2016) 012003 doi:10.1088/1742-6596/749/1/012003

Limitations to the validity of single wake superposition in wind farm yield assessment

K. Gunn[1], C. Stock-Williams[1], M. Burke[1], R. Willden[2], C. Vogel[2], W, Hunter[2], T. Stallard[3], N. Robinson[4], S. R. Schmidt[5]

[1]Uniper Engineering, Technology Centre, Ratcliffe on Soar, Nottingham, NG11 OEE, UK
[2]Department of Engineering Science, University of Oxford, OX1 3PJ, UK
[3]School of Mechanical, Aerospace and Civil Engineering, University of Manchester, Manchester, M13 9PL, UK
[4]AWS TruePower, 463 New Karner Road, Albany, NY 12205, USA
[5]E.ON Climate & Renewables, Carl Gustafs väg 1, SE-205 09 Malmö, Sweden

Kester.Gunn@uniper.energy

Abstract Commercially available wind yield assessment models rely on superposition of wakes calculated for isolated single turbines. These methods of wake simulation fail to account for emergent flow physics that may affect the behaviour of multiple turbines and their wakes and therefore wind farm yield predictions. In this paper wake-wake interaction is modelled computationally (CFD) and physically (in a hydraulic flume) to investigate physical causes of discrepancies between analytical modelling and simulations or measurements. Three effects, currently neglected in commercial models, are identified as being of importance: 1) when turbines are directly aligned, the combined wake is shortened relative to the single turbine wake; 2) when wakes are adjacent, each will be lengthened due to reduced mixing; and 3) the pressure field of downstream turbines can move and modify wakes flowing close to them.

1. Introduction

Commercially available wind farm energy yield assessment models rely on superposition of wakes calculated for isolated single wind turbines [1,2,3,4]. This is generally achieved through linear summation of momentum [5] or energy [6] deficits, although some models simply take the largest of the wake deficits present at any given point in space [1,2]. These methods of wake simulation fail to account for emergent flow physics that may affect the behaviour of multiple turbines and wakes and therefore farm yield predictions. This study investigates whether single wake superposition methods may contribute to any systematic errors in yield estimates; and whether commercial wind farm energy yield modelling software requires more explicit modelling of wake interactions. This study builds on that of Machefaux et al [7] by presenting the physical causes of discrepancies between analytical modelling and simulations or measurements.

2. Approach

2.1. Turbine Layouts

Several layouts of two turbines (126m diameter 5MW NREL reference turbine [8]) have been investigated through cross-comparisons of analytical models with numerical simulations (computational fluid dynamics, "CFD"). Additional model scale flume experiments ("tank tests") are also presented to provide comparison for several layouts. These turbine layouts, shown in Fig. 1, have been chosen to investigate the fundamental assumptions in single turbine superposition models:

1. overlapping wakes (i.e. when the inflow to the wake-generating turbines is undisturbed); and
2. interacting wakes (i.e. when the inflow to at least one of the turbines is partly or wholly in the wake of another).

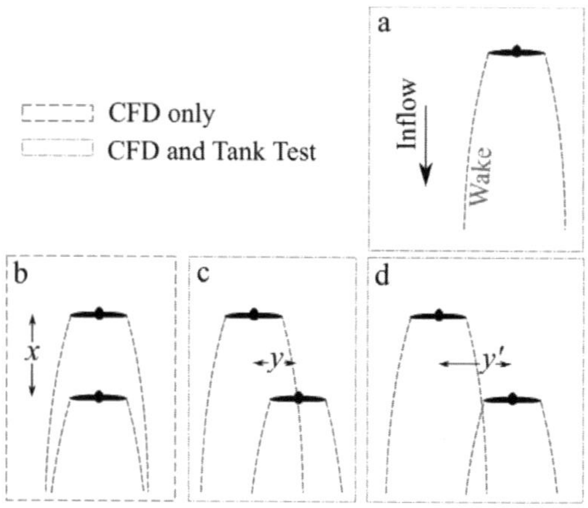

Figure 1. Summary of experimental investigations carried out with CFD and tank testing. CFD parameters at full scale are x=8D, y=1D, y'=1.75D, D=126m. Tank test parameters are x=4D, y=1D, y'=1.4D, D=0.27m.

2.2. CFD Simulations

The CFD was performed using ANSYS Fluent 15.0, solving the steady 3D incompressible Reynolds Averaged Navier-Stokes (RANS) equations using the finite volume method [9]. An embedded Blade Element actuator method was used to represent the rotors in this study, resulting in free shear emanating from the edges of the rotor disks, while avoiding the need to compute the shear caused by the rotor blade surfaces. Zero equation turbulence models, such as the Baldwin–Lomax model, calculate the eddy viscosity based on the local boundary layer profile at walls, and are therefore not well suited to actuator type simulations. Turbulence closure is therefore provided through the two-equation k-ω SST model [10], which combines the advantages of the k-ω with the k-ε turbulence closures near no-slip boundaries and through the remainder of the domain respectively; leading to it having been employed in a number of previous wind turbine studies (e.g. [11,12]). While the turbulence model is known to have an influence on the quantitative development of wake profiles, the qualitative trends presented herein are generally valid. A shear profile was generated by introducing a constant shear stress to the bottom wall of the computational domain, which results in velocity shear and a turbulence profile across the rotor height. The shear stress was adjusted to achieve a hub height turbulence intensity of around 5%.

The blades of the 126m diameter three-bladed rotor have sections varying smoothly between the DU40 and DU21 sections from 20% to 70% of the blade radius, and then the NACA64-series section out to the blade tip. The root chord is 4.6m and tapers to 1.4m at the tip. The blades are twisted through 13.1 degrees between root and tip [8]. The rotor was simulated using a RANS embedded Blade Element actuator disk model, with lift, drag and swirl injected at approximately 7000 points over each of the rotor disks in response to the simulated local flow-field [13]. Unlike analytic Blade Element Momentum methods, azimuthal averaging is not required and the effect of the shear on rotor loading and wake generation can be simulated through the embedded actuator disk.

Active torque control was implemented for the rotors, allowing the angular speed of each rotor to be independently varied. Following M. O. L. Hansen [14], the angular speed was adjusted to match the aerodynamic torque, $T_{aero} = P/\omega$, with the generator torque, $T_{gen} = K\omega^2$, where K (the "mode gain") is a property of the turbine that was determined from single rotor simulations in which ω was adjusted to achieve the peak power coefficient. The active torque control algorithm therefore ensured that the angular speed of each rotor was adjusted to achieve operation at maximum power coefficient. Rotor hubs, ellipsoids of length 1/10th of the rotor diameter, and with a radius of 1.5m, were modelled in each case but support towers were neglected. The key simulation parameters are given in Table 1.

Table 1. Summary of CFD parameters

Bulk flow wind speed	12.0 m/s
Hub height wind speed (u_0)	11.4 m/s
Hub height turbulent kinetic energy	0.538 m^2/s^2
Hub height turbulence intensity	5.26%
Domain width	1134 m
Domain length	3150 m
Domain height	1134 m
Number of cells	6-8 x 10^6
Rotor diameter	126 m
Rotor hub height	90 m

2.3. Tank tests

The tank tests were conducted in the University of Manchester wide flume with key dimensions as given in Table 2. Three-bladed rotors were employed with blade section selected for high lift to drag ratio at a chord Reynolds number of approximately 3×10^4 (typical at three-quarter radius) and with radial variation of chord length and twist selected to produce typical variation of thrust with tip speed ratio. Rotor thrust was measured through a strain-gauged supporting structure and power from rotor angular speed, measured by digital encoder, multiplied by torque supplied by a DC motor with friction subtracted. Torque was selected for equal tip speed ratio of each rotor. Flow velocity was measured with a Nortek Vectrino+ sampling at 200 Hz. The flow characteristics, rotor and instrumentation are described further by T. Stallard et al [15]. With reference to Fig. 1, the upstream rotor was located at the flume mid-span and mid-depth, 22 diameters from the inflow and the downstream rotor at x=4D, y'=1.4D. Transects of velocity were taken at hub height at 2D increments downstream.

Table 2. Summary of tank test parameters

Hub height flow speed	0.463 m/s
Hub height turbulence intensity	12%
Tank width	5 m
Length of test section	12 m
Still water depth	0.45 m
Turbine diameter	0.27 m
Turbine hub height	0.225 m

2.4. Wake combination methods

The three wake combination methods most often used in the literature are given in Eqns. 1 to 3, where $u_{norm,i}$ are the wakes at that location, from each upstream turbine i in isolation. Note that these combination methods take the effective normalisation speed for each turbine to be that turbine's inflow speed, not the free stream speed. Results from the use of these methods are presented for comparison with the CFD and tank tests (colour lines in Figs. 4 – 7).

Largest deficit: $$U_{norm} = \min(u_{norm,1}, u_{norm,2}, \dots) \quad (1)$$

Root-sum-square: $$U_{norm} = 1 - \sqrt{\sum (1 - u_{norm,i})^2} \quad (2)$$

Linear: $$U_{norm} = 1 - \sum (1 - u_{norm,i}) \quad (3)$$

3. Results and Analysis

Rotor wake data are now presented. The origin of the domain is at the centre of the upstream rotor. To account for the shear profile, velocities are normalised relative to the value at the corresponding cross-stream (y) and vertical (z) position at the furthest upstream plane of the domain, i.e.

$$u_{norm} = \mathrm{u}(x, y, z) / \mathrm{u}(-4, y, z)$$

where all distances are normalised to the rotor diameter D.

3.1. Single Turbine Analysis

In order to establish a baseline for comparison with multiple turbine data, the single turbine wake is first examined. Fig. 2a (far left column) illustrates that the location of largest velocity deficit behind a turbine shifts with downstream distance:

1. to the right, due to the swirl injected by the rotor; and
2. towards the ground, due to the velocity shear driving re-energisation of the wake region more strongly above than below the wake.

This wake evolution is not represented in commercial analytical wake models, although it does appear in Fuga simulations (a linearised CFD model) [4]. This indicates that hub height velocity may not be the most appropriate choice of measurement to characterise the wake [16].

Otherwise, as shown further in Fig. 3, the Gaussian shaped wake profile is seen in the far wake, as expected from many previous studies [17].

Journal of Physics: Conference Series **749** (2016) 012003 doi:10.1088/1742-6596/749/1/012003

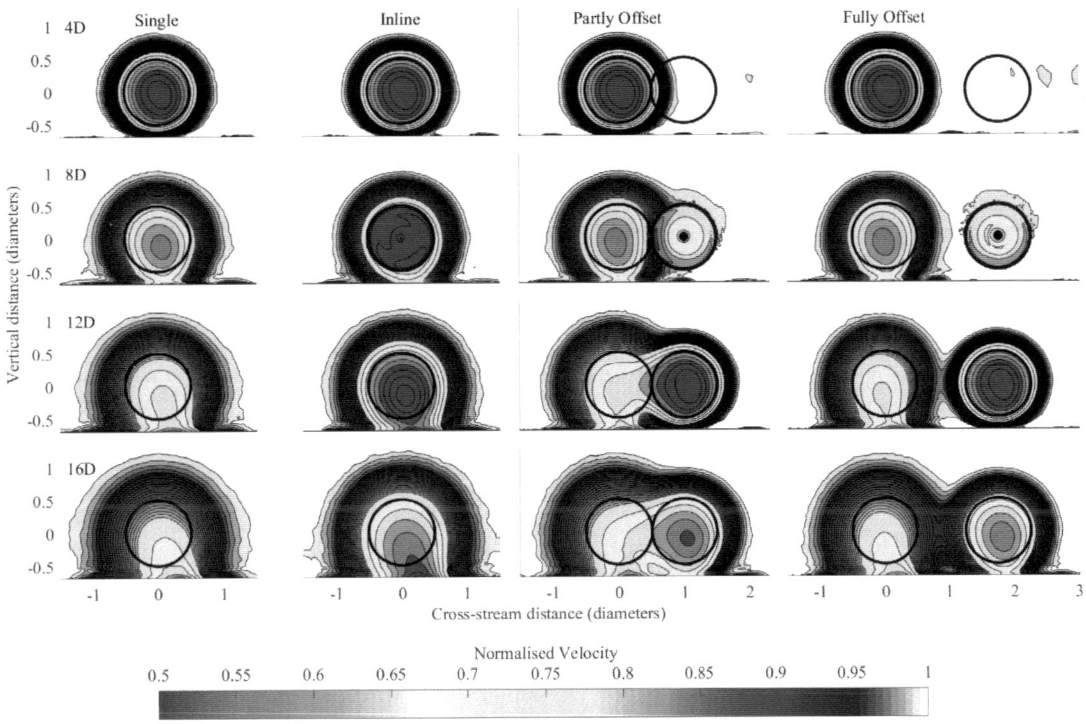

(a) Normalised velocity

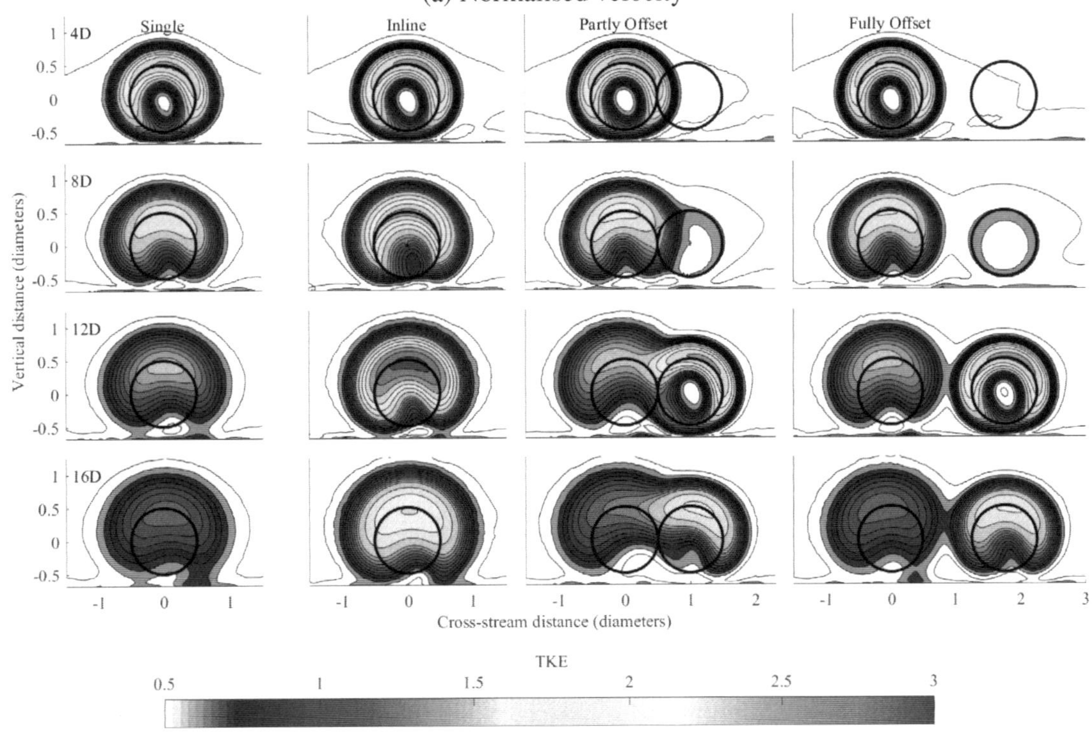

(b) Turbulent kinetic energy, TKE

Figure 2. Normalised velocity (a) and turbulent kinetic energy (b) distribution over vertical planes downstream of each rotor. Downstream position is relative to the upstream rotor. The colours indicate contours, where white is the colour of the undisturbed value. The rotor areas are marked as black outlines. (CFD simulation)

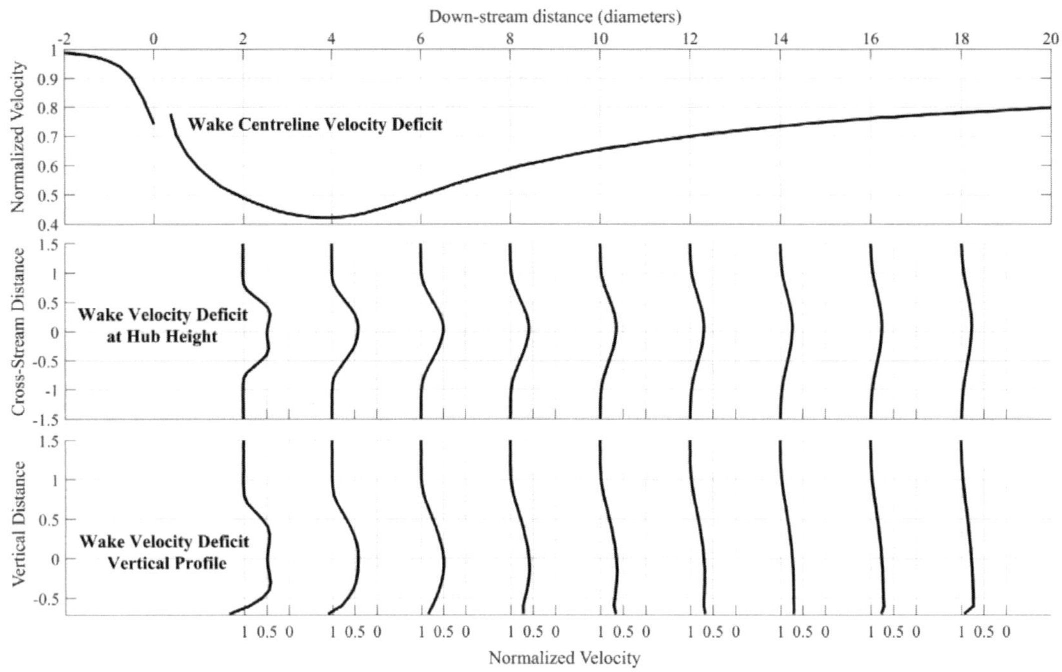

Figure 3. Normalised velocity profiles behind a single turbine (CFD simulation).

3.2. Two Turbine Analysis

Table 3 gives an overview of rotor performance from the CFD simulations.

Table 3. Summary of rotor performance. Power coefficient (C_P) and thrust coefficient (C_T) are calculated relative to the flow speed 2.5D upstream of each hub (u_i). Power and thrust of each turbine also shown relative to the single turbine case (P_0 and T_0 respectively).

Case	C_P	P/P_0	C_T	T/T_0	u_i/u_0
Single Rotor	0.536	1	0.914	1	1
Inline					
- Upstream Rotor	0.532	0.999	0.904	0.999	1
- Downstream Rotor	0.507	0.284	0.859	0.424	0.668
Partly Offset					
- Upstream Rotor	0.532	0.999	0.904	0.999	1
- Downstream Rotor	0.590	0.932	0.968	0.953	0.944
Fully Offset					
- Upstream Rotor	0.532	1.000	0.904	0.999	1
- Downstream Rotor	0.544	1.022	0.917	1.014	0.997

3.2.1. Inline Rotors. The most studied turbine configuration is the case of a downstream turbine fully contained within another's wake [18,19,20,21]. The key result from several studies of real and simulated turbines is that the flow speed recovers faster behind two turbines than behind one turbine, resulting in a third turbine - placed directly in line with the first two turbines at the same spacing - producing more power than the second turbine. This effect is shown clearly in Fig. 7 of Gaumond et al. [21].

This is also reproduced by the CFD results. Referring to Fig. 2a and comparing the single rotor cases at 4D and 8D (left-most column, top two plots) with the same distances behind the second rotor in the inline case (i.e. 12D and 16D, second-left column, lowest two plots), the wake deficit is only marginally greater. Table 2 indicates that thrust, and hence rate of momentum extraction, by the downstream turbine is lower than for an isolated turbine and thus a marginally smaller deficit is expected. The subsequent rate of wake recovery is also higher and this is explained by the higher value of turbulence observed downstream of the second rotor (Fig. 2b).

Fig. 4 further illustrates this effect and shows how the different wake combination methods perform for this case at 16D downstream of the upstream rotor. The two grey dashed lines show the individual wakes observed for the stated downstream distance for each rotor if assumed to operate in isolation. The combined wake from the CFD is shown in black. The centreline velocity deficit behind the two rotors is marginally smaller than the case for the isolated wake 8D downstream of the second turbine. The downstream turbine operates at significantly lower thrust (42.4%) and thus momentum removal than the upstream or isolated turbine, and hence the deficit is reduced relative to the isolated case. The three wake combination methods applied to the independent wakes are shown as coloured lines. It is clear that none exactly match the actual combined wake (i.e. the black line).

This demonstrates why the largest or "worst" velocity deficit (red line in Fig. 4) is used by WindFarmer and OpenWind to combine wakes together [1,2]: this method results in the best agreement in validation studies (in which cases are selected carefully, e.g. by widening or narrowing the direction sector - and hence range of onset flow angles - within which data is considered [21]). Figs. 5-7 will illustrate that this is not the case when a line of rotors is not directly aligned with the flow direction.

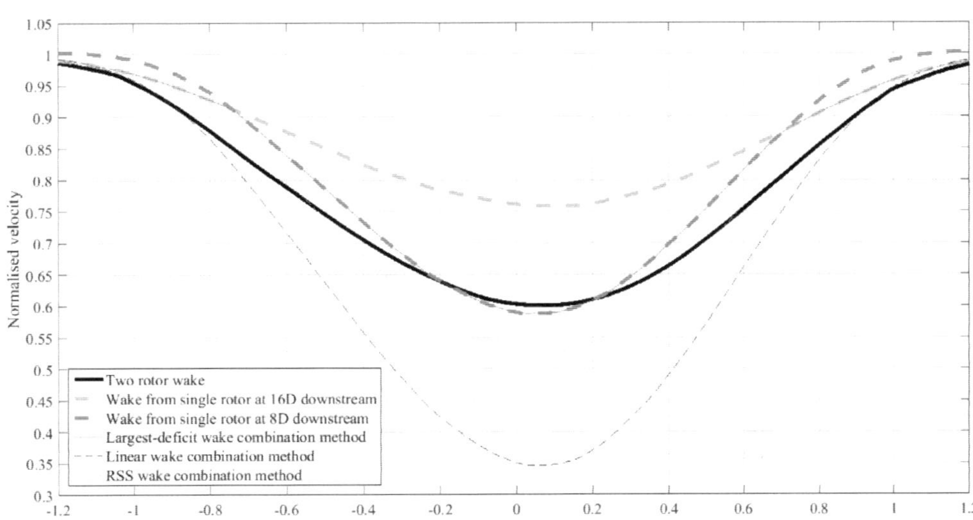

Figure 4. Normalised hub-height velocity profiles behind two in-line rotors, at 16D behind the first (8D behind the second) (black). Also shown are the wakes from the two rotors operating independently (thick lines, grey). Finally different combination methods, applied to the independent wakes, are shown (thin lines, colour). (CFD simulation).

Semi-empirical wake models could be altered to describe the enhanced recovery in the fully in-line case more accurately by using a variable near wake length model, such as that proposed by Sørensen et al [22].

3.2.2. Partly Offset Rotors. As can be seen in Table 3, the inflow to the second turbine in this case has been reduced, with a corresponding reduction in power for that turbine (93% of the single rotor). Further, as is clear from Fig. 2a, at 8D downstream of first turbine the onset flow to the second turbine is sheared horizontally (with a 21.6% difference in velocity across the rotor diameter). Fig. 5 demonstrates that the unphysical largest-deficit approach does not perform well when the swept area of the downstream rotor is only partly exposed to the wake of the upstream turbine. Indeed, the simulated effect of the two rotors is a much greater wake deficit over most of the width of the wake profile. It is postulated that this is because there is less energy available between the two wakes, resulting in recovery occurring over a longer distance as the energy must be obtained from the flow above and on either side of the aggregate wake (see Fig. 2a, second-from-right column).

A further deviation from the assumptions of commercial wake models can be observed in Fig. 2b for this configuration (second-from-right column). The TKE distribution in the downstream rotor wake is asymmetric (planes at 8D, 12D and 16D). This differs from the nearly symmetric distribution of the isolated rotor simulation, which more closely matches the standard eddy-viscosity wake models.

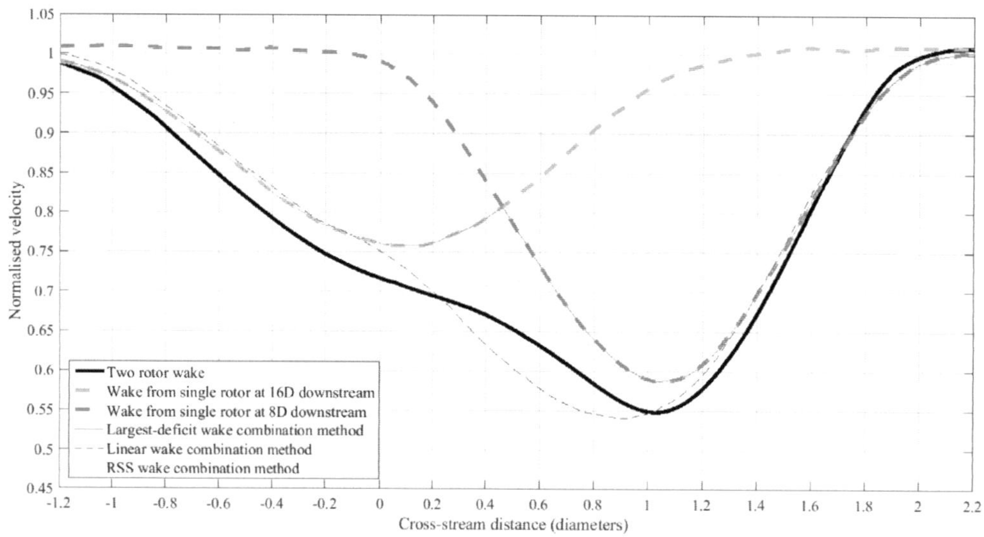

Figure 5. Normalised hub-height velocity profiles behind two partly offset rotors. Lines as described in Fig. 4 (CFD simulation).

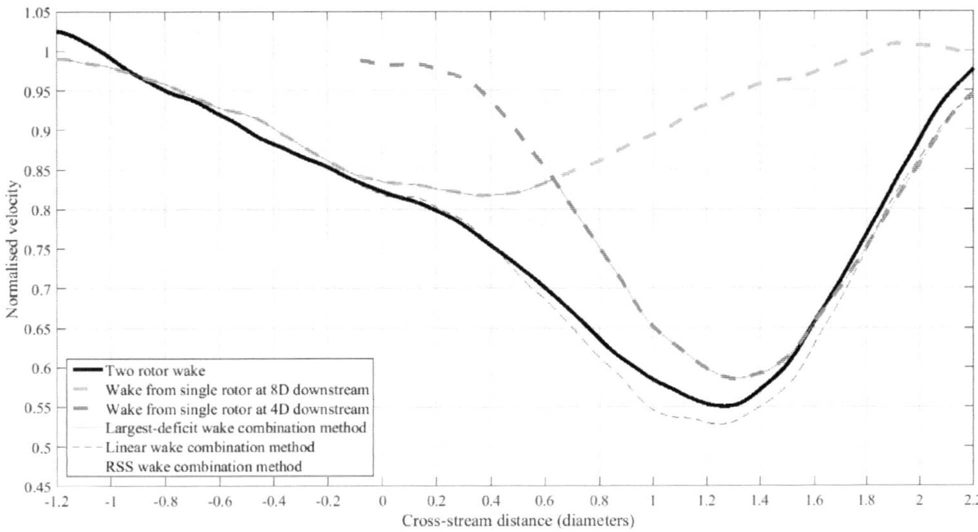

Figure 6. Normalised hub-height velocity profiles behind two partly offset rotors, at 8D behind the first (4D behind the second) (black). Also shown are the wakes from the two rotors operating independently (thick lines, grey). Finally different combination methods, applied to the independent wakes, are shown (thin lines, colour). (tank tests).

The effect is corroborated by the velocity measurements from the tank tests, as shown in Fig. 6 for the same configuration (except with a streamwise turbine separation of 4D rather than 8D, used due to the more rapid wake expansion associated with higher ambient turbulence of the tank tests).

This situation of rotors partly exposed to the wake of turbines located upwind occurs rather commonly in practice, so modification to the standard approach of superposition of single-wake models should be considered to enable better prediction of average energy yield over the life of a wind farm. If the direction sector in energy yield validation studies is chosen to be sufficiently wide (depending on the turbine layout) to average over both partly and fully in-line cases, this may result in such models providing improved long-term performance and behaving in a consistent manner when turbine positions are gradually altered for layout optimisation.

3.2.3. Fully Offset Rotors

In the fully offset case, the second rotor's wake - both velocity deficit and turbulence (right-hand columns of Figs. 2a and 2b respectively) - are only slightly altered from the single rotor case, since the two wakes do not interact strongly.

However, this case also demonstrates interesting physics. Fig. 7 shows that the upstream rotor's wake has been offset by approximately D/8, due to the pressure field created by the downstream rotor. Compared with the partly offset or fully in-line cases, the wake due to two fully offset rotors cannot therefore be expressed as a simple combination of single symmetric rotor wakes. This effect is further shown in Figs. 8 and 9.

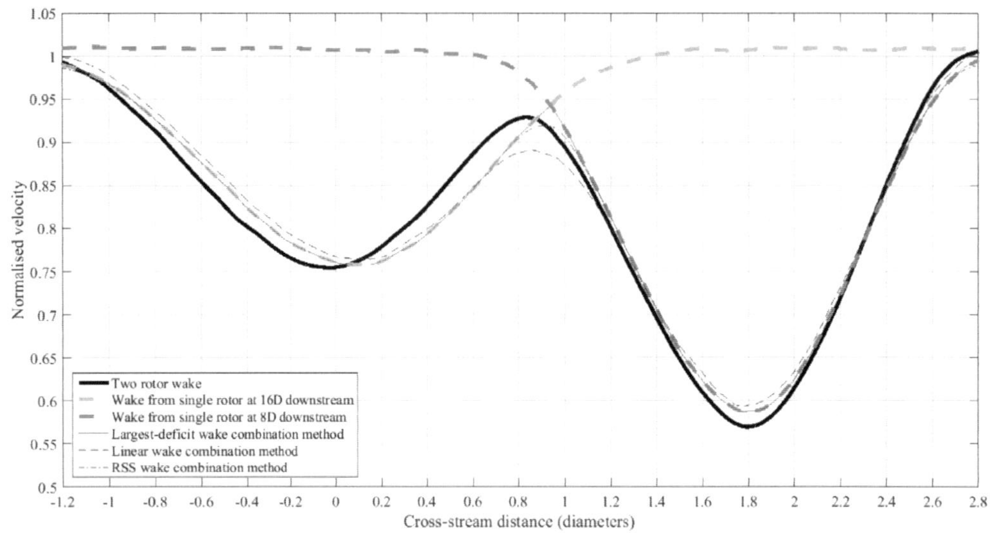

Figure 7. Normalised hub-height velocity profiles at 16D behind two fully offset rotors. Lines as described in Fig. 4 (CFD simulation).

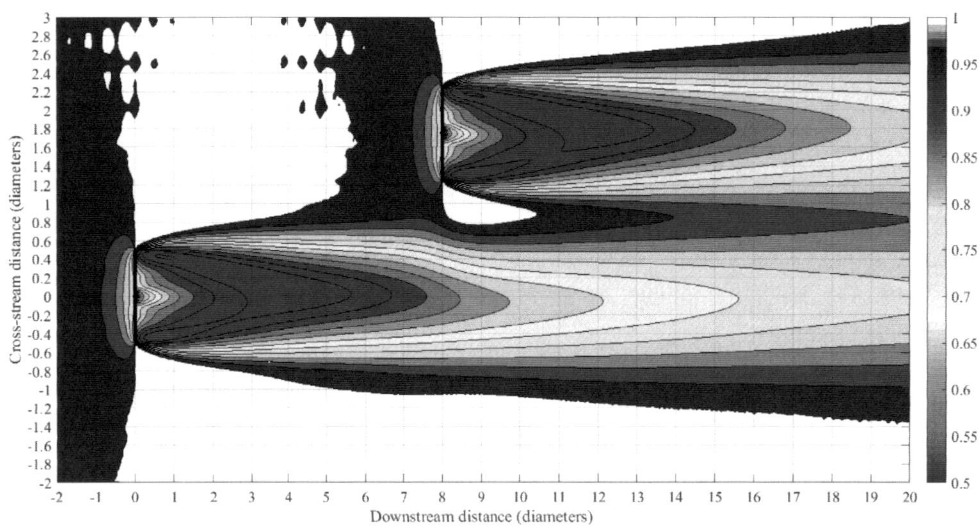

Figure 8. Map of normalised streamwise velocity at hub height for two fully offset rotors, flow incident from the left (CFD simulation).

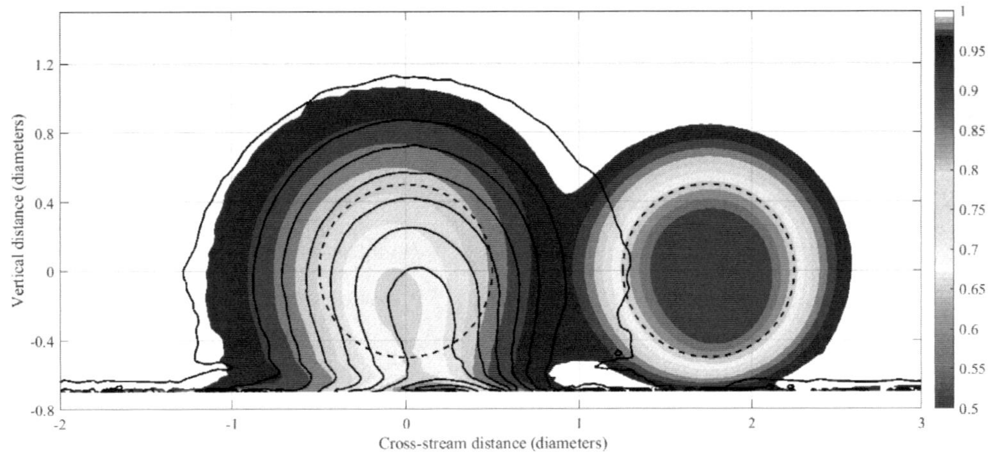

Figure 9. Normalised velocity cross-section at 12D downstream of the first of two fully offset turbines (coloured contours). The rotor areas are marked as black dashed outlines. The solid black contours indicate the locations of same velocity contours in the single turbine case, demonstrating how the upstream turbine's wake has been shifted and squeezed in the cross-stream direction (CFD simulation).

4. Conclusions

Investigations of configurations of one and two full-scale wind turbine rotors using CFD simulations, and corroborated by experiment, have shown that wake combination methods currently used in commercially available software for calculation of energy yield from wind farms neglect important aspects of the flow physics. The rate of wake recovery and rotor performance have been investigated for two rotors with the downstream rotor aligned with, partially offset and fully offset from the upstream rotor to assess typical configurations occurring within a wind farm during operation. Various wake model validation studies, typically concentrating on wind flow closely aligned with the axis of a row of turbines, have been previously used to justify a largest-deficit approach to combining individual wakes from semi-empirical or analytic models. However, this study has shown that, when the swept area of the downstream turbine is not fully enclosed in the wake of the upstream turbine, a much deeper wake deficit is produced than calculated for the same turbine in isolation, entailing significant over prediction of velocity of the two turbine wake for the largest-deficit and RSS combination approaches. Hence power output of turbines located downstream would be over predicted. When turbines are located just outside the wake of upstream turbines, this study has demonstrated emergent flow physics, particularly that the pressure field generated by the downstream rotor shifts the upstream wake laterally and compresses it.

These effects indicate that consideration should be given to implementing more than simplistic combination of pre-solved single turbine wakes when simulating wind farms for energy yield, particularly for the cases of partly offset wakes. An option would be to solve the flow equations of the Eddy Viscosity model developed by Ainslie, but with explicit consideration of the presence of upstream and nearby turbines.

A significant factor in the rate of deficit recovery is the turbulence of the surrounding flow. Improved agreement may be obtained if wake models accounted for onset wake turbulence for the in-line case, and for asymmetry between the adjacent wake turbulence and ambient turbulence for the offset wake configurations. A general approach could be to characterise the azimuthal variation of turbulence around the wake.

References

[1] AWS Truepower 2010 *OpenWind Theoretical Basis and Validation*

[2] DNV GL. 2011 *WindFarmer Theory Manual*

[3] EMD International 2011 *WindPRO Introduction to Wind Turbine Wake Modelling and Wake Generated Turbulence*

[4] Ott S, Berg J, Nielsen M. 2011 *Linearised CFD Models for Wakes*

[5] Jensen N O 1983 A Note on Wind Generator Interaction

[6] Ainslie J F 1988 Calculating the Flowfield in the Wake of Wind Turbines *J. Wind Eng. Ind. Aerodyn.*

[7] Machefaux E, Larsen G C and Murcia Leon J P 2015 Engineering models for merging wakes in wind farm optimization applications. *JPCS.*

[8] Jonkman J, Butterfield S, Musial W and Scott G 2009 *Definition of a 5-MW Reference Wind Turbine for Offshore System Development.*

[9] ANSYS Inc. 2013 *ANSYS Fluent 15.0 Theory Guide.*

[10] Menter F R 1994 Two-equation eddy-viscosity turbulence models for engineering applications. *AIAA J.* **32**

[11] Sørensen N M, Michelsen J A and Schreck S. 2002 Navier-Stokes predictions of the NREL phase VI rotor in the NASA Ames 80-by-120 wind tunnel. *In 2002 ASME wind energy symposium.* VA: American Institute of Aeronautics & Astronautics

[12] Mahu R and Popescu F 2011 NREL Phase VI rotor modeling and simulation using ANSYS Fluent 12.1. *Mathematical Modeling in Civil Engineering*

[13] Sanderse B, van der Pijl S P and Koren B 2011 Review of computational fluid dynamics for wind turbine wake aerodynamics. *Wind Energy*

[14] Hansen M O L. 2008 *Aerodynamics of Wind Turbines*, Second Edition London: Earthscan.

[15] Stallard T, Feng T and Stansby P K 2015 Experimental study of the mean wake of a tidal stream rotor in a shallow turbulent flow *J. Fluids Struct*

[16] Wagner R, Cañadillas B, Clifton A and Wagenaar J W 2014 Rotor equivalent wind speed for power curve measurement – comparative exercise for IEA Wind Annex 32 *JPCS*

[17] Vermeer L J, Sørensen J N and Crespo A 2003 Wind turbine wake aerodynamics. *Progress in Aerospace Sciences*

[18] Nygaard N G 2014 Wakes in very large wind farms and the effect of neighbouring wind farms *JPCS*

[19] Porté-Agel F, Wu YT and Chen CH 2013 A Numerical Study of the Effects of Wind Direction on Turbine Wakes and Power Losses in a Large Wind Farm *Energies*

[20] Westerhellweg A, Cañadillas B, Kinder F and Neumann T. 2012 Wake measurements at alpha ventus - dependency on stability and turbulence intensity

[21] Gaumond M, Réthoré P E, Bechmann A, Ott S, Larsen G C, Peña A, et al. 2012 Benchmarking of wind turbine wake models in large offshore wind farms. In Torque 2012

[22] Sørensen J N, Mikkelsen R, Sarmast S, Ivanell S and Henningson D 2014 Determination of Wind Turbine Near-Wake Length Based on Stability Analysis *JPCS*

Preliminary assessment of the variability of UK offshore wind speed as a function of distance to the coast

Rolando Soler-Bientz[1] and Simon Watson[1]

[1]Centre for Renewable Energy Systems Technology. Wolfson School of Mechanical, Electrical and Manufacturing Engineering, Loughborough University, LE11 3TU, UK.

R.Soler-Bientz@lboro.ac.uk, sbientz@msn.com

Abstract. In the UK, there is an interest in the expected offshore wind resource given ambitious national plans to expand offshore capacity. There is also an increasing interest in alternative datasets to evaluate wind seasonal and inter-annual cycles which can be very useful in the initial stages of the design of wind farms in order to identify prospective areas where local measurements can then be applied to determine small-scale variations in the marine wind climate. In this paper we analyse both MERRA2 reanalysis data and measured offshore mast data to determine patterns in wind speed variation and how they change as a function of the distance from the coast. We also identify an empirical expression to estimate wind speed based on the distance from the coast. From the analysis, it was found that the variations of the seasonal cycles seem to be almost independent of the distance to the nearest shore and that they are an order of magnitude larger than the variations of the diurnal cycles. It was concluded that the diurnal variations decreased to less than a half for places located more than 100km from the nearest shore and that the data from the MERRA2 reanalysis grid points give an under-prediction of the average values of wind speed for both the diurnal and seasonal cycles. Finally, even though the two offshore masts were almost the same nearest distance from the coast and were geographically relatively close, they exhibited significantly different behaviour in terms of the strength of their diurnal and seasonal cycles which may be due to the distance from the coast for the prevailing wind direction being quite different for the two sites.

1. Introduction

Wind power has seen a rapid expansion in capacity in recent years; the global capacity has more than quadrupled since the beginning of this century [1]. Offshore wind generation in particular has the potential to make a substantial contribution to future energy needs as the wind in the marine environment is generally higher than onshore and less variable. In the UK in particular, there is an interest in the expected offshore wind resource given ambitious national plans to expand offshore capacity. Since the end of 2014, the UK had the largest capacity of offshore wind generation worldwide [1].

The behavior of the wind has a critical impact on the economic viability of an offshore wind farm. However, the assessment of the wind climate in a marine environment is particularly challenging due to the difficulty and high cost of installing and maintaining meteorological masts, particularly as measurements are exceeding 100m above sea level in many cases. Observations from ships are a potential source of information but they suffer from ship motion and are sparse in space and discontinuous in time. Satellite scatterometers offer spatially well-distributed data but short records and data are irregular in time and space [2]. Measurements from anchored buoys and meteorological towers are sparse offshore in time and space and influenced by local effects. In addition, such measurements may be fine for site specific information, but it is challenging to extrapolate such data beyond the measurement site [3], [4] and [5]. In addition, meteorological masts have a high cost (>£10million). There is, therefore, an increasing interest in alternative datasets to minimize the requirement for offshore masts. These alternatives include global reanalysis datasets. Reanalysis data are available at different heights over long term time periods facilitating the interaction between wind generation models and climate models [6] and [7].

The use of reanalysis dataset as input time series to simulate wind-power applications [8] could be very useful on the aggregated level to model wake losses and to correct for systematic errors on long and short time scales [9]. In particular, the Modern-Era Retrospective analysis for Research and Applications (MERRA) reanalysis dataset has been used to produce wind-power generation time-series, which have strong correlation with the data measured in a power system [10]; to describe seasonal and inter-annual variability and to quantify the frequency of extreme events [11].

The Modern-Era Retrospective analysis for Research and Applications, Version 2 (MERRA2) is used in this paper [12][13] as a source of offshore wind speed data around the UK. The MERRA2 reanalysis replaced that of MERRA introducing among other things several improvements in the assimilation system, in the model and in the spatial resolution. MERRA2 uses the upgraded version 5.12.4 of the Goddard Earth Observing System Model, Version 5 (GEOS-5). MERRA2 is managed by the NASA Goddard Earth Sciences (GES) Data and Information Services Center (DISC) [12].

The MERRA2 grid has 576 points in the longitudinal direction and 361 points in the latitudinal direction, corresponding to a resolution of 0.625°×0.5° (a grid cell of 5/8° longitude by 1/2° latitude, known as native resolution). It also corresponds to a spatial resolution of approximately 50 km in the latitudinal direction.

Data from MERRA2 are available in a number of different file types which classify the variables according to different time and spatial scales. In particular, in this paper, the standard name of the assimilated GEOS-5 MERRA-2 data collection used was: **tavg1_2d_slv_Nx** (M2T1NXSLV) which corresponds to time averaged data with an averaging interval of 1 hour (**tavg1**) containing two dimensional fields (**2d**) in a single level (**slv**) with native resolution (**N**) just in the horizontal plane (**x**) [13].

Using the MERRA2 dataset and data from offshore masts, this paper examines the average wind speed, diurnal and seasonal cycles as a function of distance from the coast. An empirical expression is introduced to represent the average wind speed relative to the distance from the coast for the whole study period and for each season. Results for the MERRA2 dataset are compared with measurements from two offshore masts.

2. Study region

As was stated above, in this paper, we used wind speed data from the MERRA2 reanalysis dataset and data recorded from offshore mast stations. The two masts, London Array (LA) and Greater Gabbard (GG), are located offshore of the South East coast of England (see the two small white arrows in Figure 1), while the MERRA2 grid points were selected as the closest grid points to the offshore masts, plus an additional adjacent matrix of 5 x 5 points in order to cover distances to the shore up to approximately 250 km. A total of 35 MERRA2 grid points were initially used in this research which were located in Figure 1 with pins and ID numbers.

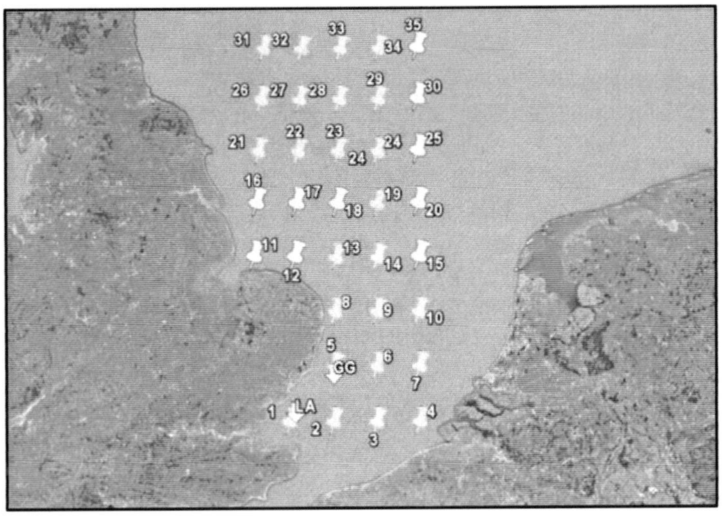

Figure 1. Map of the UK showing the study region. The small arrows indicate the location of the offshore masts: London Array (LA) and Greater Gabbard (GG). The pins represent the positions of the 35 MERRA2 grid points used in this paper.

Even though the MERRA2 datasets runs from 1980 up to the current date (~36 years), for the purpose of this research, a period of 6 years was selected as the study period, concurrent with the period of wind data available for the two offshore mast datasets. Wind speed at 50m height was used from the MERRA2 dataset and for the offshore masts the wind speed data available closest to 50m height were used. In order to reduce the number of simultaneous values to be plotted in a single chart, the results shown in the next section were analysed for the 35 MERRA2 grid points and 10 of them representing the whole range of variations of the study parameters were selected to be plotted along with the two offshore masts results. These 12 points will be referred in this paper as the "selected study points" and they are represented in Figure 1 by white pins and arrows. Table 1 shows for these selected study points their geographical coordinates, distance to the nearest coast and the direction of the nearest coast.

Table 1. Geographical coordinates, distance and direction to the nearest coast sorted by the distance to the nearest coast.

ID number	Latitude [Degree]	Longitude [Degree]	Nearest coastal distance [km]	Nearest coastal direction [Degree]
11[a]	53	0.625	3	167
12[a]	53	1.25	6	188
LA[b]	51.5940	1.38563	26	304
16[a]	53.5	0.625	28	245
GG[c]	51.9434	1.92210	29	344
17[a]	53.5	1.25	60	197
18[a]	53.5	1.875	73	209
15[a]	53	3.125	101	252
20[a]	53.5	3.125	117	113
25[a]	54	3.125	144	126
30[a]	54.5	3.125	182	140
35[a]	55	3.125	223	143

[a] Selected MERRA2 grid point
[b] London Array offshore mast
[c] Greater Gabbard offshore mast

3. Diurnal and seasonal cycles

The diurnal cycles computed over the whole study period for the selected study points have been plotted in Figure 2. In the case of the offshore masts, even though both masts are located approximately at the same distance from the coast, the London Array (LA) mast shows more influence from the land (manifested in the larger magnitude of the diurnal pattern) than the Greater Gabbard (GG) mast, particularly during the morning. This is because the coastal distance for the prevailing south-westerly wind direction is half that for the LA mast compared to the GG mast (see Figure 1 and Figure 3).

In the case of the MERRA2 grid points, the points 11, 12, 16, 17 and 18 show a clear influence from the land diurnal cycle. The points 15, 20 and 25 receive a smaller influence from the land diurnal cycle while the points 30 and 35 show almost no influence from the land diurnal cycle.

In general, it can be seen that the averages of wind speed over the diurnal cycle are smaller for the MERRA2 grid points located at similar distance of the shore than the averages recorded by the offshore mast stations. As was mentioned earlier the rest of the 35 MERRA2 grid points showed similar trends relative to their distance from the coast and are not presented for the purposes of clarity.

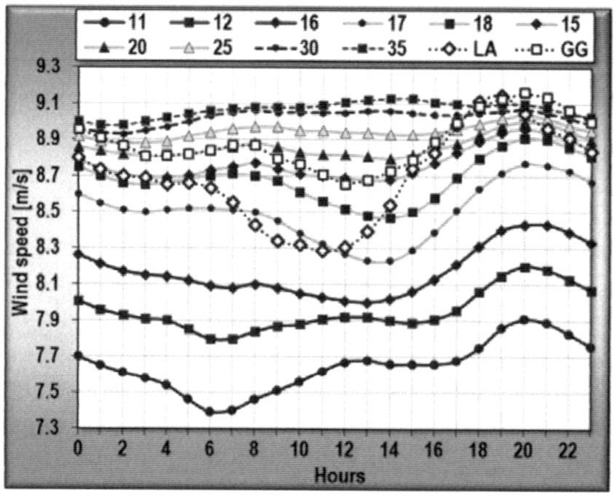

Figure 2. Average diurnal patterns of the wind speed over the whole study period of the selected MERRA2 grid points and the two offshore masts (LA and GG).

The frequency distribution of the wind direction is presented in Figure 3 below. It can be seen that the direction of the prevailing wind is around the sector located at 240 degrees. This could have some influence over the wind pattern in particular for the sites closer to the shore, as was already pointed out. This prevailing trend is even stronger for the two offshore masts which also show smaller peaks around 30-60 degrees not observed in the reanalysis data.

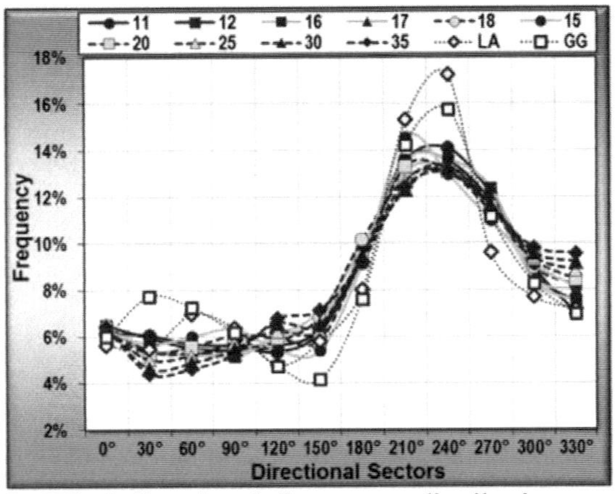

Figure 3. Wind directional frequency distribution over the whole study period for the selected MERRA2 grid points and the two offshore masts (LA and GG).

Figure 4 below presents the average wind speed for each season for every selected study point over the whole study period. As can be seen the average wind speed decreases in the Summer season and reaches a maximum over the Autumn-Winter seasons. Although the absolute values of the wind speed vary as a function of the distance to the coast, the magnitude of the seasonal cycles does not show much variation for the MERRA2 data points.

In the case of the offshore masts, there was a significant decrease in the average wind speed of the LA mast for the Autumn season which could be related with some particular local conditions around this mast location. In this case, as well as in the case of the diurnal cycles, the average values of the wind speed for the selected MERRA2 grid points are smaller than the averages wind speed recorded by the offshore mast stations.

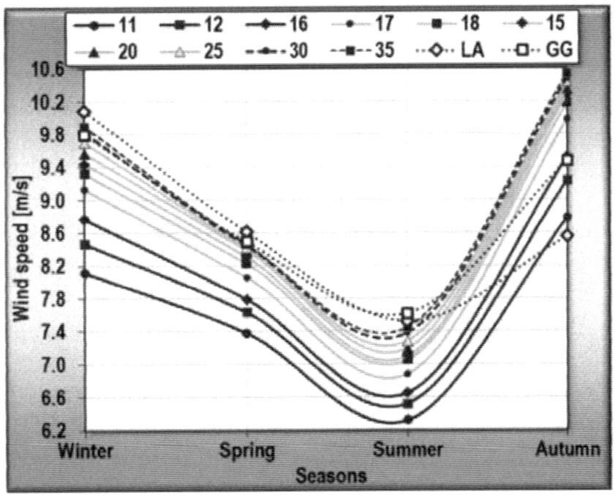

Figure 4. Average of the wind speed for each season over the whole study period of the selected MERRA2 grid points and the two offshore masts (LA and GG).

In order to better show the degree of variability of the diurnal and seasonal cycles over the whole study period, a parameter was defined as the difference between the maximum and minimum divided by the average wind speed over the study period, see equation (1) below.

$$\frac{\Delta U}{U_{avg}} = \frac{(U_{max} - U_{min})}{U_{avg}} \quad (1)$$

Figure 5 presents the parameter defined by equation (1) over the diurnal cycle for the MERRA2 grid points and the offshore masts. It can be seen that, as was mentioned before, the variations within a cycle are larger for the points located closer to the shore than those located further away. Specifically, the points located more than 150km from the nearest coast show less than the half of the variation of the sites located up to around 90km from the nearest coast. The much larger fractional variation in the diurnal cycle for LA when compared with the reanalysis data is apparent.

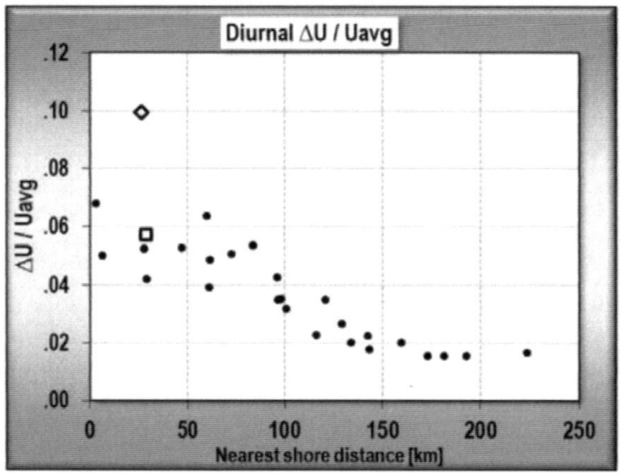

Figure 5. Variations of the average wind speed for the diurnal cycle over the whole study period for the MERRA2 grid points and for the offshore masts: LA (rhombus) and GG (square).

In the case of the seasonal patterns, Figure 6 presents the variations of the average wind speed as defined in equation (1) for the MERRA2 grid points and the offshore masts. It can be seen that other than the two points less than 30km from the coast that the seasonal cycles seem to be almost independent of the distance to the nearest shore, though interestingly there is some evidence for a peak in amplitude at around 70km from the coast What is clear is that the fractional variation seen at the offshore masts is significantly less than that of the reanalysis data.

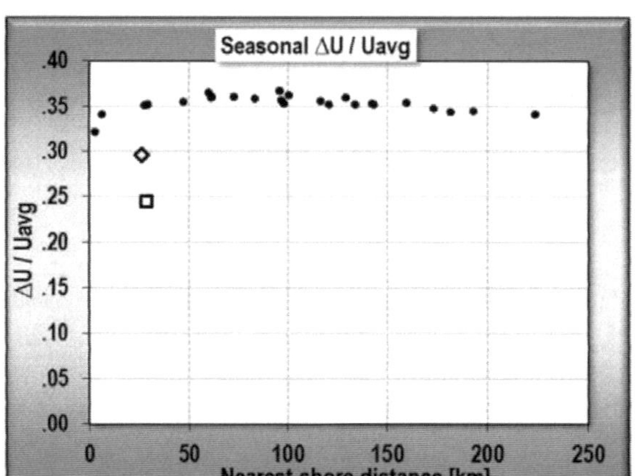

Figure 6. Variations of the average wind speed for the seasonal cycle over the whole study period for the MERRA2 grid points and the offshore masts: LA (rhombus) and GG (square).

4. Analysing average wind speed as a function of the coastal distance

The average wind speed as function of the distance to the nearest coast for the MERRA2 grid points and the offshore mast have been plotted in Figure 7 and Figure 8 for the whole period and for each season, respectively.

It can be seen that there is a relationship between average wind speed and distance. There is evidence of the average wind speed reaching an asymptotic value at large distances. Although all the points considered are offshore, clearly, the average wind speed at the coast would be expected to be non-zero. With this in mind, an empirical expression for the average long term offshore wind speed as a function of distance from the nearest coast is proposed of the form:

$$u = u_0 * \left[a - \left(\frac{x_0}{x+x_0}\right)^{1/2}\right] \quad (2)$$

Table 2. Values of $\mathbf{u_0}$ and $\mathbf{x_0}$ used to fit equation (2) to the average 50m wind speeds showed in in Figure 7 and Figure 8.

	u_0 [m/s]	x_0 [km]	R^2	Estimated bias
Whole period	2.59	55.94	0.9998	-0.0010
Winter	2.71	34.49	0.9997	-0.0021
Spring	2.14	1.72	0.9997	0.0005
Summer	1.84	1.17	0.9985	0.0020
Autumn	2.79	9.92	0.9994	0.0070

where **u** represents the average wind speed at a distance **x** from the nearest coast line, **a** is a dimensionless constant equal to 4. The empirical parameter $\mathbf{x_0}$ is a distance scale which reflects how quickly the wind speed adjusts to the offshore conditions. The value of $\mathbf{u_0}$ and $\mathbf{x_0}$ were fitted for the whole study period and by season using a least squared fit in the Mathematica programming evironment. The fits are shown as dotted lines in Figure 7 and Figure 8; and the values of $\mathbf{u_0}$ and $\mathbf{x_0}$ are given in Table 2 along with the coefficient of determination R^2 and the estimated bias for each fitted value.

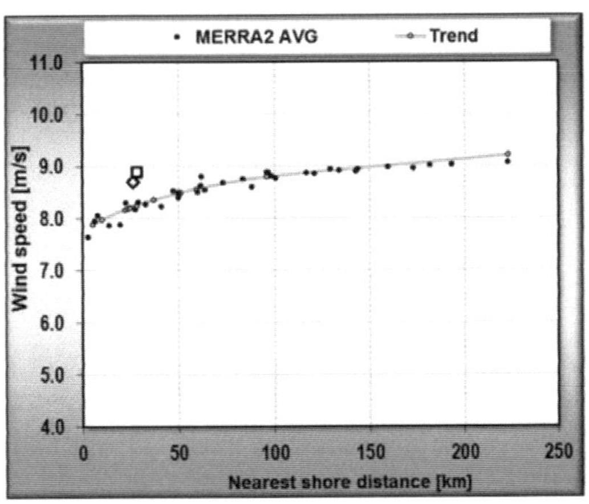

Figure 7. Distribution of the averages wind speed relative to the nearest shore distance and the interpolation function, as represented by equation (2) and table 2, over the whole study period. The average wind speed for the offshore masts is represented for LA by a rhombus and for GG by a square.

The average wind speed would equal $\mathbf{u_0}$ at the coast and would asymptote at large distances to $4\mathbf{u_0}$, however, it should be stressed that the fit has only been made between the closest site at 3km and the

Journal of Physics: Conference Series **749** (2016) 012004 doi:10.1088/1742-6596/749/1/012004

furthest at 223km, so would only strictly be valid in this range. Although it can be seen that there is some degree of scatter for the values at each site around the trend lines, the empirical expression in equation (2) does provide a reasonably good fit to the data over a range of distances between ~5km and ~220km from the coast, with the possible exception of the summer season.

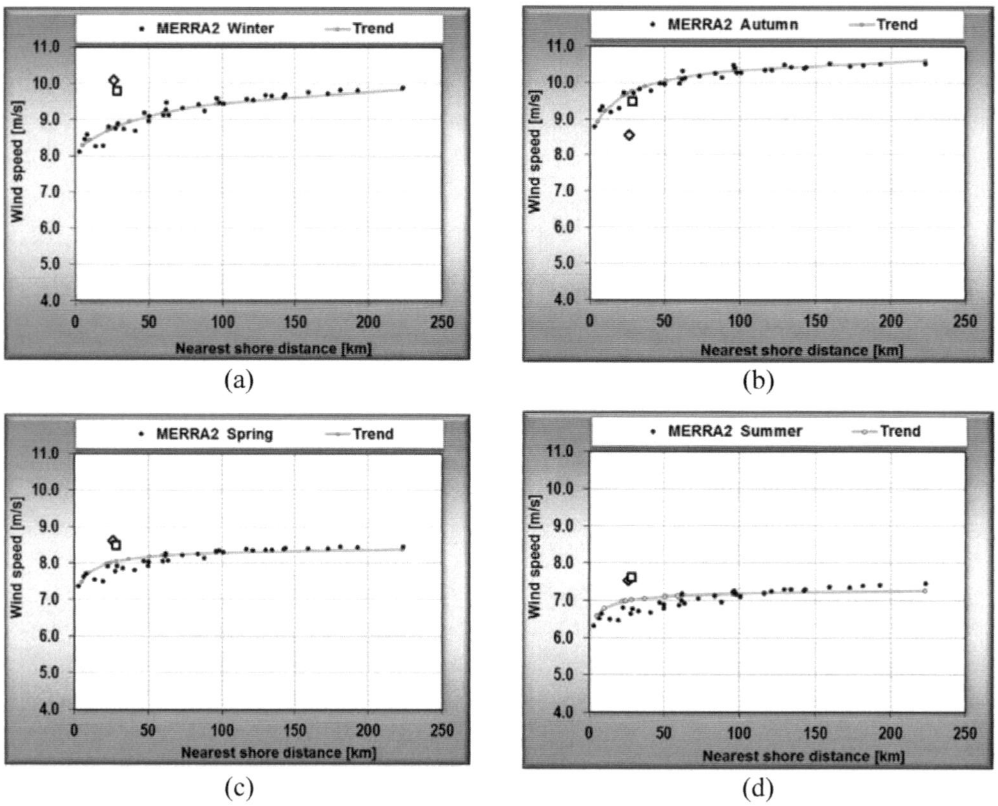

Figure 8. Distribution of the average wind speed relative to the nearest shore distance and the proposed interpolation curves, following equation (2) with the values listed in table 2, over each particular season (a) Winter, (b) Autumn, (c) Spring and (d) Summer. The average wind speed for the offshore masts is represented for LA by a rhombus and for GG by a square.

Figure 7 shows clearly that the reanalysis data significantly under-predicts the overall average wind speed at both LA and GG. Looking at Figure 8, this is also true by season with the exception of the autumn where there is an over-prediction.

5. Conclusions

This paper has used 50m long term reanalysis data from MERRA2 in order to study the spatial characteristics of the offshore wind speed close to the UK's development sites. A comparison was made between these data and measurements from two offshore masts. Trends in mean wind speed, diurnal and seasonal variation were studied as a function of distance from the coast. A relatively simple two-parameter empirical expression was identified to estimate wind speed based on distance from the coast giving a reasonable fit to the data, though this needs to be validated further by analysing data from a larger number of sites and a greater range of distances from the coast.

From the results presented in this paper, it can also be concluded that:

- The position of the coast for the prevailing wind direction seems to have a dominant effect over the magnitude of the diurnal cycle This was clear from looking at the results for London Array and

Greater Gabbard where the distance to the coast for the prevailing wind direction is significantly different but the nearest distance to the coast is almost the same.

- Data from the MERRA2 grid points located more than 150 km from the nearest coast show a magnitude in the fractional diurnal variation which is less than half of that for grid points located 90km form the nearest coast.
- The data from the MERRA2 grid points used in this research consistently seem to under-predict the measured of wind speed overall and for the diurnal and seasonal cycles, with the exception of the autumn season.
- The fractional variation in magnitude of the seasonal cycles seem to be independent of the distance to the nearest shore for distances above the 30km, though there is some evidence for a peak at around 70km.

References

[1] Global Wind Energy Council (2015). Global Wind Statistics 2014. Global Wind Energy Council, Brussels, Belgium. Available from http://www.gwec.net/wp-content/uploads/2015/02/GWEC_GlobalWindStats2014_FINAL_10.2.2015.pdf, last accessed 12th March 2016.

[2] Bentamy, A. and D. Croize-Fillon, Spatial and temporal characteristics of wind and wind power off the coasts of Brittany. Renewable Energy, 2014. 66(0): p. 670-679. 0960-1481. http://dx.doi.org/10.1016/j.renene.2014.01.012.

[3] Foley A M, Leahy P G, Marvuglia A and McKeogh E J 2012 Current methods and advances in forecasting of wind power generation *Renewable Energy* **37** 1-8.

[4] Jung J and Broadwater R P 2014 Current status and future advances for wind speed and power forecasting *Renewable and Sustainable Energy Reviews* **31** 762-77.

[5] Okumus I and Dinler A 2016 Current status of wind energy forecasting and a hybrid method for hourly predictions *Energy Conversion and Management* **123** 362-71.

[6] Pryor S C, Barthelmie R J and Schoof J T 2006 Inter-annual variability of wind indices across Europe *Wind Energy* **9** 27-38.

[7] Rose S and Apt J 2015 What can reanalysis data tell us about wind power? *Renewable Energy* **83** 963-9.

[8] Göransson L and Johnsson F 2013 Cost-optimized allocation of wind power investments: a Nordic–German perspective *Wind Energy* **16** 587-604.

[9] Olauson J and Bergkvist M 2015 Modelling the Swedish wind power production using MERRA reanalysis data *Renewable Energy* **76** 717-25.

[10] Kubik M L, Brayshaw D J, Coker P J and Barlow J F 2013 Exploring the role of reanalysis data in simulating regional wind generation variability over Northern Ireland *Renewable Energy* **57** 558-61.

[11] Cannon D J, Brayshaw D J, Methven J, Coker P J and Lenaghan D 2015 Using reanalysis data to quantify extreme wind power generation statistics: A 33 year case study in Great Britain *Renewable Energy* **75** 767-78.

[12] The MERRA-2 data are available online through the Goddard Earth Sciences (GES) Data and Information Services Center (DISC) at http://disc.sci.gsfc.nasa.gov/mdisc/).

[13] Global Modeling and Assimilation Office (GMAO) (2015), *tavg1_2d_slv_Nx: MERRA-2 2D single level, time averaged 1-hourly (p-coord, 0.625x0.5), version 5.12.4*, Greenbelt, MD, USA: Goddard Space Flight Center Distributed Active Archive Center (GSFC DAAC), Accessed **11/07/2016** at doi: 10.5067/VJAFPLI1CSIV.

[14] April 2016, The Crown Estate; *The Marine Data Exchange*; at http://www.marinedataexchange.co.uk/.

Journal of Physics: Conference Series **749** (2016) 012005 doi:10.1088/1742-6596/749/1/012005

Orientation correction of wind direction measurements by means of staring lidar

Michael Schmidt, Juan José Trujillo, Martin Kühn

University of Oldenburg, ForWind – Institute for Wind Energy Research, Küpkersweg 70, 26409 Oldenburg, Germany

E-mail: `michael.schmidt@uni-oldenburg.de`

Abstract. In spite of the efforts made at the time of installation of wind vanes or ultrasonic anemometers (Sonic), there is always a remaining uncertainty of several degrees in the absolute north of such sensors. In this research a method is presented to reduce the azimuthal orientation error of wind direction sensors by means of Doppler Lidar measurements. The method is based on the comparison between the conventional sensor and a distant long range lidar pointing to it in staring mode. By comparing their line-of-sight wind speeds any misalignment between both systems can be estimated more accurately. This method was applied in an measurement campaign in the offshore wind farm alpha ventus next to the meteorological mast FINO 1. The maximum alignment error of a Sonic was reduced to below $\pm 1^{\circ}$. This accurate alignment has asserted, that no bias exists between Lidar and Sonic wind speed measurements.

1. Introduction

At present the wind energy community is experiencing a need for more accurate orientation of wind direction sensors. For instance for testing new wind farm control techniques with active wake control as proposed by [1, 2] . Lately, Trujillo et al. have evidenced for instance the need of higher wind direction accuracy to study the effects of yaw misalignment on wake deviation [3]. Furthermore, Lidar measurements need accurate wind direction information, as they only measure the projection of the wind speed onto the Lidar's line of sight: the "line-of-sight speed" V_{los}. This speed has to be transferred to wind speed V using the wind direction α. The last is usually measured by wind vanes or ultrasonic anemometers (hereinafter referred to as "Sonic"). For some time, also "Velocity-Azimuth-Display-" (VAD-) Lidar's provide – next to the vertical wind profile – wind direction information.

Typically, wind direction data are subject to biases of several degrees. Nevertheless, this is often overlooked or in the most cases accepted as a practical limitation consequence of lack of very accurate means at installation time. Unfavourable mast structures as well as other local (e.g. weather) conditions often impede the orientation of devices exactly to North. Particularly difficult is the installation of measurement devices in offshore conditions, where no object is given to serve as reference point.

It seems, that little attention has been paid so far on the negative impact that poor wind direction sensor alignments have not only, but also on the quality of Lidar data. It is almost impossible to find reports on efforts to align direction data in literature. An exception is the use of remote sensing techniques to assess alignment errors. For instance A. Westerhellweg et al. [4], have used ground-based Lidars to estimate the wind direction using the VAD technique. This relies on a sort of average on a circle around the sensor in study, which in typical setups has a diameter equal to the height at which the sensor is installed. Due to their volumetric character, VAD measurements depend on high uniformity of the flow in that area and any non-uniformity – especially occurring onshore – will generate some bias. A direction data bias occurs as well, of course, if the VAD Lidar itself is misaligned.

In order to circumvent such alignment issues, we propose – in contrast to the VAD technique and other sensor alignment attempts – a long-range Lidar. Operated in staring mode, this Lidar is pointing from a far distance to the sensor to be aligned making quasi one-point-measurements at the location of the sensor.
With this the accuracy of both, the orientation of direction sensors, and the estimation of the staring mode Lidar wind speed bias can be improved substantially.

At first, we show the method, then we apply it to real data and finally we demonstrate it's performance in terms of wind speed biases between Sonic and Lidar data.

2. Method

2.1. Staring velocity azimuth display

We propose a method based on the analysis of the magnitude of the velocity measured in the beam direction of a staring lidar compared to the wind direction. This resembles the classical velocity azimuth display (VAD) technique, therefore we call the method staring velocity azimuth display (SVAD). In the following we explain the VAD technique to illustrate our approach.

A ground-based VAD-lidar interrogates the wind by describing a vertical cone with vertex at the lidar system. At a defined height the measurement domain is given by discrete points on a circle. These are covered sequentially in some period of time, ranging from some seconds to some minutes. The mean wind vector on that circle is estimated with the VAD technique, where two variables are plotted against each other: on the horizontal axis the magnitude of the horizontal projection of the measured line-of-sight wind speed at each point, and on the vertical axis the azimuthal position of the laser beam. In an uniform flow this plot reveals a sinusoidal shape from which the magnitude and direction of the wind can be derived. In effect, these are obtained from the amplitude and the phase shift of a fitted cosine function, respectively. See an example of the conical scan and the resulting sinusoidal function as presented by C. Werner [5] in Fig. 1.

If we let the lidar pointing fixed into one azimuthal direction, then it is not possible to estimate neither the wind speed, nor the wind direction. However, if at least a wind direction signal is given we can reconstruct the instantaneous wind vector. The relation of these two signals can be analysed also in long time periods. Then, the behaviour of the line-of-sight wind speed can be observed with respect to all wind directions. If then data are selected for a constant magnitude of the wind vector, then the plot of these two signals shows also a sinusoidal shape:

$$V_{los}(\alpha) = V \cdot cos(\alpha) \tag{1}$$

A comparison of two measurement techniques, that measure the wind at the same point, but having different orientations, would show a phase shift γ between their V_{los}-sinusoidals (Fig. 2). The angle Φ is necessary and sufficient to transfer the cosine function from the ("mathematical") Cartesian to the ("meteorological", wind-direction providing) compass system (0° means "North"). The origin of both systems is the measurement point (here the centre of

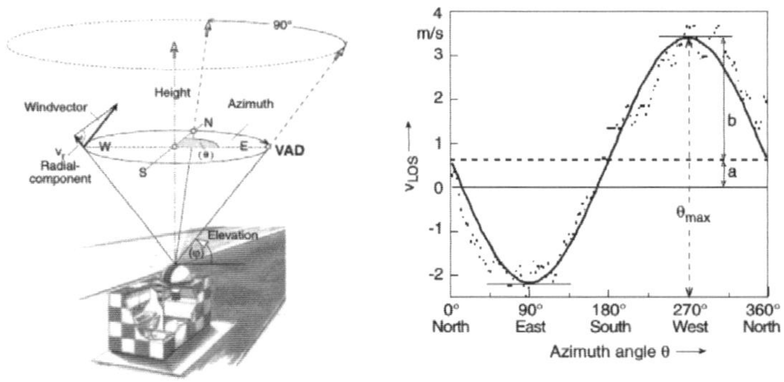

Figure 1. Left: sketch of the VAD-Lidar scan technique. Right: the measured line-of-sight speeds are arranged on a sinusoidal curve [5].

the Sonic device, equal to the centre of that Lidar beam range gate, that is the closest one to the Sonic). To the same time, Φ is used here to shift the V_{los}-maximum to the given line-of-sight-direction. It is easily seen, that one angle is sufficicient for both intentions: a cosine function can only be shifted to the left ($\Phi > 0$) or to the right ($\Phi < 0$), according to: the direction sensor is turned to the left (North-west), or right (North-east).

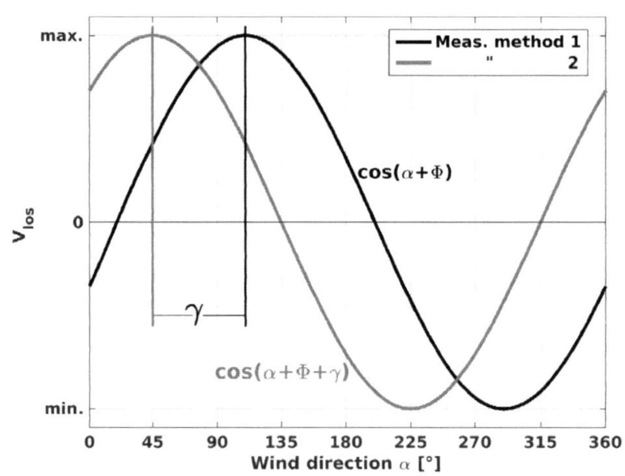

Figure 2. The V_{los} of two measurement devices, orientated to different direction, are shifted by the angle γ. The angle Φ is explained in the text.

If the position of the V_{los}-curve of one of the techniques is accepted as to be true (because the line-of-sight is accurately known), this curve serves as a reference. Then, a shift γ denotes the misalignment of the other instrument. We mention, that the described method would also work, if wind speed data from one or both measurements are biased. Not the maximum V_{los}, but only their phases are of interest here.

In our configuration a long-range Lidar serves as reference, as the line-of-sight direction can be estimated very accurately. It points to a second instrument with an unknown azimuthal bias, which is a Sonic in our case. For other configurations the second direction measurement method could also be a combined cup anemometer and wind vane, or even a ground-based VAD-Lidar.

2.2. Determination of the phase shift

As already mentioned in section 2.1, the V_{los}-data of both instruments (in our example staring mode Lidar and Sonic) have to be categorized ("binned“) by wind speed magnitude to reveal their direction-depending sinusoidal character. In other words: there is one sinusoidal for each bin.

To have satisfying numbers of V_{los}-data in each wind speed bin, we choose the binning $0.5 - 1.5m/s, 1.5 - 2.5m/s, ..., 30.5 - 31.5m/s$ (number of bins is n=30). Then, for each bin, the sinusoidal function

$$V_{los} = \overline{V} \cdot cos(\alpha + \Phi + \gamma) \quad (2)$$

has to be fitted to the V_{los}-data of method 2.

In equation 2, $\overline{V}$ denotes the centres of the wind speed bins, Φ is fixed (was introduced to shift the expected maximum V_{los} to the line of sight in compass system), and γ is the fit parameter. Non-zero-γ's indicate, that sinusoidals (and maxima) of method 2 are not at the expected position, but shifted with respect to the reference method's sinusoidals. The reference method's V_{los}-data don't have to be fitted, as it's sinusoidal is fixed for all bins: maximum is at line-of-sight, so γ of the reference method is 0 always.

We get slightly different γ's for the wind speed bins, because the V_{los} data must not exactly lie on the sinusoidals, as each wind speed bin has a finite width (e.g. bin width $1m/s$ allows distances of V_{los}-data to their sinusoidals of $< \pm 0.5m/s$). In this way, the finite width of wind speed bins introduces a statistical uncertainty to the method. We propose to calculate the weighted average of the γ's using

$$\overline{\gamma} = \frac{\sum_{i=1}^{n} w_i \cdot \gamma_i}{\sum_{i=1}^{n} w_i}. \quad (3)$$

As weights w_i we propose the number of V_{los}-data in the n wind speed bins. The weight-averaged $\overline{\gamma}$ is stated as misalignment of sensor 2 with respect to sensor 1 (reference, precisely aligned by the well-known line-of-sight, here: direction of the long-range Lidar pointing to the Sonic).

2.3. Re-calculating Sonic and Lidar data

Now that the size of sensor 2's misalignment is found, $\overline{\gamma}$ is used to correct the data. To align Sonic data, a rotation matrix (fixed coordinate system, vector (u,v) rotates counter-clockwise, if $\gamma > 0$) is used:

$$\begin{pmatrix} u \\ v \end{pmatrix}_{\substack{Sonic, \\ aligned}} = \begin{bmatrix} cos(\overline{\gamma}) & -sin(\overline{\gamma}) \\ sin(\overline{\gamma}) & cos(\overline{\gamma}) \end{bmatrix} \begin{pmatrix} u \\ v \end{pmatrix}_{\substack{Sonic, \\ not\, aligned}} \quad (4)$$

Of course, the wind speed ($\sqrt{(u^2 + v^2)}$) is unchanged, but the wind direction α has to be re-calculated. Knowing α precisely, we are able to compute the right V_{los} of the Sonic as well as the right Lidar wind speed V. Equation 1 is applicable for both operations.

Journal of Physics: Conference Series **749** (2016) 012005 doi:10.1088/1742-6596/749/1/012005

2.4. SVAD procedure

The accurate alignment of a sensor's wind direction data by means of a staring mode Lidar requires the following steps:

(i) Pre-process all data as usual to ensure high quality data,

(ii) Apply a filter method to the Lidar data to remove periods of low signal-to-noise ratio,

(iii) Average all data to 10-min means,

(iv) Project the sensor's data to the Lidar's line-of-sight to get $V_{los,sensor}$,

(v) Classify all data into direction and wind speed bins,

(vi) Fit cosine-functions (Eq.2) to the Lidar data to find the misalignment γ for each wind speed bin. Here data have to be rejected for wind directions with non-free flow conditions such as mast and wind turbine wakes, etc. Moreover, data in directions near the perpendicular to the Lidar line-of-sight $\pm 10°$ should be rejected.

(vii) Average the γ's to get the weight-averaged misalignment angle $\overline{\gamma}$'s: this is the estimated angle to align the sensor's direction data,

(viii) Use the matrix equation given in section 2.3 for aligning, and finally re-calculate the sensor's wind direction data α as basis for the correction of all V_{los}- and wind speed data.

2.5. Theoretical estimation of Lidar wind speed errors

The usefulness of corrected, i.e. aligned direction data is illustrated here by an estimation of the – systematic – wind speed error $\Delta V(\alpha)$, that arises with a misalignment $\Delta\alpha$ (identical with a misalignment γ). The wind speed is simply calculated by:

$$V(\alpha) = \frac{V_{los}}{cos(\alpha)}. \tag{5}$$

For the error estimation the first derivative of equation 5 is used as customary:

$$\Delta V(\alpha) \sim \frac{\partial V(\alpha)}{\partial \alpha}\Delta\alpha \tag{6}$$

To derive the secant function $1/cos(\alpha)$ the quotient rule is used to get

$$\frac{\partial(1/cos(\alpha))}{\partial\alpha} = \frac{sin(\alpha)}{cos^2(\alpha)}. \tag{7}$$

By combining equations 6 and 7 we obtain finally:

$$\Delta V(\alpha) \sim \frac{sin(\alpha)}{cos^2(\alpha)}\Delta\alpha. \tag{8}$$

The factor $sin(\alpha)/cos^2(\alpha)$ is mathematically fixed, it's behaviour is shown in Fig. 3. At directions 90° and 270°, identified as to be perpendicular to the Lidar's laser beam (where $V_{los}=0$), the wind speed error is inevitable infinite. For directions close to 0° and 180° (parallel and anti-parallel to the Lidar beam, resp.) errors are small.

In the end, the fast growth of the secant functions derivative ($\sim 1/cos^2(\alpha)$) urges us to minimize the misalignment $\Delta\alpha$ to keep the wind speed error ΔV as small as possible.

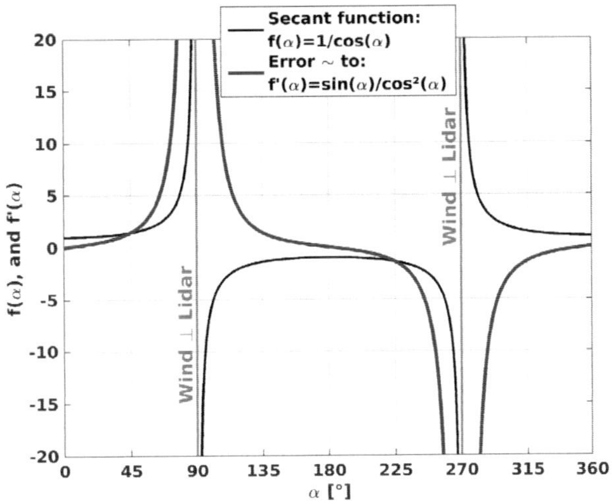

Figure 3. Closer to angles $\alpha = 90°$ and $270°$ (wind direction is perpendicular to Lidar beam), the derivative of the secant function is growing faster than the secant function itself.

3. Experimental setup

We have applied the method to align Sonic data by using staring Lidar data. Sonic and Lidar data have been recorded during a four week measurement campaign in the offshore wind farm alpha ventus next to the meteorological mast FINO 1 in the North Sea in December 2013 and January 2014, when South-Easterly to South-Westerly wind directions were predominant.

A Gill R3-50 Sonic anemometer was mounted on the North-western boom of FINO 1 at height of 41.5 m LAT (lowest astronomical tide). A Windcube 200S Lidar was located 2864 m apart from the Sonic on the substation at height of 23.5 m LAT South-west of alpha ventus.

The direction from the Sonic towards the Lidar (the line-of-sight expressed as wind direction) was estimated as 306.47° as result of a positioning of the Lidar and the meteorological mast through GPS (Fig. 3). The wind direction perpendicular to the Lidar beam, where Lidar errors are expected to be highest, is at 216.47° (of course, 90° apart from line-of-sight).

3.1. Lidar and Sonic data

The Lidar was operated in staring mode only for wind directions between 140° and 300°, the sampling rate was about 1.5 Hz. We have used the Lidar data filter of H. Beck et al. [6], basically to filter out data having a low signal-to-noise ratio. Subsequent averaging has left 1243 10-min mean $V_{los,Lidar}$-values for the comparison with the Sonic data.

The 10 Hz Sonic wind speed data were averaged to 10-min mean values as well. Mast distortion effects on the data (up to 2.7m/s in the mast shadow sector) were corrected using the mast correction function provided by the German Wind Energy Institute DEWI [7]. The "Sonic-line-of-sight wind speed" $V_{los,Sonic}$ was obtained by projecting the wind speed onto the Lidar beam direction.

The Lidar $V_{los,Lidar}$ data were classified by the magnitude of the Sonic wind speed using 1m/s-bins. For each wind speed bin, all the V_{los} data lie close to sinusoidal curves. An example is shown in Fig. 5 for the wind speed bin 12.5–13.5m/s. The maximum $V_{los,Sonic}$ appears at 306.47°.

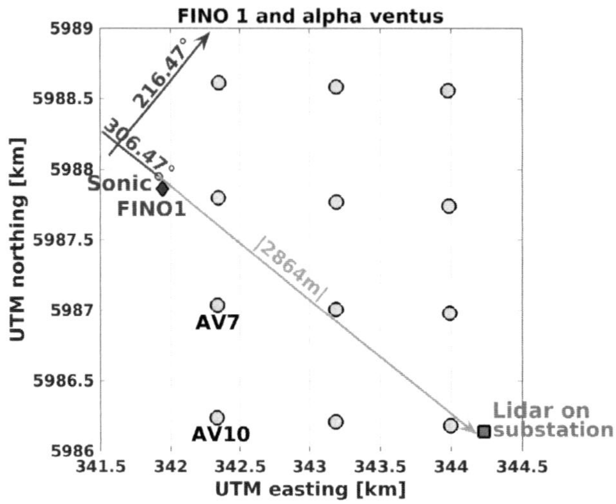

Figure 4. Measurement setting in wind farm alpha ventus, consisting of 12 5MW-turbines. We'll see, that turbine AV7 has influenced the measurements, but AV10 did not.

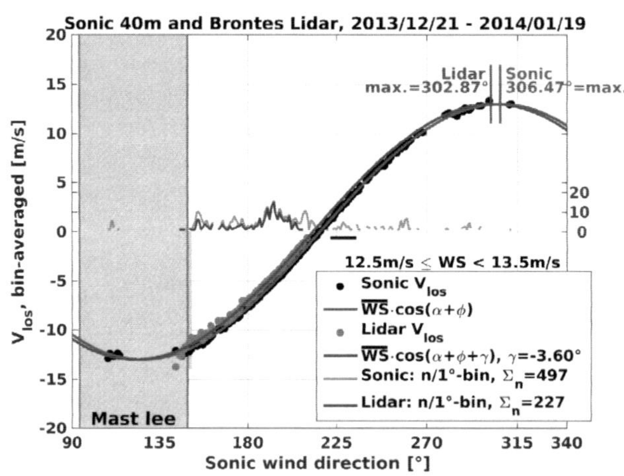

Figure 5. The shift between the sinusoidal curves for the Sonic and the Lidar in the wind speed bin 12.5−13.5m/s is $\gamma = -3.60°$. The amplitude of the sinusoidal is equal to the bin centre (13m/s).

The $V_{los,Lidar}$ data are located apart from the given sinusoidal function for $V_{los,Sonic}$. The fitting procedure using Eq. 2 – executed only for the direction range $170° - 210°$ to exclude mast shadow and farm wake distorted data – yields for the example in Fig. 5 the phase shift $\gamma = -3.60°$. This involves the maximum $V_{los,Lidar}$ being at 302.76°. The explanation for not finding the Lidar's maximum at the line-of-sight: the x-axis of Fig. 5 – labeled with "Sonic wind direction" – is shifted by γ: this is the Sonic's misalignment.

3.2. Error estimation

Assuming maximum geographical position errors of $\pm 5\,m$ (according to typical GPS inaccuracies) for the Sonic as well as for the Lidar, the direction error is in the range of $\pm 0.28°$ – this is simply due to the large distance. Caused by the Windcube 200S setting options for the beams azimuth and elevation angles it is not possible to target for example a Sonic at a met mast with higher accuracy than $\pm 0.1°$. From both, position and targetting errors, the maximum distance from the Sonic to the centre of the Lidar range gate at $2864\,m$ is estimated to be shorter than $8\,m$.

As explained in chapter 2.2, the $1m/s$- binning fixes the V_{los} not entirely, giving statistical uncertainty some space. For the 12 wind speed bins $10.5 - 11.5m/s, ..., 21.5 - 22.5m/s$ (we have ignored low weighted wind speed bins containing less than 30 10-min mean values), we got by fitting the γ-values $2.98°, 3.21°, 3.60°, 3.71°, 3.68°, 3.66°, 3.88°, 3.91°, 4.00°, 4.13°, 3.55°$, and $4.04°$. The weighted average according to Eq. 3 is $\overline{\gamma} = -3.69°$, the standard deviation of the γ's is $0.34°$.

The height difference of the Sonic and the Lidar results in the elevation angle $0.36°$. This allows to assume that the measurement setting was 2-dimensional, because the horizontal wind components are almost fully captured. The vertical wind component w is – except for extreme situations like in tropical cyclones or downbursts in thunderstorms – smaller than $\pm 0.1m/s$ (usually 10 to 100 times smaller than the horizontal components u and v). Furthermore, the Lidar setting here measures only the horizontal projection of w. Therefore, w is negligible in this experiment.

To estimate the overall-error of the aligning method for our data set, we sum up the location errors of Sonic and Lidar (together $0.28°$ at maximum), the Lidar's hard targeting error ($0.1°$ at maximum), and the standard deviation of the distribution of the γ's for each wind speed bin ($0.34°$) using

$$E_r = \sqrt{0.28^2 + 0.1^2 + 0.34^2}\ [°], \tag{9}$$

resulting in $E_r \approx 0.45°$. The SVAD method determines the misalignment of FINO 1's Sonic at 41.5 m LAT height to

$$\overline{\gamma} = -3.69° \pm 0.45°. \tag{10}$$

4. Evaluation of the SVAD method

The alignment of the Sonic on 41.5 m LAT at FINO 1 by applying $\gamma = -3.69°$, reduces the bias between the V_{los} of Sonic and Lidar in the sector $170° - 210°$ from $-0.95m/s$ to $+0.02m/s$ (Fig. 6).

Comparing the wind speed measurements of the Lidar and the Sonic after the alignment, we have to distinguish three direction sectors (Fig. 7). Around 155°, in the wake of wind turbine AV 7, we notice differences up to $1m/s$ (Lidar wind speed is reduced). We explain these differences with an "integrating effect" of the Lidar. It measures the reduced wind speed along a path (length is estimated to be 60 m) in line-of-sight, but the Sonic measures the wind only in a small volume. The difference between a remote sensing- and a in-situ technique is clearly seen only here. We could not find such a wake effect of turbine AV 10 in our data set. Presumably, the distance between this turbine and FINO 1 ($1626\,m$) is too far for this (the distance of AV 7 to FINO 1 is 872 m).

In the direction sector $170° - 210°$, the Lidar-to-Sonic bias is $-0.02m/s$. The bias error of the Sonic alone should be 0, if it has been calibrated properly, but we do not know. However, the small Lidar-to-Sonic bias in our experiment might be explaind by statistical uncertainty due to the restricted amount of Lidar data (around 900 10-min mean values in this sector).

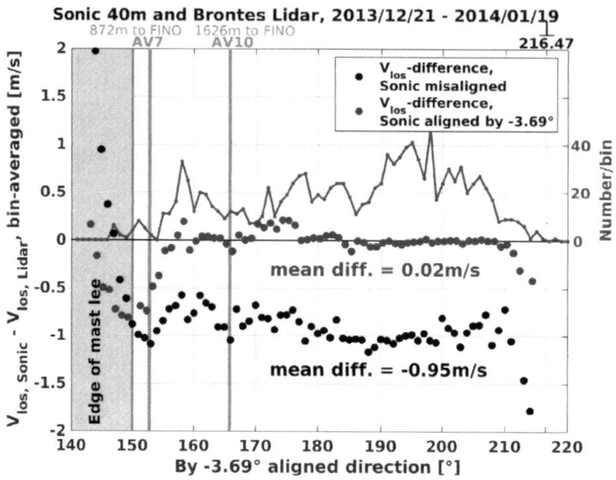

Figure 6. The aligning procedure reduces the V_{los}-difference considerably. The Sonic data suffer from the mast shadow more than the Lidar data do.

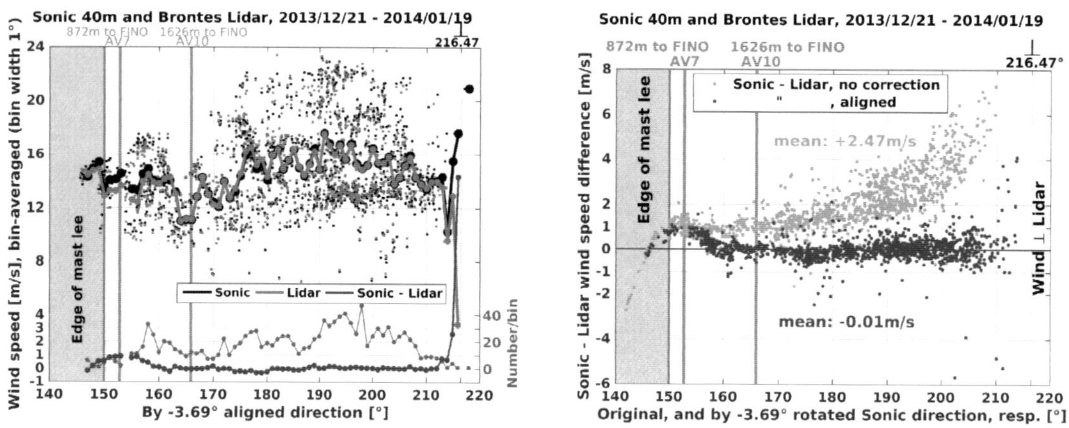

Figure 7. Left: red and black dots, small: 10-min mean values, big: these data direction-bin averaged. Blue dots: Sonic-to-Lidar differences. Left: reasons for Lidar-to-Sonic differences for three direction sectors (wake region of turbine AV 7, around 155°, wind almost perpendicluar to Lidar beam close to 216°, and the "undisturbed" sector in between) are explained in the text.

The left part of Fig. 7 shows big Lidar-to-Sonic differences only for wind directions close to 216.47° (wind perpendicular to Lidar beam). In this region, $V_{los,Lidar}$ is due to the projection of the wind speed onto the line-of-sight small (exactly 0, if wind is exactly perpendicular to the Lidar beam). This leads to big relative V_{los}-errors, and therefore (see Equation 1) to large Lidar wind speed errors. In this way, the measurement data confirm the theoretical error estimation for wind speed that we made in section 2.5 (please see Fig. 3 again). Even after a proper aligning procedure, Lidar wind speed data from the sector ±5° around the 'perpendicular direction' still are not reliable – fortunately only in this small direction sector.

Just to emphasize the importance of having highly accurate wind direction data just once more, we look at the the Sonic-to-Lidar wind speed differences before and after the aligning procedure in Fig. 7. Without alignment, the Lidar's wind speed error starts to become bigger than $2m/s$ for wind directions 35° apart from the 'perpendicular' direction. Aligning the Sonic wind direction reduces Lidar wind speed errors in a wide direction range: the Lidar-to-Sonic bias has almost vanished.

An alternative aligning method, suggested by the left part of Fig. 7: if there is some confidence in the similarity of Lidar and Sonic wind speed data, also a minimisation method may used to detect that alignment angle, for which the cost function, given by the Lidar and Sonic wind speed differences, is minimal.

An advantage of the alternative method (not checked up to now): the wind direction binning is obsolete, so, a relatively short measurement period might produce a sufficient amount of data for wind direction alignment. Nevertheless, measurements for a wide range of wind directions will improve the alignments accuracy.

5. Conclusions

This research illustrates that even small misalignments ($<4°$) of wind direction sensors produce fatal Lidar wind speed errors. It shows that long-range Lidar's, operated in staring mode technique, can be applied to improve the estimation of the absolute orientation of wind sensors, in fact after installation of the measurement system. Moreover, it can be applied from the ground without need for access to the sensor or the meteorological mast.

We presented the application of this method for comparisons of Lidar and ultrasonic anemometer measurements. The staring velocity azimuth method (SVAD) achieves direction information with previously unknown accuracy.

Line-of-sight- and wind speed differences between the ultrasonic anemometer and the Lidar measurement techniques have revealed to be negligible after the alignment of direction data.

We expect this method to be useful also for aligning wind vane or VAD-Lidar data. Directing a long-range Lidar to met mast sensors at several heights may provide more accurate insight in the height-depending rotation of wind (veer).

6. References

[1] Jimenez A, Crespo A and Migoya E 2010 *Wind Energy* **13** 559–572 ISSN 1099-1824 URL `http://dx.doi.org/10.1002/we.380`
[2] Fleming P A, Gebraad P M, Lee S, van Wingerden J W, Johnson K, Churchfield M, Michalakes J, Spalart P and Moriarty P 2014 *Renewable Energy* **70** 211 – 218 ISSN 0960-1481 special issue on aerodynamics of offshore wind energy systems and wakes URL `http://www.sciencedirect.com/science/article/pii/S0960148114000950`
[3] Trujillo J J, Seifert J K, Wrth I, Schlipf D and Khn M 2016 *Wind Energy Science* **1** 41–53 URL `http://www.wind-energ-sci.net/1/41/2016/`
[4] Westerhellweg A, Canadillas B, Beeken A and Neumann T 2010 *10th German Wind Energy Conference – DEWEK 2010*
[5] Werner C 2005 *Doppler Wind Lidar* chap 12, pp 325–354
[6] Beck H Schmidt M K M 2016 *18th International Symposium for the Advancement of Boundary-Layer Remote Sensing (ISARS 2016)* oral presentation
[7] Westerhellweg A, Neumann T and Riedel V 2012 *DEWI Magazin* 60–66 URL `http://www.dewi.de/dewi_res/fileadmin/pdf/publications/Magazin_40/09.pdf`

Short time ahead wind power production forecast.

Alla Sapronova[1], Catherine Meissner[2], and Matteo Mana[2]
1. Uni Research, Bergen, 5008, Norway
2. WindSim, Tonsberg, 3125, Norway

alla.sapronova@uni.no

Abstract. An accurate prediction of wind power output is crucial for efficient coordination of cooperative energy production from different sources. Long-time ahead prediction (from 6 to 24 hours) of wind power for onshore parks can be achieved by using a coupled model that would bridge the mesoscale weather prediction data and computational fluid dynamics. When a forecast for shorter time horizon (less than one hour ahead) is anticipated, an accuracy of a predictive model that utilizes hourly weather data is decreasing. That is because the higher frequency fluctuations of the wind speed are lost when data is averaged over an hour. Since the wind speed can vary up to 50% in magnitude over a period of 5 minutes, the higher frequency variations of wind speed and direction have to be taken into account for an accurate short-term ahead energy production forecast. In this work a new model for wind power production forecast 5- to 30-minutes ahead is presented. The model is based on machine learning techniques and categorization approach and using the historical park production time series and hourly numerical weather forecast.

1. Introduction

Currently, the wind energy plays a significant role in total electricity production in Europe. Therefore it is vital to predict the wind energy output timely and accurate. The variability of the wind is the primary challenge for obtaining an accurate forecast minutes to hours ahead.

Many studies have been devoted to the improvements of wind forecasting techniques with a number of models developed and launched. These models are based on physical [1,2], statistical [3,4] or hybrid approaches [5,6].

The physical approach focuses on integrating well-known physical aspects into the model, such as information about surrounding terrain and properties of the wind turbines. Statistical approaches relies more on historical observations and their statistical relation to meteorological predictions as well as measurements from Supervisory Control And Data Acquisition (SCADA). Statistical models are usually built around Numerical Weather Prediction (NWP) at tens-of-kilometers resolution.

But NWP-based models are failing to provide site-specific short time ahead forecasts at satisfactory accuracy due to a relatively coarse resolution of the NWP model.

For site-specific forecasts, the mesoscale-microscale coupling model was proposed in [7], where the artificial intelligence methods were used to issue 1 to 3 hours ahead wind speed forecast from NWP data, and later the CFD module was used to calculate the flow at finer scales [8]. The accuracy of the coupled model [7] was found to be superior comparing to that based on polynomial fittings as well as AutoRegressive Moving Average (ARMA) models, although the accuracy for shorter-than-one-hour forecast remains unsatisfactory.

Very short-term models (less than 9 hours) used for wind power forecasting usually consists of statistical methods like Kalman Filters, ARMA, Auto-Regressive with Exogenous Input, Box-Jenkins etc. Inputs to these models are historical observations of wind speed, wind directions, temperature, etc.

Machine learning methods used in this area include Neural Networks [9], Support Vector Machines [10], Nearest Neighbour Search [11], Random Forests, etc. Summarizing the reports for very-short time ahead forecast the error of 12% of the total rated power can be stated as the state-of-the-art. [12,16]

2. Model selection

In this work, the model proposed in [7] has been adapted to forecast a wind power output for less than one hour ahead by utilizing time series of historical power production in addition to NWP data. Adapted model also uses a data categorization approach, as described in [13]. There the information obtained from the categorization of a single variable (wind speed, for instance) is supplied as an additional input to the model.

The model for short-time ahead prediction of the wind power is based on time series analysis of the historical park production and NWP data. The quality of the forecast depends on the ability of the model to predict the wind flow near the ground in the complex terrain, where the roughness and complexity affect the flow at microscale. Therefore the model employs machine learning techniques as those are proven to be efficient for nonlinear multivariable functions approximation when explicit physical-based models have a limited application or not available.

In this work, a combination of artificial neural networks (ANN) module and kernel module (support vector machine, SVM) has been used to forecast the total wind park power production 5- to 30-minutes ahead. The modules are trained, tested and validated on 4-months historical data from several onshore wind parks from Sweden and Norway.

To forecast the wind power production 5- to 30-minutes ahead a feed-forward (FF) ANN with a single hidden layer and single output neuron has been built.

Two different sets of the input variables were used for ANN training. One ANN received the historical power production values as inputs and obtained the total power production minutes ahead as output. Another ANN received time-series of multiple environmental variables as inputs (registered wind speed and direction time-series and hourly NWP data) in addition to historical power production time-series.

The length of a time window was two times longer than the prediction horizon. For example, for 5-minutes ahead prediction, the last 10-minutes park power production records (averaged over 1 minute) were used as inputs.

The pre-processing included data cleaning, normalization, and time averaging. The data was averaged by 1-minute for 5- to 10-minutes ahead prediction, and 5-minutes averages were used for 15- to 30-minutes ahead prediction. The dataset split in the ratio 70:20:10 for training, testing, and validation correspondingly. After cleaning, the total number of records available for the model training was 129600 for 1-minute averaging intervals and 25920 for 5-minutes averaging intervals.

To measure the differences between values predicted by the model and the values observed the normalized root-mean-square error (root-mean-square percentage error, RMSPE) had been calculated:

$$RMSPE=\sqrt{\frac{1}{n}\sum_{i=1}^{n}\left(\frac{y_i-y_{ipredict}}{y_i}\right)^2} ,$$

where y denotes the observed value and $y_{predict}$ denotes the corresponding prediction.

As in RMSPE the negative and positive errors don't cancel out each other and the higher weight is given to larger errors it makes an excellent general purpose error metric for numerical predictions: the smaller the RMSPE, the better fit of the model. RMSPE for ANN models for validation data sets is shown in Table 1.

Backpropagation training method as provided by Encog Workbench open source software package [15] has been used.

Table 1. Root-mean-square percentage error of wind power production forecast performed by ANN-module.

Prediction time-window	Hourly weather forecast and historical power production time series	Historical power production time series
5 minutes ahead prediction	12.2	8.2
15 minutes ahead prediction	14.9	9.7
30 minutes ahead prediction	15.1	9.5

Because of the lower prediction error, the ANN trained only on historical power production time series has been selected for further modification.
From a theoretical point of view, one can approximate almost any function with one layer neural network. Therefore, the most of the literature suggests that a single layer neural network with a sufficient number of hidden neurons will provide a good approximation for most problems and that adding a second or third layer yields little benefit. Here a new double-module model has been suggested to improve the model's accuracy. One single hidden layer FF ANN has been used to issue a "coarse" prediction and the output from the first module has been submitted to the second machine learning unit (a "correcting" module) to issue a more accurate forecast. With the suggested double-module approach when the output of the coarse predictor has been used as additional input to the fine predictor, RMSPE has been lowered to 7.9 and 8.9 for 5 minutes ahead and 30 minutes ahead predictions respectively.
The accuracy of the "correcting" module based on single hidden layer FF ANN is compared to the one based on kernel method (SVM). RMSPE for different "correcting" module architectures is provided in Table 2.

Table 2. Root-mean-square percentage error of wind power production forecast performed by "correcting" module.

Prediction time-window	ANN module	SVM module
5 minutes ahead prediction	7.9	6.4
30 minutes ahead prediction	8.9	7.9

The performance of double-module model has been compared to data categorization approach based model. This method is described in [13] and suggests that the entire dataset should be grouped into several discrete categories. Here, the wind speed data has been split into nine categories according to the turbine power curve characteristics (with categories' names similar to Beaufort scale) as shown in Table 3. Obtained numerical attribute of the category has been fed as an additional input to "coarse" module FF ANN.

Table 3. Wind speed ranges and corresponding numerical attributes used for categorization.

Wind speed range, m/s	<0.5	0.5-1	1-3	3-7	7-10	10-15	15-20	20-25	>25
Category	Failed record	Wind calm	Light breeze	Gentle breeze	Fresh breeze	Strong wind	Near gale	Gale	Cut off
Numerical attribute	1	2	3	4	5	6	7	8	9

Previously it was shown [14] that the selection of methods for categorization is not critical, so in this work, the wind speed was categorized. RMSPE on validation data set for all the models mentioned above summarized in Table 4.

Table 4. Root-mean-square percentage error of 30 minutes ahead power output forecast for different model architectures.

Single-module, ANN model	Double-module, ANN based model	Double-module, ANN and SVM based model	Double-module model with data categorization approach
14.9	8.9	7.9	7.5

3. Results

The suggested double-module model comprise of (1) ANN that uses park power production time-series to predict power output 5- to 30-minutes ahead, and (2) SVM that utilizes output from ANN as one of the inputs along with NWP data and the wind power production time-series. Schematic description of the model shown in Figure 1.

Figure 1. Double-module model: "coarse"-predicting module is using longer time-series of a single variable; the output from "coarse" module is utilized by "correcting" module along with shorter time-series of multiple variables.

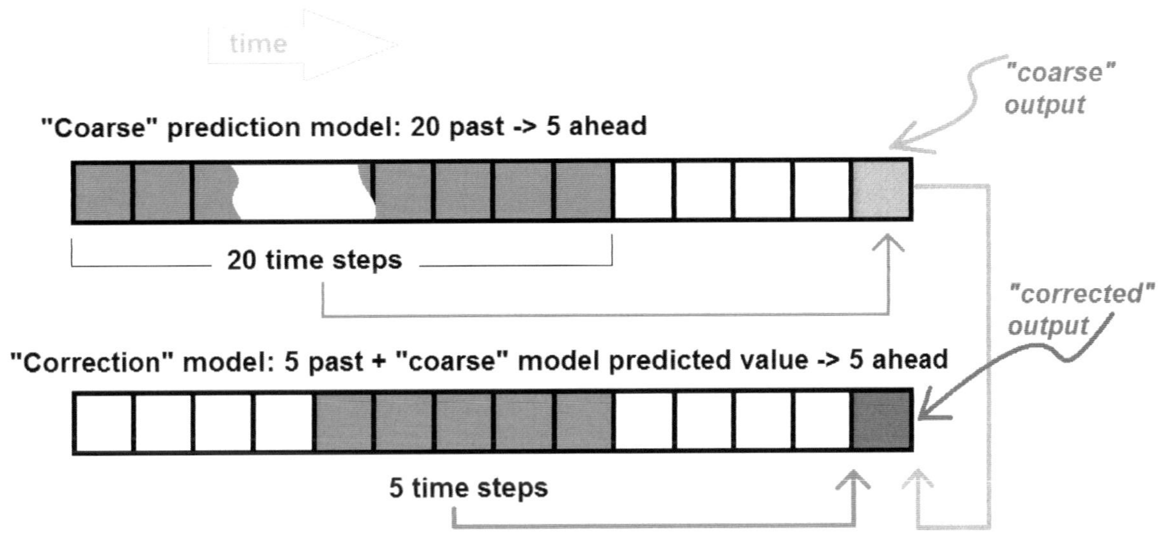

ANN architecture and performance for various length of time-windows and prediction horizons is summarized in Table 5.

Table 5. ANN modules' architectures and root-mean-square percentage error of wind power production forecast performed by corresponding modules.

Prediction horizon	5 minutes			10 minutes			20 minutes			30 minutes		
Time-series window interval, minutes	10	20	30	20	30	40	30	40	60	40	60	90
Number of hidden neurons	10 to 30	20 to 60	30 to 90	20 to 60	30 to 90	40 to 100	30 to 90	40 to 100	60 to 100	40 to 100	60 to 100	90 to 100
Validation error (RMSPE)	14.9	9.7	7.9	15.1	12.2	8.1	18.1	12.4	8.6	16.2	8.9	8.9

The output from the ANN module along with other inputs has been submitted to SVM to predict the power output same time ahead. With the double-module approach, RMSPE for 10 minutes ahead forecast has been lowered from 7.9 to 5.9. The SVM module performance is shown in Table 6.

Table 6. SVM modules' architectures and root-mean-square percentage error of wind power production forecast performed by corresponding modules.

Prediction horizon	5 minutes				10 minutes				30 minutes					
Power production	5	5	10	10	5	5	10	10	5	5	10	10	20	20

time-series length, minutes														
NWP data used	Yes	No	Yes	No	Yes	No	Yes	No	Yes	No	Yes	No	Yes	No
Wind speed and direction time-series, minutes	5	5	5	5	5	5	5	5	10	10	10	10	10	10
Validation error (RMSPE)	6.6	6.7	5.9	5.9	6.9	6.8	6.1	6.2	9.1	9.4	8.6	9.2	8.2	8.5

SVM module performance is not significantly affected by the lack of NWP data for very short term ahead prediction, yet for longer prediction horizons, like 30-minutes ahead, using NWP data as input helps to lower RMSPE.
Finally, the model has been modified to employ the categorization data (as described in [4]). The wind speed data were categorized and supplied as one of input variables to SVM module. The model that is using categorization approach shows better performance, as shown in Table 7.

Table 7. SVM module performances for the short time ahead wind power forecast with and without categorization approach.

Prediction horizon	5 minutes		10 minutes		30 minutes		
Power production time-series length, minutes	5	10	5	10	5	10	20
Wind speed and direction time-series, minutes	5	5	5	5	10	10	10
RMSPE for model without categorization approach	6.7	5.9	6.8	6.1	9.1	8.6	8.2
RMSPE for model with categorization approach	4.5	4.6	5.1	5.1	6.5	6.2	6.2

4. Conclusions

In this work a new model to forecast the wind power production 5- to 30-minutes ahead is presented.
The model uses NWP, historical wind power production data and employs machine learning methods and categorization approach. The best performance is observed for the combination of ANN and kernel methods. The proposed model provides 10-minutes ahead prediction with RMSPE 4.5 and 30-minutes ahead prediction RMSPE 6.2 which is much better than reported state of the art for less than one hour ahead prediction [12, 16].
It is shown that use of NWP data does not significantly improve the accuracy of very short time ahead forecast (5-10 minutes ahead), yet it is valuable for 30-minutes ahead forecasts.
If categorization approach is used, better accuracy of the forecast can be achieved even with smaller number input variables. It is observed that model's generalization arises from the model's ability to find similarity in the training data that usually consists of continuous numeric data. Since numbers are rarely the same from one example to the next, the model can fail in selecting the margins for identical properties. In this case, the generalization can be improved by classification. As shown in [14] the choice of methods for categorization is irrelevant to the generalization improvement. Therefore, the generalization approach can be adapted to various input parameters.

5. Aknowledgement

The work is sponsored by the Norwegian Research Council ENERGIX programm, 2015-2016.

6. References

[1] Landberg L and Simon J W 1994 Short-term prediction of local wind conditions. *Boundary-Layer Meteorology*, **70(1-2)** 171–195

[2] Gaertner M A, Gallardo C, Tejeda C, Martínez N, Calabria S, Martínez N, and Fernández B 2003 The casandra project: results of wind power 72-h range daily operational forecasting in spain. In *European Wind Energy Conference*

[3] Rodrigues A, Peças Lopes J A, Miranda P, Palma L, Monteiro C, Bessa R, Sousa J, Rodrigues C, and Matos C 2007 Eprev–a wind power forecasting tool for portugal. In *Proceedings of the European Wind Energy Conference EWEC* **7**

[4] Gonzalez G, Diaz-Guerra B, Soto F, Lopez S, Sanchez I, Usaola J, Alonso M, and Lobo M G 2004 Sipreólico-wind power prediction tool for the spanish peninsular power system. *Proceedings of the CIGRÉ 40th general session and exhibition*, París, France

[5] Giebel G, Brownsword R, Kariniotakis G, Denhard M, and Draxl C 2011 The state-of-the-art in short-term prediction of wind power: A literature overview. *Technical report ANEMOS plus*

[6] Focken U, Lange M, and Waldl H P 2011 Previento-a wind power prediction system with an innovative upscaling algorithm. *Proceedings of the European Wind Energy Conference 2011* **276**

[7] Sapronova A, Meissner C, and Mana M 2014 Mesoscale-microscale coupled model based on artificial neural network techniques for wind power forecast. *Poster at EWEA Offshore 2014*

[8] Mana M 2013 Short-term Forecasting of wind energy production using CFD simulations. *Poster at EWEA 2013*

[9] Kusiak A, Zheng H, and Song Z 2009 Wind farm power prediction: a data-mining approach. *Wind Energy*, **12(3)** 275–293

[10] Fugon L, Juban J, and Kariniotakis G 2008 Data mining for wind power forecasting. *In European Wind Energy Conference & Exhibition* EWEC 2008

[11] Jursa R and Rohrig K 2008 Short-term wind power forecasting using evolutionary algorithms for the automated specification of artificial intelligence models. *International Journal of Forecasting*, **24(4)** 694–709

[12] Wang X, Guoc P, and Huanga X, A 2011 Review of Wind Power Forecasting Models, *Energy Procedia* **12** 770–778

[13] Sapronova A, Meissner C, and Mana M 2014 Improving an Accuracy of ANN-Based Mesoscale-Microscale Coupling Model by Data Categorization: With Application to Wind Forecast for Offshore and Complex Terrain Onshore Wind Farms, *Lecture Notes in Computer Science* **8817** 61-66

[14] Li L, Pratap A, Lin H T, and Abu-Mostafa Y S 2005 Improving generalization by data categorization. In: Jorge A M, Torgo L, Brazdil P B, Camacho R, and Gama J. (eds.) PKDD 2005. LNCS (LNAI) *Springer, Heidelberg* **3721** 157–168.

[15] http://www.heatonresearch.com/wiki/Encog_Workbench [online; assessed 10-08-2016]

[16] Svensson M 2015 Short-term wind power forecasting using artificial neural networks Master's Thesis at School of Computer Science and Communication Royal Institute of Technology KTH Machine Learning Programme

Investigation and validation of wake model combinations for large wind farm modelling in neutral atmospheric boundary layers

E Tromeur, S Puygrenier, S Sanquer

Meteodyn France, 14bd Winston Churchill, 44100, Nantes, France
Email: eric.tromeur@meteodyn.com

Abstract. This study is focused on assessing the ability of two refined large wind farm models to describe the disturbance of the neutral atmospheric flow caused by large offshore wind farms. Sensitivity studies of internal boundary layer parameters are carried out. An optimum large wind farm correction is then proposed and combined with two different standard single wake models, the Park and EVM models. The large wind farm wake effect is evaluated and validated against measurements of two offshore wind farms at Horns Rev and Nysted and four standard wake models by computing velocity deficit and normalized power. All large wind farm models proposed were able to capture wake width to some degree and the decrease of power output moving through the wind farm. Despite some uncertainties, this very promising model combinations allows us to take into account the slowdown in large wind farms.

Keywords: wake model, atmospheric boundary layer, wind farm

1. Introduction

When air under neutral conditions flows from one surface through a wind turbine with a different roughness, the air is slowed [1] [2], reducing the wind velocity (Figure 1, left and side). The region in the flow behind the turbine is called the wake of a wind turbine. Its effects are seen as wake effects.

Generally, these effects can be neglected when the turbines are spaced apart by more than 10 rotor diameters. However, the turbines are increasingly clustered in large wind farms, the spacing between turbines being usually smaller because of the very favorable wind conditions on some sites. In that case, an internal boundary layer grows downwind from the roughness change [3][4][5] (Figure 1, right and side), the turbines influencing each other. A large wind farm wake effect then occurs by combination of single wake effects and this internal boundary layer modification (Figure 1).

Wake models in wind resource softwares like WindFarmer [6], ECN-Wakefarm [7], WAsP [8][9], NTUA [10] or Meteodyn WT [11] were evaluated for small wind farms [12] or single wakes [13]. However, it has become apparent that standard single wake models as Park [14][15] and EVM [16] models tend to underestimate wake losses in large wind farms as offshore arrays [17]. It is thus important to compute correctly the velocity deficit by taking into account the large wind farm wake effect.

Nowadays, two types of models are used to estimate velocity deficit and power losses due to wind turbine wakes. One is a wind farm model using a wake model that has been simplified or parameterized. The second is a CFD-type k-ε model which solves basic equations of the atmosphere and produces results on a fine mesh in space and time [18]. In particular, Schlez and Neubert [6] described the disturbance of the atmospheric flow caused by the wind farm with an empirical large wind farm correction. However, there are uncertainties in their extrapolation which need to be further explored [6].

The goal of this paper is to assess the ability of two refined large wind farm models based on [6] to describe the disturbance of the neutral atmospheric flow caused by the wind farm. Sensitivity studies of internal boundary layer parameters are first carried out. An optimum large wind farm correction is then proposed and combined with two different standard single wake models, the Park [14][15] and EVM [16] models. The large wind farm wake effect is finally evaluated and validated against measurements of two offshore wind farms and four standard wake models as in [18] by computing velocity deficit and normalized power.

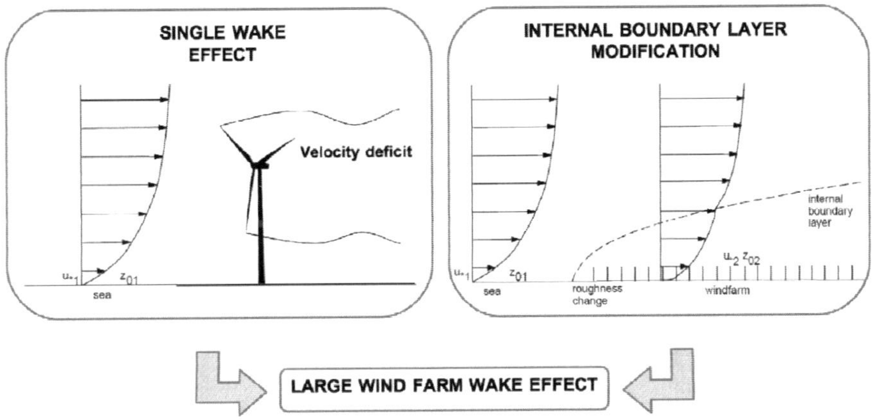

Figure 1: Sketch of large wind farm wake effect

2 . Methodology

Single wake models don't consider the change of the atmospheric boundary layer by the additional roughness associated with wind turbines. More wind turbines are far from the windward side of the park, more the velocity deficit increases due to speed slowing down over the park. Therefore, the boundary layer profile is a function of the equivalent roughness z'_0 and the wind position relative to the upstream turbine.

Three sensitivity studies are performed to optimize the large wind farm correction:

1. equivalent roughness z'_0 computation
2. internal boundary layer profile estimation
3. large wind farm correction activation

The large wind farm correction is parametrized according to Horns Rev offshore wind farm data and validated against production data at Nysted offshore wind farm as in [18]. The velocity deficit is then calculated in each point of the wind farm by combining a single wake effect from a wind turbine with this boundary layer modification. Two different standard single wake models are considered: the Park model [14][15] and the fast algorithm for EVM model [19] implemented in Meteodyn WT software [11]. Thereafter, these two large wind farm models are named WT Park+IBL and WT Fast EVM+IBL.

A model intercomparison is finally performed at the two offshore wind farms. These two combined models are compared with four different wake models: WindFarmer [6], ECN-Wakefarm [7], WAsP [8][9] and NTUA [10] as in [18].

An overview of the main features of the models used in this intercomparison is given in [20]. Some of the models applied are industry standard models, for example, WAsP, WindFarmer, WT Park+IBL and WT Fast EVM+IBL, whereas others are primarily research models. In WasP, the Park model is used as a wake model while a large wind farm correction is combined with this single wake model in the WT Park+IBL model.

Besides, both large wind farm models WT Fast EVM+IBL and WindFarmer are based on the same single wake model, namely the EVM model. However, the main differences are in the boundary layer modification parameterization which is the goal of this paper.

There are a number of issues in comparing model simulations and wind farm observations of velocity deficit and power losses in wakes that were detailed in [20]. It is difficult to make exactly the same simulations with models of different types even after the main variables, such as thrust coefficient, wind speed at hub height, free-stream wind profile, etc., have been set. Examples of this are that it is not possible to run WAsP for extremely narrow wind speed and direction bins because WAsP relies on a Weibull fit to the wind speed observations. For CFD, one issue is to accurately determine the turbulence profile and to recall that, for narrow sectors, the wake is centered on the given direction and no directional variability is included [18].

There are also practical issues relating to computing resources. Running a full wind farm simulation in WAsP or Meteodyn WT takes of the order of minutes, while in the NTUA model requires a time scale of days to make even one simulation limiting the number of runs performed.

To provide a quantitative evaluation of the refined large wind farm model performances versus the observations and other wake models, the normalized power and its root-mean-square-error (RMSE) are compared for each case.

This normalized power is defined in the next section presenting briefly the two offshore wind farms.

3. Measurements

The observations used to parameterize and validate the large wind farm correction are taken from the large offshore wind farms at Horns Rev [21] and Nysted [22][23] in Denmark. The turbine spacing at Horns Rev is 7 D_{rotor} in both north–south and west–east directions (Figure 2), whereas at Nysted the turbine spacing is 5.8 D_{rotor} in the north–south direction and 10.5 D_{rotor} in the east–west direction (not shown).

To examine the single wake, the average power at each turbine is calculated in each column of the wind farm for seven wind direction sectors centered on an exact wind farm row (ER) (270° +/- 2.5° at Horns Rev and 278° +/- 2.5° at Nysted), and for mean wind directions of +5°, +10°, and +15° and -5°, -10°, and -15° from ER. Flow down at ER thus represents the likely maximum wake effect, while the wind directions that are slightly offset from ER assist in assessing the wake width.

Finally, The power in each column is normalized to the power in the first column (1 ... 8 at Horns Rev). In both cases, wake effects is evaluated for a free-stream velocity mainly coming from the west (not shown) and equal to 8 $m.s^{-1}$ as in [18].

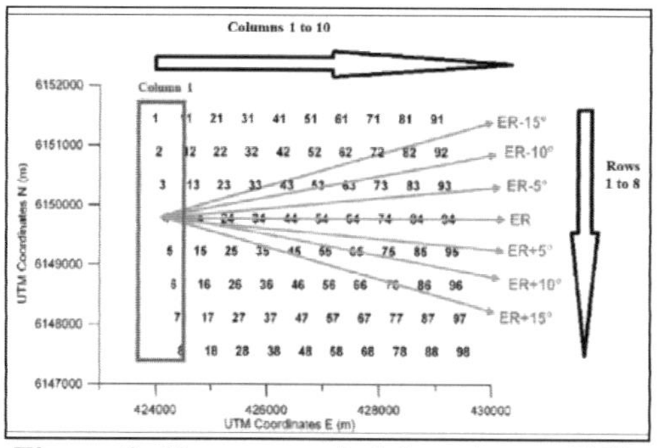

Figure 2: Horns rev wind farm layout [18].

4. Large wind farm wake effect: parametrization and activation

4.1. *Roughness z'_0 influence*

The equivalent roughness z'_0 is calculated with the method of Frandsen [24][25] for each wind direction and wind speed at each turbine. It depends on the spacing between two rows of wind turbines along the wind direction S_d and the crosswind direction S_c. S_c has a huge influence on the roughness (example on Figure 3 for the wind turbine number 74 (WT74) at the Horns Rev with $S_d = 7$). It impacts directly the velocity deficit coefficient correction and the normalized power with respect to WT04, the last one going down to 10% if $S_c = 3$ and the wind speed is equal to 8 $m.s^{-1}$ (not shown).

An algorithm is developed to calculate S_c and S_d whatever the type of wind farms and the wind directions. Figure 4 presents an example of S_c and S_d evolutions at Horns Rev for ER incidence (other incidences not shown here). The number of upstream wind turbines for a specific position is increasing for a wind turbine going far away from the first column of the array. S_c and S_d are homogeneous over the all wind farm considering at least one wind turbine is detected upstream. Finally, S_c and S_d has been found equal to 7 for both wind farms in Denmark.

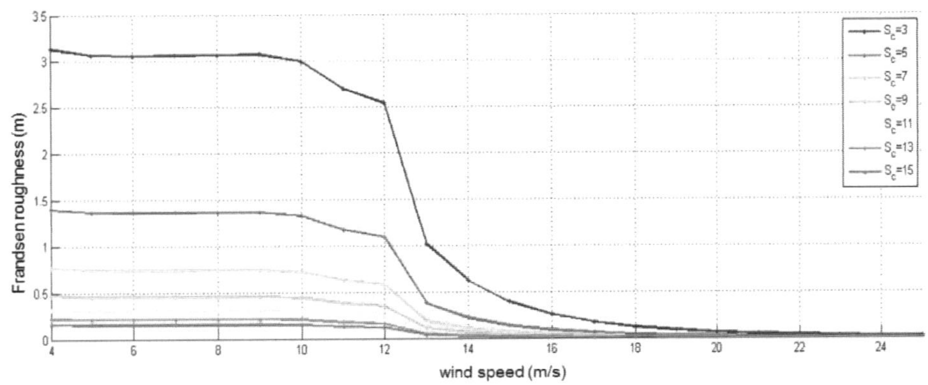

Figure 3: Frandsen roughness function of wind speed and S_c with S_d=7 at ER incidence and wind turbine WT74

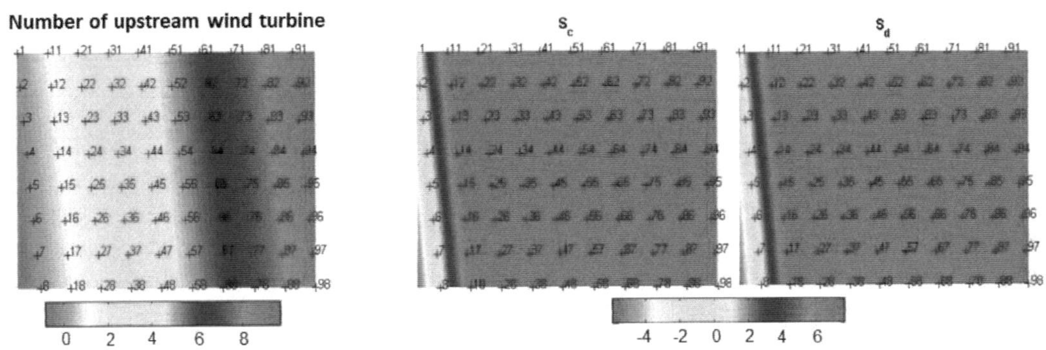

Figure 4: Evolution of S_c and S_d at ER incidence at Horns Rev

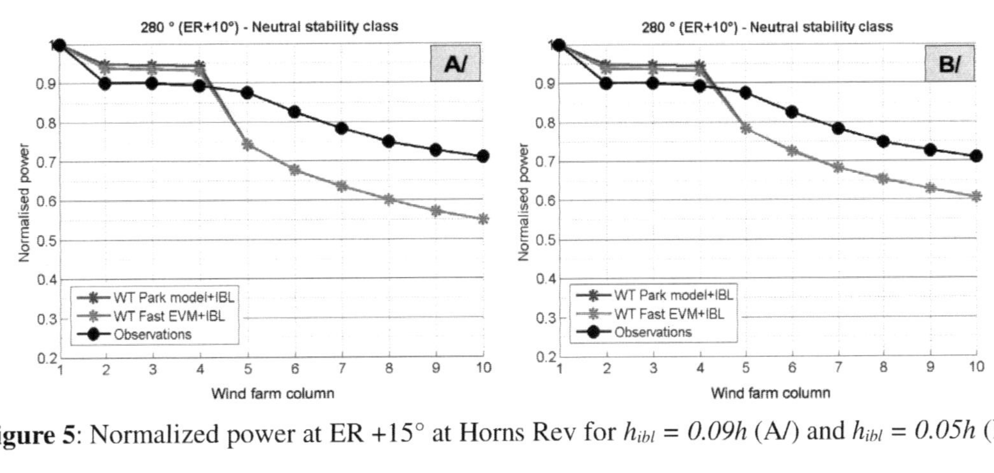

Figure 5: Normalized power at ER +15° at Horns Rev for h_{ibl} = $0.09h$ (A/) and h_{ibl} = $0.05h$ (B/).

4.2. *Internal boundary layer influence*

The velocity deficit coefficient correction is the ratio between the wind speed in the IBL and the wind speed taken at the same height before the roughness change. However, an offset H_{start} (function of the fetch and z'$_0$.) from which the boundary layer starts and the IBL height h_{ibl} influence it. Sensitivity studies of H_{start} and h_{ibl} are then carried out at Horns Rev with the two combined wake models in order to optimize wind speed and power corrections:

- According to [26], $0.05h \leq h_{ibl} \leq 0.09h$, where h is the boundary layer height. Comparisons between both combined models and observations in Figure 5 show a better agreement for $h_{ibl}=0.05h$ (case B/) against 9% of h in [6]. The same is observed for all other directions, except for ER-15° (not shown).
- As shown in Table 1, the more H_{start} is low, the more velocity and power deficits are low. On the contray to [6] proposing H_{start} = 2/3 h_{hub} (with h_{hub} the hub height), the optimum H_{start} is equal to zero, meaning the inner boundary layer influence starts from the ground.

All these optimized parameters are considered by default in the next validation section 5.

Table 1: Evolution of wind speed and power correction function of H_{start} for the wind turbine WT74 at incidence ER at Horns Rev (WT Park+IBL model). D_{rotor} is the rotor diameter.

H_{start}	Coefficient Correction WT74	Power Correction WT74
0	0.83	0.57
1/4 H_{hub}	0.78	0.47
1/3 H_{hub}	0.76	0.45
H_{hub}-D_{rotor}/2	0.75	0.41
2/3 H_{hub}	0.66	0.27

4.3. *Large wind farm correction activation*

A geometric measure of turbine density is used to activate the large wind farm correction. For each small direction sector, the horizon is scanned and the presence of upstream turbines detected. On the contrary to [6], an optimum turbine density for 5° sectors is parameterized rather than 30° sector. Several angular sectors between 5° and 30° have been tested (not shown) to find this optimum sector. The large wind farm correction to ambient wind speed is then applied if there is at least one turbine in the selected sector.

Figure 6 shows a schematic diagram for wind turbine identification. Two wind turbines are in a sector [*Dir-dα ; Dir+ dα*] with *Dir* being the wind direction and *dα* the half turbine density sector. In that case, the large wind farm correction is combined to the single wake model to estimate the large wind farm wake effect. This model is always activated from the fourth wind farm column.

Finally, the velocity deficit is computed as the velocity deficit minimum taken between the large wind farm model and the single wake Park or Fast EVM models.

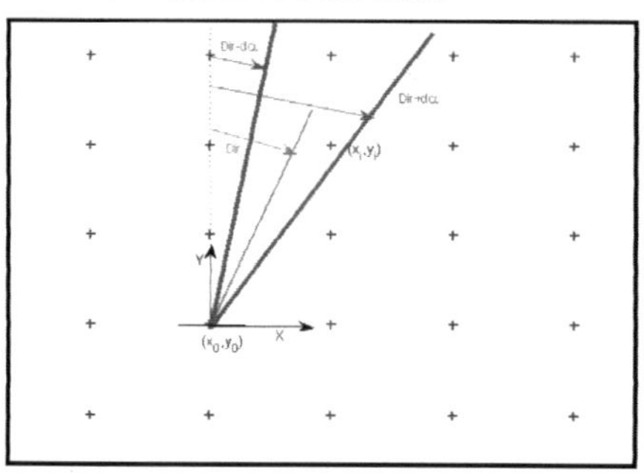

Figure 6: Schematic diagram for wind turbine identification

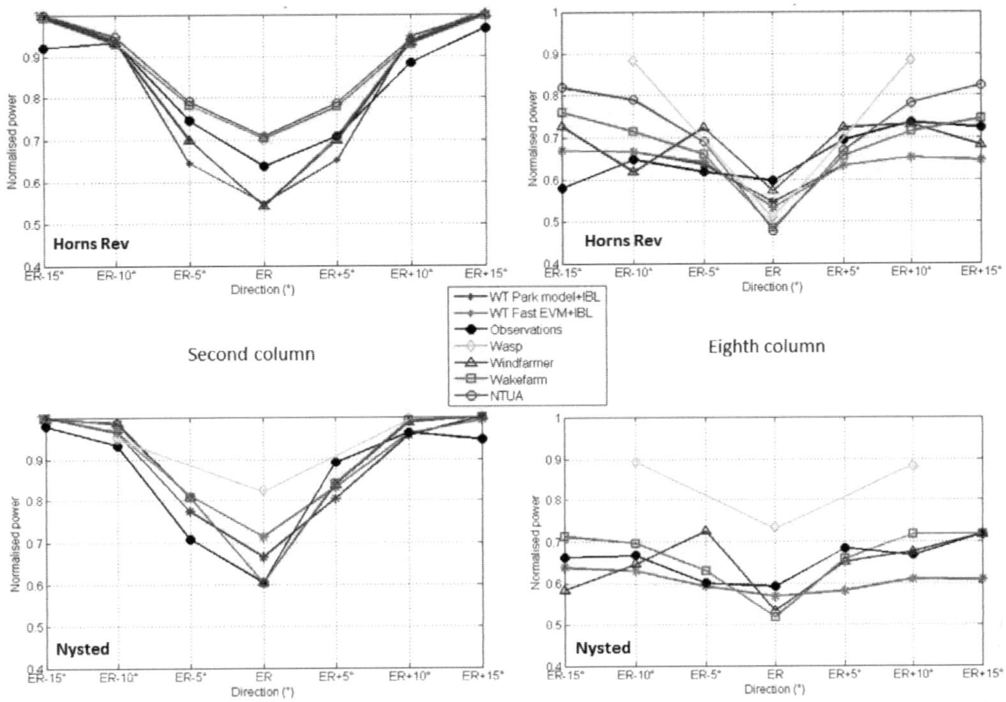

Figure 7: Mean normalized power from Horns Rev (top), Nysted (down) and model simulations for the second (left) and the eighth (right) columns of wind turbines.

5 . Model comparisons with offshore wind farm data

A model intercomparison is performed at the two offshore wind farms for four different wake models as in [18] and the two combined models.

5.1 . Wake width

As for other models, WT Park+IBL and Fast EVM model+IBL capture well the wake width at the second column of wind turbines (Figure 7) and show greater agreement with the observed wake depth than WAsP though both overestimate (respectively underestimate) the magnitude of the wake width at Horns Rev (Nysted).

For the entire wind farm (column 8), normalized powers of both combined models fit better with observations than other models even if they tend to overestimate (underestimate) the power for sectors less (greater) than ER.

In general, the root-mean-square error (RMSE) of normalized power shown in Table 2 indicates that WT Park+IBL and WT Fast EVM+IBL models perform better (i.e., exibit lower RMSE) for direct flow down the row (i.e, ER) than for oblique angles.

5.2 . Power deficit by downwind distance

Averaged normalized power as a function of downwind distance for a freestream wind speed of 8 m.s^{-1} for the seven wind directions are shown in Figure 8 (Horns Rev) and Figure 9 (Nysted). Both combined models appear to capture the shape of power deficit as a function of distance into both wind farms. In particular, this model has a very good agreement with observations at incident wind directions of ER-10°, ER-5°, ER+5° for Horns Rev and ER-15°, ER-10°, ER+10° for Nysted.

Moreover, WT Fast EMV+IBL model results have the same order of magnitude than data from Wakefarm and WindFarmer models, being even better at incident wind directions of 255°, 260°, 275° for Horns Rev and 263°, 268°, 273°, 288° for Nysted.

Overall the performance of a single wake model and the large wind farm correction combination is very promising.

Table 2: RMSE of normalized power from the models vs observations at Horns Rev (top) and Nysted (bottom).

Horns Rev						
Direction (°)	**WT Park+IBL**	**WT Fast EVM+IBL**	**WindFarmer**	**Wakefarm**	**WAsP**	**NTUA**
ER-15°	0.11	0.11	0.15	0.17		0.07
ER-10°	0.04	0.03	0.05	0.07	0.17	0.11
ER-5°	0.05	0.03	0.05	0.03		0.03
ER	0.06	0.06	0.04	0.08	0.06	0.08
ER+5°	0.07	0.05	0.04	0.05		0.04
ER+10°	0.08	0.08	0.03	0.03	0.11	0.03
ER+15°	0.07	0.07	0.07	0.03		0.10

Nysted					
Direction (°)	**WT Park+IBL**	**WT Fast EVM+IBL**	**WindFarmer**	**Wakefarm**	**WAsP**
ER-15°	0.02	0.02	0.09	0.05	
ER-10°	0.04	0.04	0.05	0.04	0.21
ER-5°	0.04	0.06	0.13	0.10	
ER	0.04	0.07	0.04	0.06	0.09
ER+5°	0.08	0.07	0.04	0.03	
ER+10°	0.03	0.03	0.04	0.03	0.12
ER+15°	0.07	0.06	0.03	0.04	

6 . Discussion and conclusion

A new correction to the classic wind farm wake model is presented that allows the disturbance of the ambient flow field caused by large wind farms to be modelled. Sensitivity studies of internal boundary layer parameters are carried out in order to assess the combination of single wake models (Park or Fast EVM) with a refined version of boundary layer model based on [6] and [26].

The two new combined model simulations were evaluated comparing wake width and normalized power output by turbine moving through the wind farm. The large wind farm models were able to capture wake width to some degree and the decrease of power output moving through the wind farm. Root-mean-square errors indicate generally better model performance for direct down the row flow than for oblique angles.

The large wind farm correction compared with four standard wake models is then validated with success for two offshore wind farms (Horns Rev and Nysted), solving a complex flow problem with little computational cost. This suggests that both combined wake models proposed assess well the power losses in those wind farms.

The main strenght of this new correction is the automatic estimation of the spacing between two rows of wind turbines along the wind direction S_d and the crosswind direction S_c. Indeed, the new algorithm developed to calculate S_c and S_d allows us to estimate directly the equivalent roughness z'_0 for each wind direction and wind speed at each turbine whatever the type of wind farms.

However given the limited set of validation cases there is still uncertainties in such extrapolation. The large wind farm model based on fundamental physics of the boundary layer is designed to scale to offshore wind farm layouts. In particular, the internal boundary layer parametrization depends on these wind farms and their equivalent roughness z'0.

Further validation cases are then needed and the large wind farm correction should be also tested on onshore wind farms where the topography is more complex.

In the future, a linear combination of single wake models with the boundary layer modification will be assessed to compute velocity and power deficits. Moreover, the influence of thermal stratification of the boundary layer will be also evaluated.

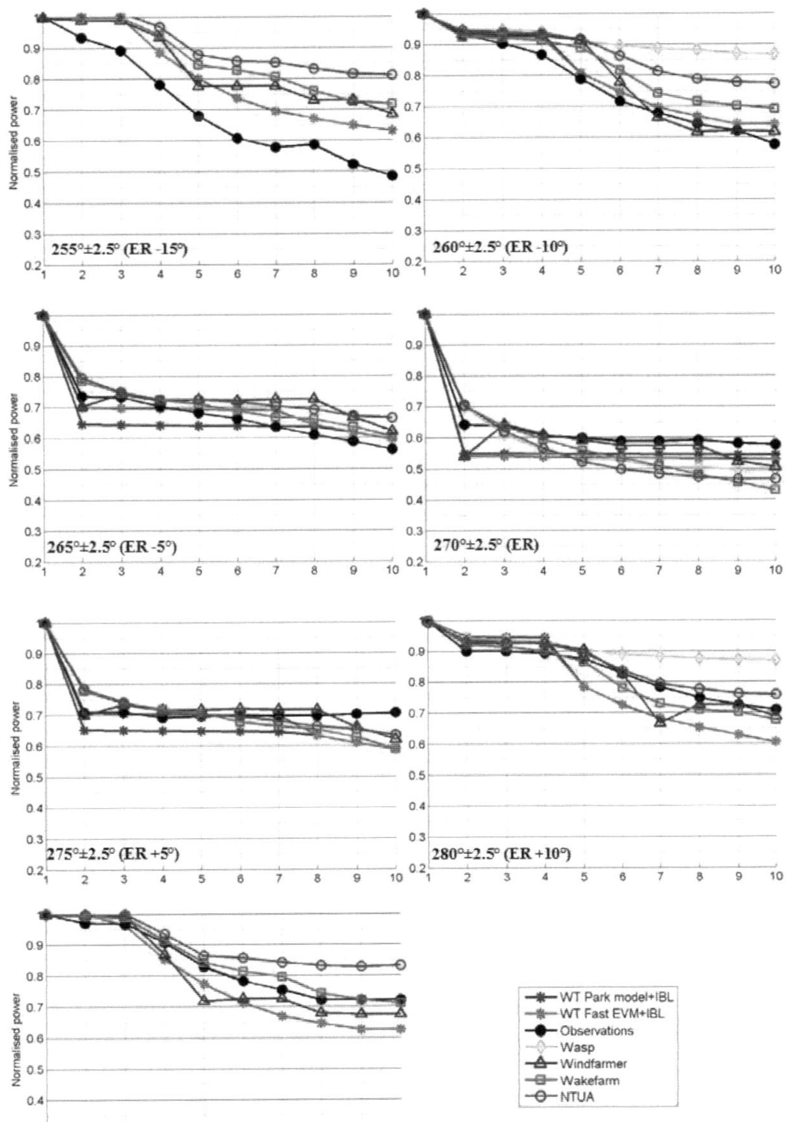

Figure 8: Normalized power at Horns Rev.

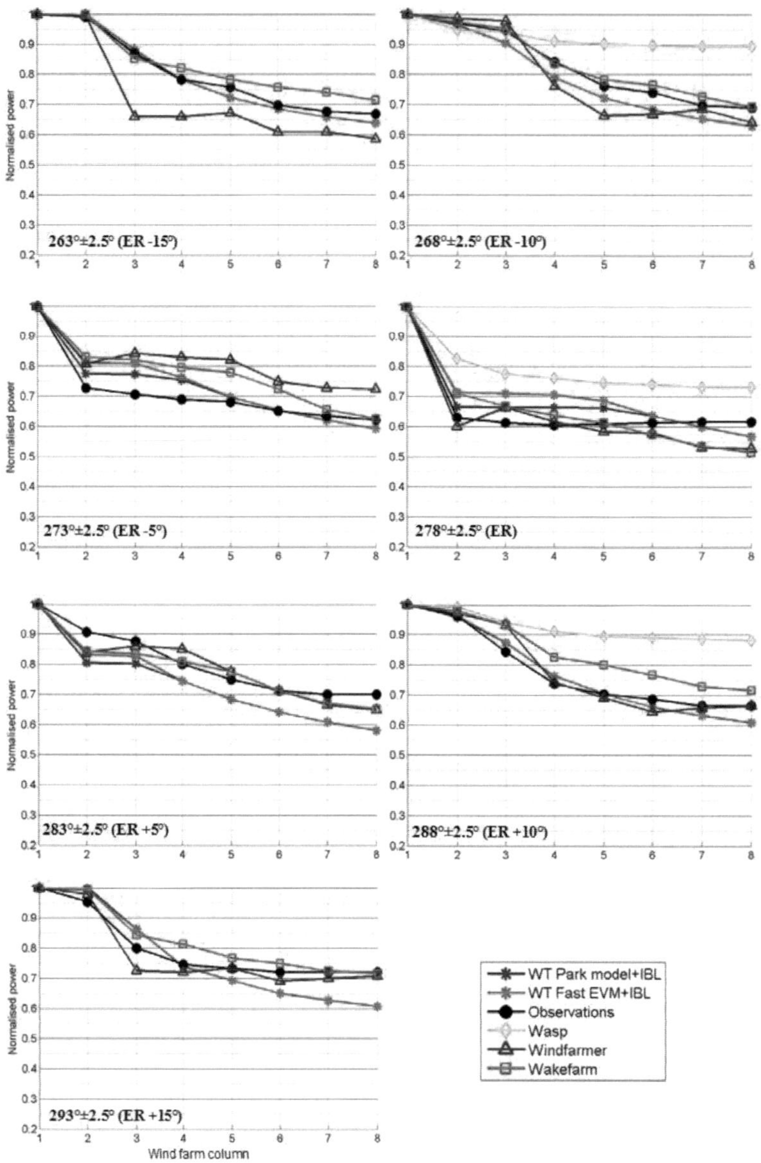

Figure 9: Normalized power at Nysted.

Nomenclature

$d\alpha$	Half turbine density sector (°)
Dir	Wind direction (°)
D_{rotor}	Rotor diameter *(m)*
ε	Turbulent dissipation rate *(J/(kg.s))*
ER	Exact wind farm row (°)
h	Boundary layer height *(m)*
h_{hub}	Hub height *(m)*
h_{ibl}	Internal boundary layer height *(m)*
H_{start}	Boundary layer offset *(m)*
k	Turbulent kinetic energy *($m^2.s^{-2}$)*
S_c	Spacing between two wind turbine rows along the crosswind direction*(m)*
S_d	Spacing between two wind turbine rows along the wind direction *(m)*
z'_0	Equivalent roughness *(m)*

Acronyms

CFD	Computational Fluid Dynamics
ECN	Energy research Center of the Netherlands
EVM	Eddy Viscosity Model
IBL	Internal Boundary layer
NTUA	The National Technical University of Athens
RMSE	Root-mean-square error
WasP	Wind Atlas Analysis and Application Program

References

[1] Crespo A, Hernandez J, Frandsen S. "Survey of modelling methods for wind turbine wakes and wind farms". *Wind Energy* 1999; 2:1-24.

[2] Vermeer LJ, Sørensen JN, Crespo A. "Wind turbine wake aerodynamics". *Progress in Aerospace Sciences* 2003; 39:467-510.

[3] Bradley EF. "A micrometeorological study of velocity profiles and surface drag in the region modified by a change in surface roughness". *Quart. J. R. Met. Soc.*1968; **94**:361-379.

[4] Jensen NO. "Change of surface roughness and the planetary boundary layer", *Quart. J. R. Met. Soc.* 1978; 104:351-356.

[5] Rao KS., Wyngaard JC, Coté DR. "The structure of the two-dimensional internal boundary layer over a sudden change of surface roughness". *J. Atmos. Sci.*, **26**, pp. 432-440, 1974.

[6] Schlez W, Neubert A. "New developments in large wind farm modelling". *Proc. European Wind Energy Conf.*, Marseille, France, 2009; EWEA PO.167, 8 p.

[7] Schepers, JG. "ENDOW: Validation and improvement of ECN's wake model". *Energy Research Center for the Netherlands rep.* ECN-C-03-034, 2003; 113 p.

[8] Mortensen NG., Heathfield DN., Myllerup L, Landberg L, Rathmann O. "Wind atlas analysis and application program: WAsP 8 help facility". *Risø National Laboratory*, Roskilde, Denmark, 2005.

[9] Rathmann O., Barthelmie RJ, Frandsen ST. "Turbine wake model for wind resource software". *Proc. European Wind Energy Conf.*, Athens, Greece, EWEA 2006, BL3.313.

[10] Magnusson M., Rados KG, Voutsinas SG. "A study of the flow downstream of a wind turbine using measurements and simulations". *Wind Eng*. 1996; 20:389-403.

[11] Li R., Delaunay D, Jiang Z. "A new Turbulence Model for the Stable Boundary Layer with Application to CFD in Wind Resource Assessment". *EWEA Proceedings*, Paris, France, 17-20 November, 2015; 9 p.

[12] Barthelmie RJ and Coauthors. "Efficient development of offshore windfarms (ENDOW): Modelling wake and boundary layer interactions". *Wind Energy* 2004; 7, 225-245.

[13] Barthelmie RJ, Folkerts L, Rados K, Larsen GC, Pryor SC, Frandsen S, Lange B, Schepers G. "Comparison of wake model simulations with offshore wind turbine wake profiles measured by sodar". *J. Atmos. Oceanic Technol.* 2006; 23, 88-901.

[14] Jensen NO. "A note on wind generator interaction". *Technical report from the Risø National Laboratory (Risø-M-2411)*, Roskilde, Denmark, 1983; 16 p.

[15] Katic I, Hojstrup J, Jensen NO. "A simple model for cluster efficiency". *EWEC Proceedings*, Rome, Italy, 1986; 5 p.

[16] Ainslie JF. "Calculating the flowfield in the wake of wind turbines". *I. Wind Eng. And Ind Aero.*, 1988; 27:213-224.

[17] Beaucage P, Robinson N, Brower M, Alonge C. "Overview of six commercial and research wake models for large offshore wind farms". *EWEA Proceedings*, Copenhagen, Germany,16-19 April, 2012;10 p.

[18] Barthelmie RJ, Pryor SC, Frandsen ST, Hansen KS, Schepers JG, Rados K, Schlez W, Neubert A, Jensen LE, Neckelmann S. "Quantifying the Impact of Wind Turbines Wakes on Power Output at Offshore Wind Farms". *Journal of Atmospheric and Oceanic Technology*, 2010; 27: 1302-1317.

[19] Anderson M. "Simplified solution to the Eddy-Viscosity wake model". RES document 01327-202, Issue 02, 10 July 2009.

[20] Barthelmie RJ, and Coauthors. "Modelling and measuring flow and wind turbine wakes in large wind farms offshore". *Wind Energy*, 2009, 12, 431–444, doi:10.1002/we.348.

[21] Jensen, L. "Wake measurements from the Horns Rev wind farm. Proc". European Wind Energy Conf., London, United Kingdom, European Wind Energy Association, 2004.

[22] Barthelmie RJ, Rethore, P. E. and Jensen L. "Analysis of atmospheric impacts on the development of wind turbine wakes at the Nysted wind farm". Proc. European Offshore Wind Conf., Berlin, Germany, European Wind Energy Association, PO.36., 2007c [Available online at http://www.eow2007proceedings.info/ allfiles2/132_Eow2007fullpaper.pdf.]

[23] Cleve, J., M. Grenier, P. Enevoldsen, B. Birkemose, and L. Jensen. "Model-based analysis of wake-flow data in the Nysted off- shore wind farm". Wind Energy, 2009, 12, 125–135, doi:10.1002/we.314.

[24] Frandsen S, Barthelmie R, Pryor S, Rathmann O, Larsen S, Højstrup J, Thøgersen M. "Analytical modelling of wind speed deficit in large offshore wind farms". *Wind Energy*, 9, 2006.

[25] Frandsen ST. "Turbulence and Turbulence-Generated Structural Loading in Wind Turbine Clusters". *Technical report from the Risø National Laboratory* (Risø-R-1188), Roskilde, Denmark, 2007; 130p.

[26] Larsen S, Mortensen E, Gylling N, Sempreviva AM, Troen, Ib. "Response of neutral boundary layers to changes of roughness". *Meteorology and Wind Energy Department*. Annual Progress Report. 1 January - 31 December 1987; pp. 15-43.

WindEurope Summit 2016 IOP Publishing
Journal of Physics: Conference Series **749** (2016) 012008 doi:10.1088/1742-6596/749/1/012008

3-D Time-Accurate CFD Simulations of a Multi-Megawatt Slender Bladed HAWT under Yawed Inflow Conditions

M. Sayed, Th. Lutz and E. Krämer

Institute of Aerodynamics and Gas Dynamics (IAG), University of Stuttgart, Pfaffenwaldring 21, D-70569 Stuttgart

E-mail: sayed@iag.uni-stuttgart.de

Abstract. In the present study numerical investigations of a generic Multi-Megawatt slender bladed Horizontal-Axis Wind Turbine (HAWT) under yawed inflow conditions were conducted. A three-dimensional URANS flow solver based on structured overlapping meshes was used. The simulations were conducted at wind speeds of 7 m/sec, 11 m/sec and 15 m/sec for different yaw angles ranging from $+60°$ to $-60°$. It was concluded that, for below rated wind speeds, under small yaw angles (below $\pm 15°$) the magnitudes of the blade forces are slightly increased, while under high yaw angles (above $\pm 15°$) there is a significant decrease. Moreover, the load fluctuations, for the different yaw angles, have the same frequency but different amplitude and oscillation shape. It was concluded that at the above rated wind speed of 15 m/sec, the blade aerodynamic loads are significantly affected by the yaw inflow conditions and the magnitude values of the loads are decreased with increasing yaw angle. It can be concluded that the angle of attack and the tower interference are the utmost variables affecting the yawed turbines.

1. Introduction

As the wind direction is always random and continuously changing, HAWTs are basically exposed to yawed flow conditions most of their operating time. Under these conditions, highly unsteady flow conditions over the wind turbine blades is introduced due to a parallel velocity component to the rotor disk. This unsteadiness results from strong radial flow along the span-wise as well as dynamic stall on the blade and also leading edge separation bubbles [1]. As a result, the rotor performance is strongly affected due to the blade load fluctuations which will also affect the operational life of the wind turbine. Consequently, it is important to understand the flow unsteadiness under yawed turbine operation conditions for better further development in the future.

Different aerodynamic methods are used to solve the flow field around the wind turbine. These methods vary in their degree of complexity and accuracy ranging from Blade Element Momentum theory (BEM) and lifting line theory [2] to the solution of Navier-Stokes equations using Computational Fluid Dynamics (CFD). Simplified methods such as the lifting line theory were used in [1] and [2] to investigate the aerodynamic loads of the wind turbine under yaw conditions. Simulations for yaw angles of 10°, 30° and 60° were performed at a wind speed of 10 m/sec. The results showed that the method cannot capture the dynamic stall and stall delay and therefore, a dynamic stall model should be included to improve the accuracy of the results. CFD models of wind turbines have been widely used, as in [3], [4], [5], [6], [7], [8] and [9], to capture the flow phenomena that cannot be resolved using simplified numerical methods. CFD simulations of a compressible Euler flow using unstructured meshes were performed in [6]. This aerodynamic model was used because of an inviscid flow assumption at this wind speed. Only one flow case was investigated at a yaw angle of 30° and a low wind speed of 7 m/sec.

Simulations of yawed wind turbines based on the solution the Navier-Stokes equations were examined in literature. A structured mesh flow solver combined with a BEM model were used to investigate the wind turbine at a yaw angle of 45° for a low wind speed of 7 m/sec [7]. It was concluded that the classical inflow model used in the BEM should be improved to accurately capture the downwash flow of the rotor. Moreover, a hybrid Navier-Stokes/Vortex-wake method coupled to a turbulence model of BaldwinLomax was used to investigate the blade loadings at different flow cases [8]. But there were no detailed studies on the time-accurate blade loadings.

In the present paper, 3-D time-accurate CFD simulations were conducted on a multi-megawatt HAWT rotor under yawed inflow conditions. The simulations were performed at wind speeds of 7 m/sec, 11 m/sec and 15 m/sec for different yaw angles ranging from +60° to −60°. The blade loads are used to present and discuss the time-accurate results and detailed flow features over the blades are also observed.

2. Methodology

2.1. Wind Turbine configuration

The wind turbine examined in the present study is the generic reference turbine of the ongoing European InnWind project [10] designed by DTU. It is a large HAWT with 178.3 m rotor diameter, a hub height of 119 m and FFA-W3 airfoil series. The rotor is installed at 5° tilt angle and a cone angle of −2.5°. The solid modelling of the complete wind turbine is presented in Figure 1, in which the used coordinate system, wind direction and yaw and azimuthal directions are defined. For the present simulations, all the wind turbine components are assumed to be rigid. A cylindrical tower is considered with 8.3 m outer diameter at the ground and 5.5 m outer diameter at the top. The wind turbine rotor rotates clockwise as viewed from the inlet boundary. Pre-bended blades are used with about 3.73%R at the blade tip. The blade at the root region is equipped with wedge shaped Gurney flaps to increase the aerodynamic performance [10].

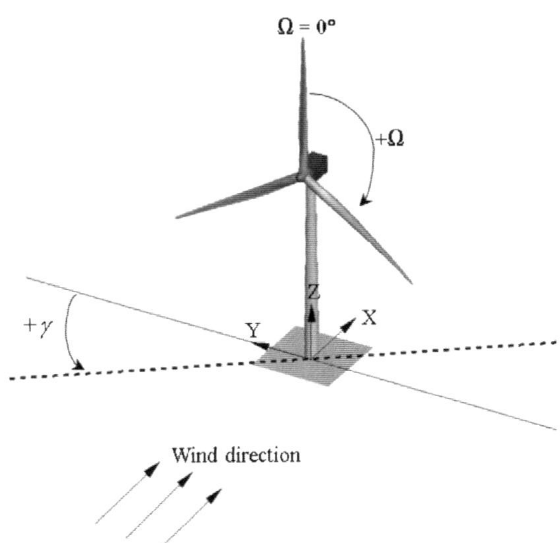

Figure 1: Wind turbine configuration and definitions of global coordinate system, wind direction, positive yaw and positive azimuthal directions

2.2. Process Chain for Wind Turbine Simulations

The simulations performed here have been achieved with FLOWer [11] which is a block structured flow solver. It is provided by the German Aerospace Center DLR with specific extensions implemented at the IAG. In order to achieve the proposed simulations of the wind turbine, a process starting from generating the geometries of the used model to the final discretized model must be performed. In the recent years, a process chain of the wind turbine simulations was developed at the IAG. This process chain includes a user friendly script for blade grid generation in purpose of minimizing the effort in the blade meshing process. Hence the blade mesh is the most important key feature in the turbine simulations, one must carefully set the different parameters in the script to insure the quality of the generated mesh. Several CFD simulations of different offshore and onshore wind turbines have been conducted using that process chain at IAG [12], [13] and [14]. These CFD simulations proof the quality of the used process chain at the IAG.

2.3. Computational Domain

In this paper, a structured overset mesh technique is used to generate the computational domains. This overset feature allows more mesh control over all the wind turbine parts, as every component was meshed separately and finally the domains are placed together and overlapped using CHIMERA overlapping mesh technique. The computational domain of the whole turbine consist of three blades, hub, nacelle and tower. The meshes were generated using Gridgen®. The blade mesh was generated and refined by the developed script at IAG. For each sub-domain, an independent grid was created with adopted refinement. Finally a background mesh was created to extend the computational domain to the far field [14]. The grid statistics of the different domains used in the simulations are as shown in Table 1 and Figure 2.

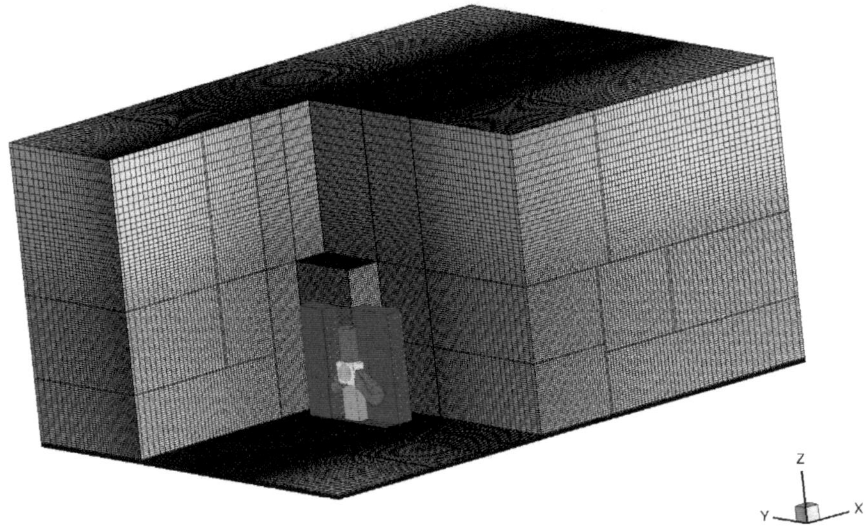

Figure 2: The assembled domains, Red are blades domains, Orange is the tower domain, Green represents the hub domain, Cyan shows the nacelle domain, Blue is the fine background domain and Black is the extended background domain

Two different background meshes were used. The first one is the coarse background mesh with cell size of 5 m in all directions around the turbine and cell size of 20m at the far field boundaries. This domain has a cubic shape (as show in Figure 2) with dimensions of 6R in the upstream direction, 9R in the downstream direction, 8R in the far field direction and 6R in each of the horizontal direction. These values are used based on studies done by [14]. At the ground, the mesh was refined to account for the boundary layer at the ground. The second

Table 1: Grid statistics

Domain	Number of cells [*million*]
Blade	6.5
Tower	1.15
Hub	2.2
Nacelle	1.75
Fine Background	0.9
Extended Background	3.1
Blade-Hub Connector	0.8
Total Number of Cells	31

one is the fine background mesh with 2m cell size in all directions. This domain has a cubic shape with dimensions of 1R×3R×3R in the $x-$, $y-$, $z-$direction respectively. It is used only to avoid having the whole background mesh as a fine mesh which is too costly and also to get good overlapped regions. Both are non-rotating meshes. The tower mesh has an outer cubic shape with 0.5R in x- and y-direction. The grid size at the outer boundary is 0.8m and refined close to the surface of the tower to fulfill the $y^+ \leq 1$ condition at the first grid cell. A no-slip boundary condition is applied to the tower surface. The nacelle mesh was also generated taking into account the $y^+ \leq 1$ condition at the nacelle wall and also the no-slip boundary condition is applied. The cell size at the nacelle surface is 0.3 m and it is the last non-rotating grid. The hub mesh was generated and it is the first rotating mesh in this setup. It was connected to a rotating blade-hub-connector mesh and the non-rotating nacelle mesh. The hub rotation is around the rotor axis as defined before at 5° tilt angle.

Finally the blade mesh was generated (blade sectional mesh is shown in Figure 3) using the proven script at the IAG [13] and [15]. 300 cells were used in chord-wise direction and 100 cells in (span-wise) radial direction. 35 cells were introduced to resolve the boundary layer with $y^+ \approx 1$ for the first boundary layer cells. The time-step was equivalent to 3° azimuthal-step. These values were chosen based on studies performed at IAG [14]. The turbulence model used in the simulations is the Wilcox k-ω turbulence model [16] with fully turbulent flow state. The k-ω model has been successfully used to simulate the flow over airfoils and wind turbines. Moreover, it has been used in many wind-turbine CFD studies such as [1], [3] and [14]. All simulations were performed on the Cray XE6 (HLRS) named HORNET using 448 cores

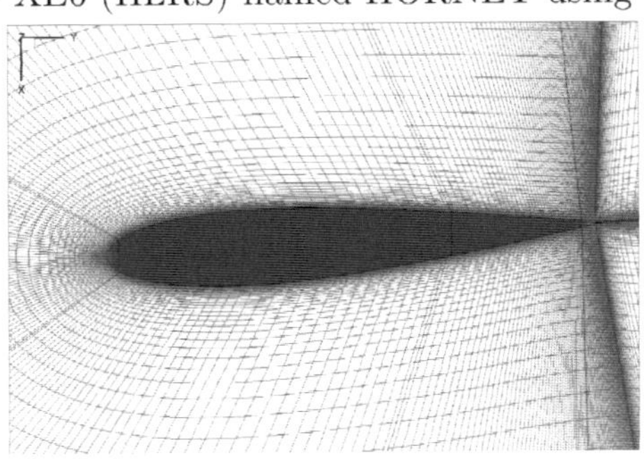

Figure 3: Blade profile meshing at $r = 0.8R$

3. Results

In the present study, the simulations were performed for wind speeds of 7, 11 and 15 m/sec at the yaw angles (γ) of −60°, −45°, −30°, −15°, 0°, 15°, 30°, 45° and 60°. As designed, rotor rotational speeds of 6, 8.83564 and 9.6 RPM were used respectively for the used wind speeds. The reason of choosing these wind speeds was to investigate the wind turbine at minimum, rated and maximum rotational speed. Analyzing yawed inflow conditions can be done either by yawing the wind turbine and keeping the wind direction constant (as in this article) or by keeping the turbine fixed and yawing the wind direction. As the wind turbine is assembled, it was easier to rotate the wind turbine around the $z-$axis by rotating the tower. Because all other components are in the tower coordinate system, all other parts will be rotated as the tower rotates.

For each case, the simulations were performed until periodic unsteady blade aerodynamic loadings were obtained. 25 revolutions were simulated in order to ensure the convergence of the load values, then the results of the last rotor revolution are evaluated. The time-averaged and unsteady time-varying results for the last rotor revolution are presented in form of the blade surface pressure and the normal and tangential force coefficients. Detailed analyses of the flow physics are presented and analyzed to study the dynamic nature of the rotor behavior at yaw misalignment.

3.1. *Effects of Yawed Inflow on Wind Turbine Power*

As the most important factor in the design of the wind turbine is the power output, the effect of the yawed inflow conditions on the wind turbine power and thrust is studied. The effect of the yawed inflow on the power coefficient C_p and the thrust coefficient C_a is shown in Figure 4. The presented values are a result of time-averaging over the last rotor revolution and then normalized by the results for 0° yaw angle (Baseline). As shown, for small yaw angles between +15° and −15° there is a slight increase in both values of C_p and C_a with respect to the baseline. The reason for that might be that under these small yaw angles there is stall delay which means more attached flow over the blade in the advancing half of the rotor plane. Then for the higher yaw angles, a sudden decrease can be observed of about ≈70% of the C_p and ≈40% of the C_a of the baseline at ±60° yaw angles. These values are for the wind speeds of 7 m/sec and 11 m/sec as shown in Table 2. In addition severe reduction is predicted in case of 15 m/sec where negative values are obtained at the ±60° yaw angles. As the CFD simulations were performed without controller, i.e constant RPM of 9.6 and constant pitch angle, negative power values were predicted due to negative blade forces in the advancing half of the rotor as a result of geometrical tilt, precone and twist angle and yaw angle.

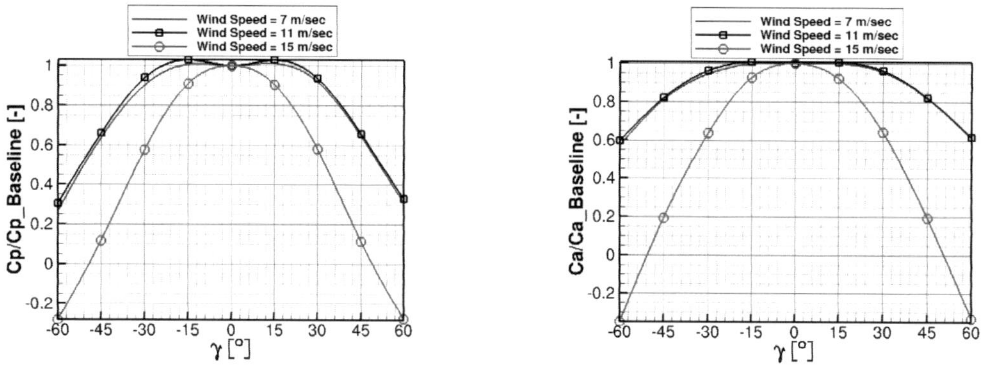

Figure 4: Effect of yawed inflow on the power output at different wind speeds

It was suggested in the literature that the power is reduced by yawed inflow by a cosine-cubed factor (equation 1) and the thrust is reduced by a cosine-squared factor (equation 2) [7] and [17].

$$C_p(\gamma) = C_p(0)cos^3\gamma \tag{1}$$

$$C_a(\gamma) = C_a(0)cos^2\gamma \tag{2}$$

Table 2: Percentage value of the reduction in Power and Thrust

γ [°]	Wind speeds [m/sec]					
	7		11		15	
	C_p [%]	C_a [%]	C_p [%]	C_a [%]	C_p [%]	C_a [%]
−60	73	43	69	40	128	135
−45	36	19	34	18	88	81
−30	10	5	6	4	42	36
−15	−1	0	−3	−1	9	7
0	0	0	0	0	0	0
+15	−1	0	−3	−1	9	8
+30	8	5	6	4	42	36
+45	35	18	34	18	89	81
+60	69	39	67	38	128	133

It was also suggested to change the cosines power value in order to find a good matching between the suggested equations and the results. As shown in Figure 5 for the case of 7 m/sec and 11 m/sec, the reduced power and thrust coefficients are not following the reduction factors suggested by [7] and [17]. The reason for that is the slight increase in the C_p and C_a coefficients at small yaw angles. Moreover, in case of 15 m/sec, the suggested factors can be used for yaw angles between ±30° with less than 2% error with cosine power of 3.5 for the power coefficient C_p as shown in Figure 5.

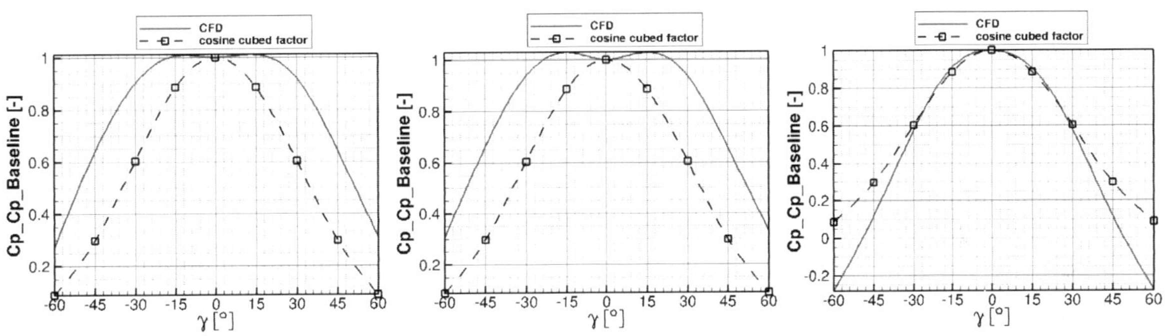

(a) 7 [m/sec] (b) 11 [m/sec] (c) 15 [m/sec]

Figure 5: CFD power coefficients compared to the suggested reduction cosine-cube factor

In addition, the azimuthal variations of the power coefficient C_p are presented for all wind speeds and all yaw angles as shown in Figure 6. In addition to the power reduction, which is also clear in this figure, the power oscillation amplitude is sensitive to the yaw angle as shown in Figure 7. For wind speeds of 7 m/sec and 11 m/sec, the amplitude is minimum at the baseline operating condition of zero yaw. Moreover, larger amplitudes for the positive yaw than for the negative one at the same yaw angle are observed. The oscillation amplitude is smoothly increased till 45° yaw angle and then starts to decrease again. In case of a wind speed of 15 m/sec, the amplitude is 5% which is more than the double for the other wind speeds. The reason for that is the strong tower interference with the rotated blades as there is more reduction in the inflow velocity.

3.2. Effects of Yawed Inflow on Blade loads

3.2.1. Blade loads at a Wind speed of 11 m/sec

First, the simulations were performed for rated wind speed. The loads resulting from the different yawed conditions are also discussed. The azimuthal variation of the normalized thrust and torque is shown in Figure 8. The presented values are normalized by the results for 0° yaw

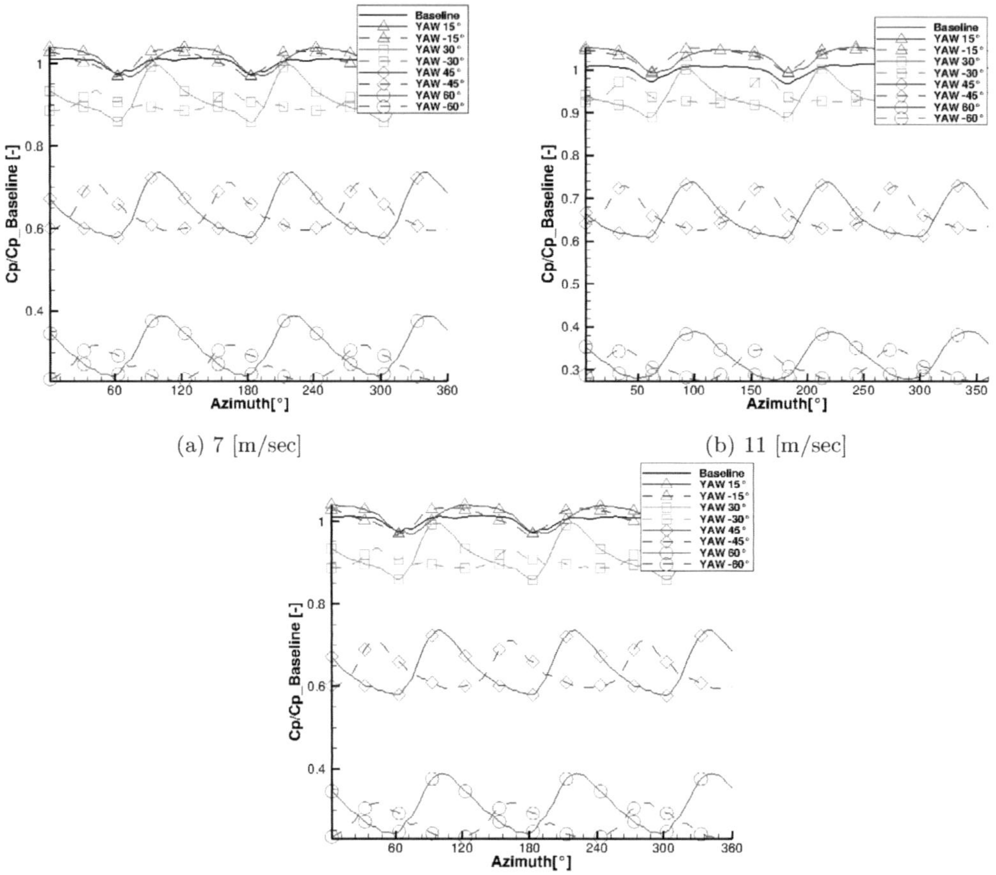

(a) 7 [m/sec] (b) 11 [m/sec]

(c) 15 [m/sec]

Figure 6: Effect of yawed inflow on the power output at different wind speeds

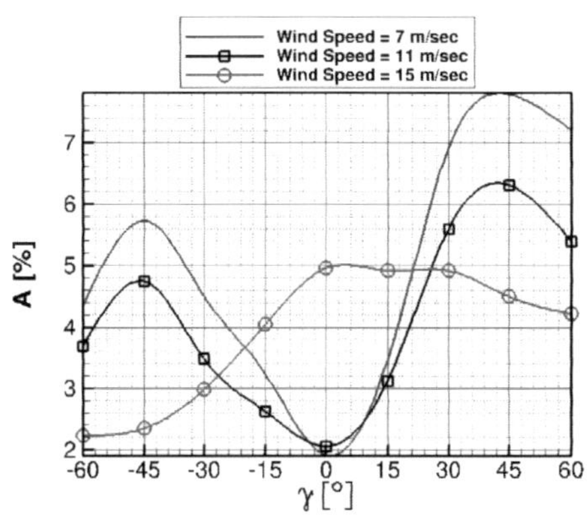

Figure 7: Amplitude variation over the different yaw angles

angle (Baseline). It can be observed that, for small positive yaw angles of less than 30°, there is a slight variation in the thrust about the mean value. There is an increase in the advancing half of the rotor plane (azimuth angle between 90° and 270° for the positive yaw angles), and a slight decrease in the retreating half of the rotor plane (azimuth angle between 270° and 90° for the positive yaw angles).

At 45° there is a dominant increase in the advancing half or the rotor with about 13.5% maximum value at an azimuth angle of 115°. Further increase can be seen at 60° which about 35% higher thrust values were predicted in the advancing half of the rotor at azimuth angle of 120°. For the torque (Figure 8 right), at yaw angles of 15° there is about 20% increase at the

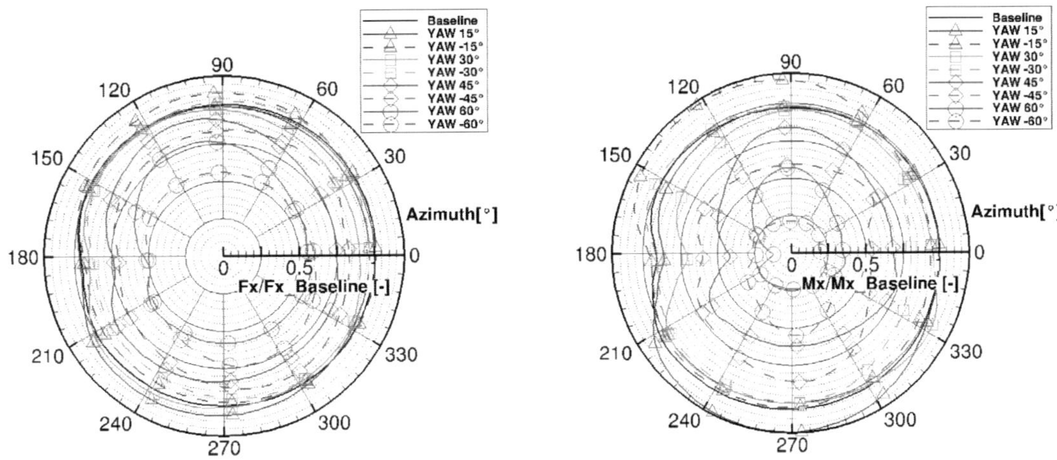

Figure 8: Normalized Thrust (left) and Torque (right) at different yaw angles.

end of the advancing half of the rotor, this is because of the increase in the edge-wise force in the advancing half of the rotor as shown in Figure 9a. At yaw angles of 60° there is about 120% increase in the advancing half of the rotor. The reason behind that is that in the advancing half of the rotor plane the angle of attack increases and hence the blade thrust is increased reaching the maximum at about 120° as shown in Figure 8. In case of negative yawed inflow angles the behavior is inversed.

At an azimuth angle of 180°, the tower interference with blade 1 is seen in Figure 8. The impact increases with increasing yaw angle, the reason for that is the inflow velocity reduction due to the yaw angle. Moreover, it can be concluded that there is more inflow velocity reduction at positive angles than at negative ones. The effect of yaw on the edge-wise force and moment are presented in Figure 9. As shown in Figure 9a, the tower effect is dominant at the baseline. The normalized edge-wise force is reduced by about 15% at the azimuth angle of 180°. Moreover, it is more affected by the tower interference at yaw angle of 45° which is reduced by about 73%. It was expected that in the advancing half of the rotor the edge-wise force will be also increased as in the case of flap-wise forces because of the angle of attack increase. As shown, a slight increase in case of +45° of about 6% is observed. For the edge-wise moment, less tower influence at ±60°, but it is increased by about 60% in the advancing half as shown in Figure 9b.

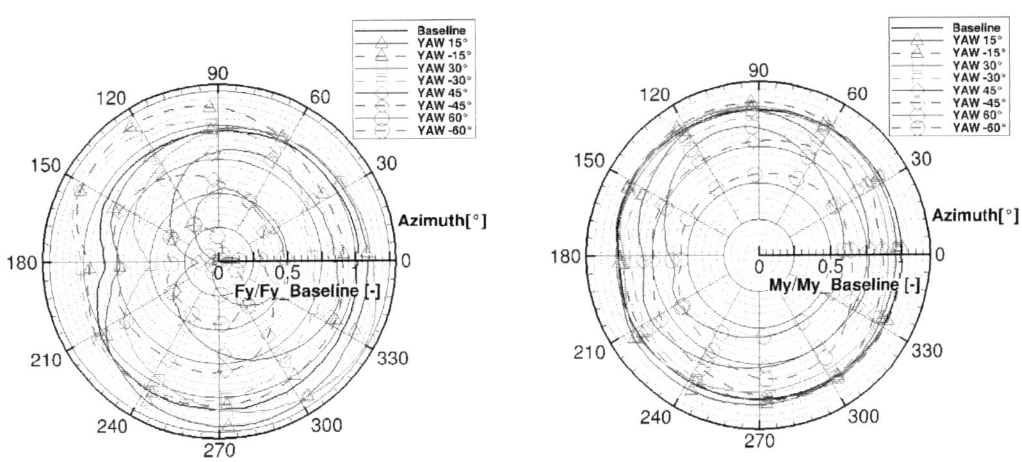

(a) Normalized edge-wise force (b) Normalized edge-wise moment

Figure 9: Normalized forces and moments of blade 1

Journal of Physics: Conference Series **749** (2016) 012008 doi:10.1088/1742-6596/749/1/012008

These changes in the load values might affect the structural vibration of the blades. Therefore, a Fast Fourier Transform (FFT) was performed. As shown in Figure 10, for the forces, the baseline has the smallest amplitude and as the turbine is yawed the amplitude is increased. It can be concluded that the amplitude is higher for negative yaw compared to the positive one for the same yaw angle except for 60° where both have the same amplitude. In contrast, for the moments, at positive yaw angles the amplitude is slightly higher than at the negative ones. The highest amplitudes are in the advancing half on the rotor, which is expected from the load distribution over one revolution. These high amplitudes will have strong influence on the fatigue and aero-elastic results and should be taken into account.

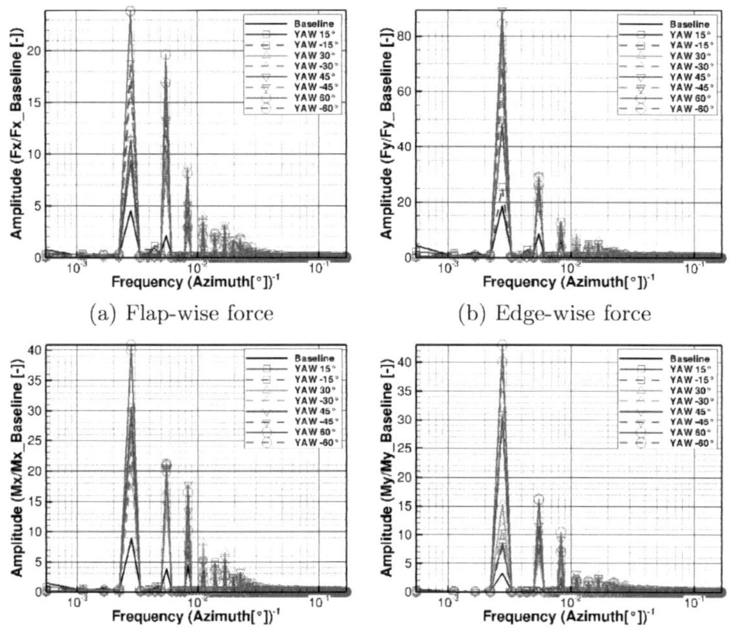

(a) Flap-wise force (b) Edge-wise force

(c) Flap-wise moment (d) Edge-wise moment

Figure 10: FFT of the normalized forces and moments of blade 1

3.2.2. Blade loads at a Wind speed of 7 m/sec

For the wind speed of 7 m/sec, the azimuthal variation of the normalized forces and moments are shown in Figure 12. As described previously, the presented values are normalized with the results of the baseline at the wind speed of 7 m/sec. As shown in Figure 12, it can be concluded that, the yawed inflow at this wind speed has almost the same effect as in the case of 11 m/sec. The reason behind that is that the flow is still attached and the blade is not pitched. Therefore, a slight change occurs for small yaw angles of less than +30°. At yaw angle of +45° there is an increase by about 12.5% of the thrust as shown in (Figure 11a). For +60° an about 33% higher thrust value was predicted in the advancing half of the rotor. For the torque (Figure 11b), at yaw angles of +15° there is about 22% increase in the advancing half of the rotor, this is because of the edge-wise force increase in the advancing half of the rotor as shown in Figure 11c. At yaw angles of +60° there is about 140% increase in the advancing half of the rotor. The reason behind that is that in the advancing half of the rotor there is an increase in the angle of attack and hence the blade thrust is increased reaching the maximum at about 120°. So it can be concluded that for wind speeds less than the rated wind speed, the wind speed has almost no effect on the load distribution and the inflow yawed condition is utmost. The FFT results confirm the same predicted frequencies with very small decrease in the amplitudes compared to the case of 11 m/sec wind speed.

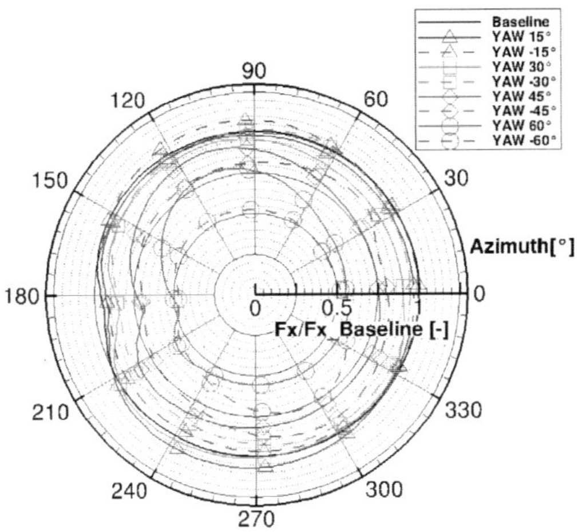

(a) Thrust force

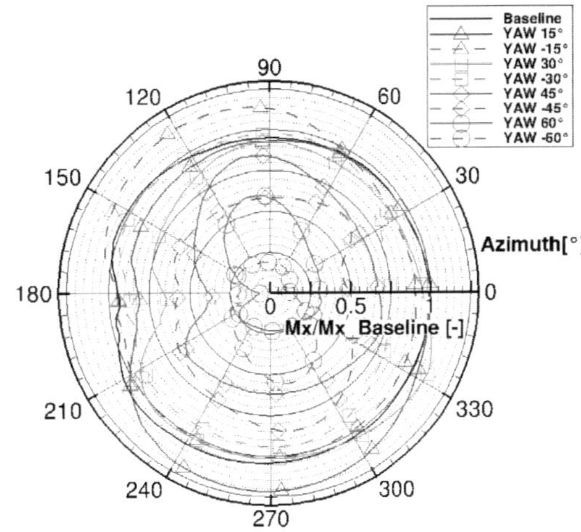

(b) Torque

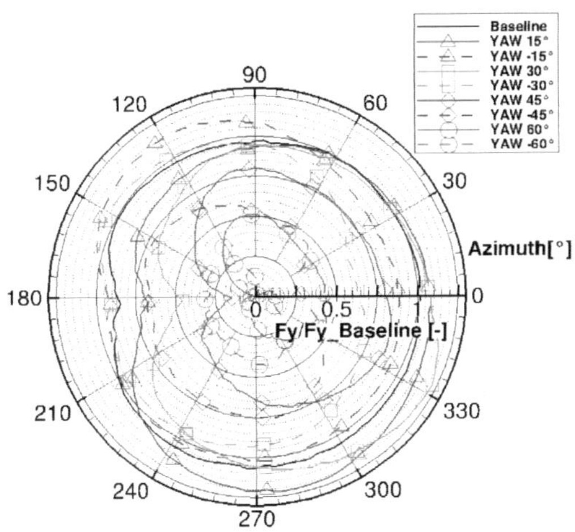

(c) Edge-wise force

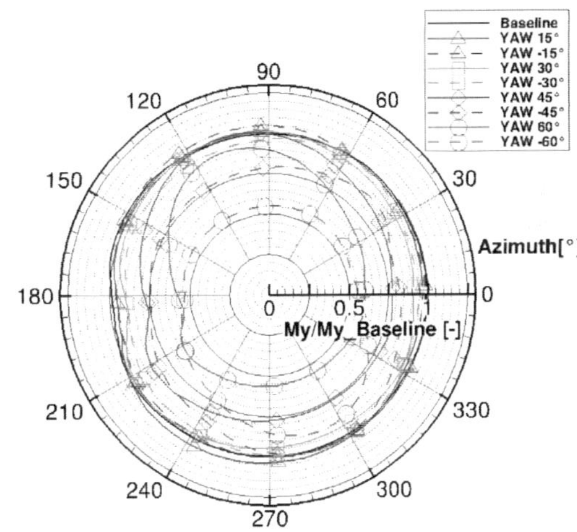

(d) Edge-wise moment

Figure 11: azimuthal variations of the normalized forces and moments of blade 1

3.2.3. Blade loads at a Wind speed of 15 m/sec

Finally, the simulations were performed for a wind speed of 15 m/sec. The blades were pitched with 10.95805° in order to keep the flow attached to the surface of the blade by means of decreasing the angle of attack. The azimuthal variation of the normalized forces and moments are shown in Figure 13. The presented values are normalized by the results of the baseline at the wind speed of 15 m/sec. As shown in Figure 13, for the case of the baseline, the load distributions over one revolution is not symmetric. That is the first difference between this wind speed and the previous ones. The reason for that is the pitch angle in addition to the twist angle which varies from $-14.5°$ to $3.428°$ at the tip. In addition to that is the cone and the tilt angle. All these angles finally affect the angle of attack at the blade sections which might change the load distribution over one revolution for the baseline case. Moreover, for all yaw angles from $+60°$ to $-60°$ the loads are strongly affected by the tower at azimuth angle of 180° for blade one. There is decrease of about $15\% - 30\%$ for all loads when blade 1 is crossing the tower. This is because of more decrease in the inflow velocity due to higher wind speed and also because of the blade pitch. Between yaw angles of $+15°$ and $+45°$, there is increase in the load values in the retreating half of the rotor and decrease in the advancing half. Starting from $+45°$, severe tower effect is predicted and the load values becomes negative reaching the minimum value at the azimuth angle of 180°.

Journal of Physics: Conference Series **749** (2016) 012008 doi:10.1088/1742-6596/749/1/012008

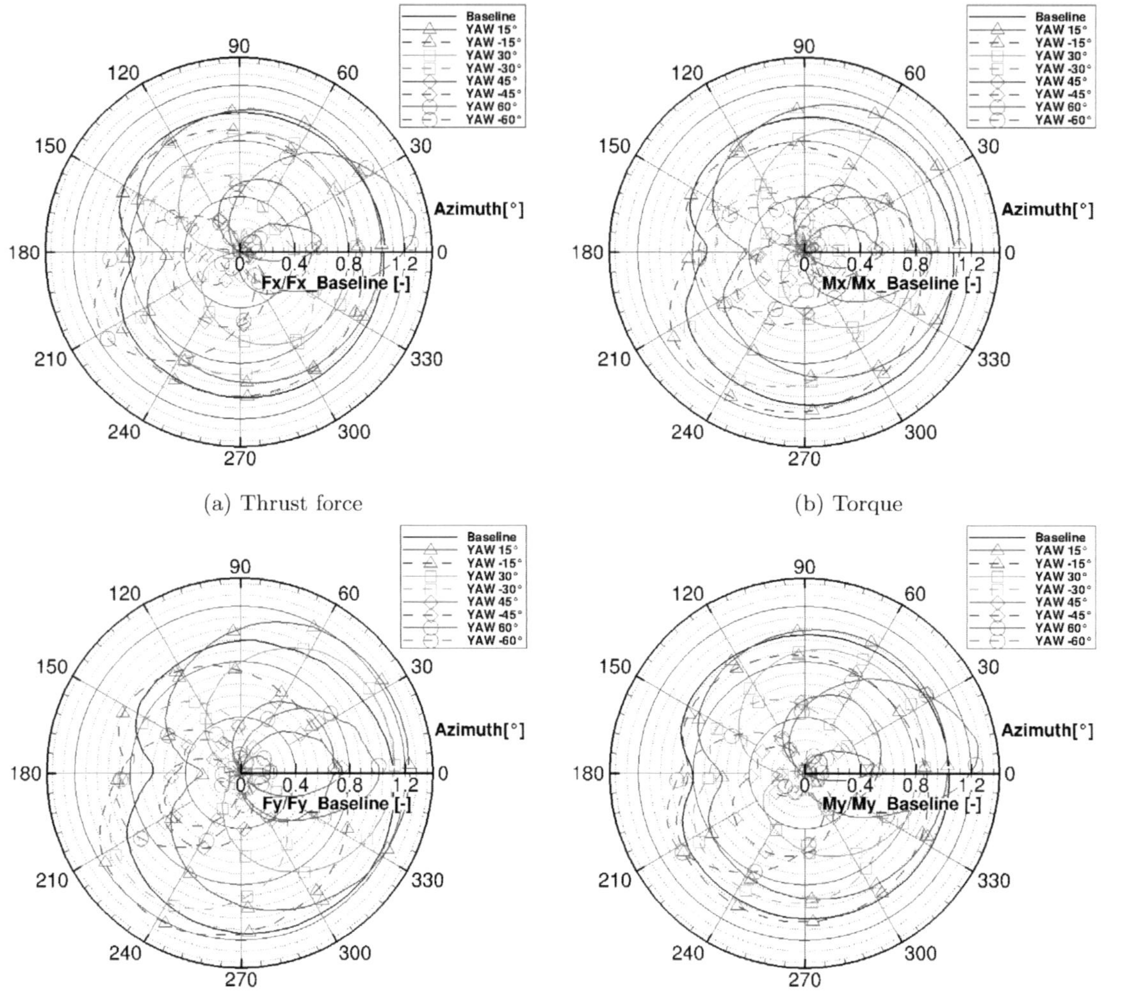

(a) Thrust force (b) Torque

(c) Edge-wise force (d) Edge-wise moment

Figure 12: azimuthal variations of the normalized forces and moments of blade 1

These utmost changes in the load values might be critical on the structural vibration of the blades therefore FFT was performed. As shown in Figure 13, all load components have greater amplitude values compared to the previous two cases of low and rated wind speeds. Moreover, for all load components, the amplitude at positive yaw is greater than at negative one. For the normalized flap-wise force and the torque, the largest amplitude is at $\pm 45°$ yaw angle which decreases again at $\pm 60°$. In addition, for the blade twisting moment the largest amplitude is at $\pm 30°$ yaw angle. These ultimate amplitudes will have strong influence on the fatigue and aero-elastic results and should be taken into account.

3.3. Effects of Wind speed on Blade loads

The effect of the wind speed on the distribution of the predicted loads at different yaw angles was investigated. As shown in Figure 14, the below rated wind speeds of 7 m/sec and 11 m/sec have almost the same distribution of the load over one revolution. The presented values are the normalized thrust by the mean value over the blade radius at every wind speed. It is clear that at wind speeds of 7 m/sec and 11 m/sec the distribution is the same for the all positive yaw angles. At 15 m/sec, it is again clear that strong tower influence on the inflow reduction is predicted. The reason for that as discussed before is that the blade is pitched which strongly affect the distribution of the angle of attack over the blade sections.

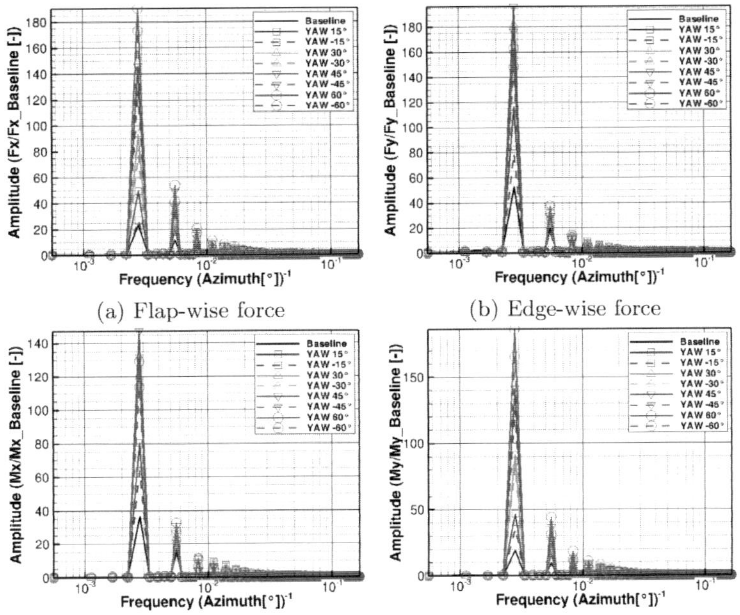

(a) Flap-wise force (b) Edge-wise force

(c) Flap-wise moment (d) Edge-wise moment

Figure 13: FFT of the normalized forces and moments of blade 1

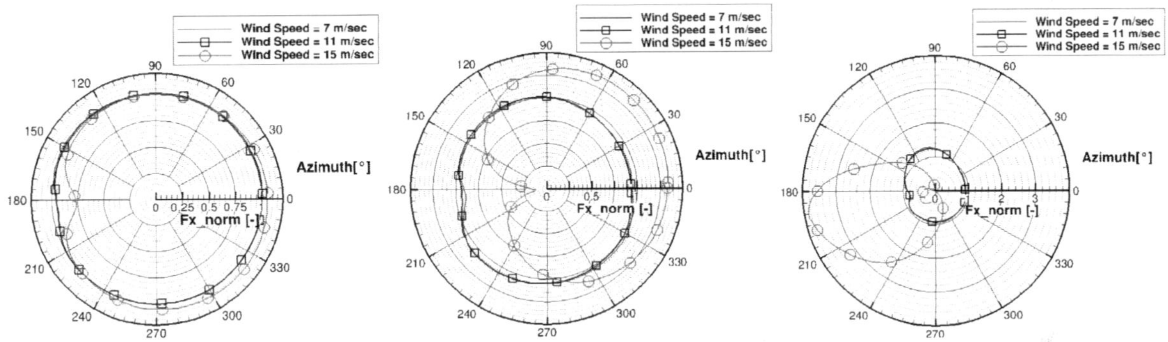

(a) Baseline (b) +30° yaw angle (c) +60° yaw angle

Figure 14: The normalized thrust forces of blade 1 at different wind speeds

4. Conclusion

In this paper, CFD simulations of a yawed large slender bladed HAWT were conducted using a 3D URANS flow solver based on structured overlapped meshes. The Wilcox k-ω turbulence model has been used for the all simulations. Different yawed conditions have been examined ranging from $+60°$ to $-60°$. At the rated wind speed of 11 m/sec and the below rated wind speed of 7 m/sec, time-accurate simulations were conducted. In both cases, as the yaw angle increases to $\pm 15°$, the aerodynamic loads over the blades are slightly increased. This might be because of the stall delay results from the reattachment of the flow over the blade surface. As the yaw angle is further increased above $\pm 15°$, the aerodynamic loads are significantly reduced. The reason for that is that the wind component becomes more aligned tangent to the rotor disk plane. Larger load fluctuations were predicted as the yaw angles were increased.

There is a massive increase in the amplitude at higher yaw angles and the highest amplitudes are in the advancing half on the rotor. These high amplitudes will definitely affect the structural design of the blade and must be taken into account. At wind speeds less than rated wind speed, the yaw angle is the most dominating factor affecting the blade load distributions over one revolution more than the wind speed. It was predicted that the load distributions were the same with different absolute values. In contrary, at above rated wind speed of 15 m/sec, the wind speed has an utmost effect on the load distributions over one revolution. The reason behind that is the change in the pitch angle which has a great influence on the distribution of the angle of attack along the blade radius.

Acknowledgments
The DAAD (Deutscher Akademischer Austauschdienst) is acknowledged for financial support.

References

[1] Tsalicoglou C, Jafari S, Chokani N and Abhari R 2013 Rans computations of wind turbine near-wake aerodynamics in uniform and yawed inflow *Proceedings of the ASME Turbo Expo* vol 8 (New York, N.Y.: The American Society of Mechanical Engineers) pp 1–12
[2] Duque E P, Burklund M D and Johnson W 2003 *Journal of Solar Energy Engineering* **125** 457–467
[3] Sørensen N N, Michelsen J and Schreck S 2002 *Wind Energy* **5** 151–169
[4] Hansen M O L, Sørensen J N, Voutsinas S, Sørensen N and Madsen H A 2006 *Progress in aerospace sciences* **42** 285–330
[5] Zahle F and Sørensen N N 2007 On the influence of far-wake resolution on wind turbine flow simulations *Journal of Physics: Conference Series* vol 75 (IOP Publishing) p 012042
[6] Sezer-Uzol N and Long L N 2006 *AIAA paper* **394** 2006
[7] Madsen H A, Sørensen N N and Schreck S 2003 Yaw aerodynamics analyzed with three codes in comparison with experiment *ASME 2003 Wind Energy Symposium* (American Society of Mechanical Engineers) pp 94–103
[8] Tongchitpakdee C, Benjanirat S and Sankar L N 2005 *Journal of solar energy engineering* **127** 464–474
[9] Kim D H, Yu D O, Kwon O J, Fletcher T M, Scheurich F and Brown R 2010 Predicting unsteady blade loads of a wind turbine using rans and vorticity transport methodologies *European Wind Energy Conference* pp 33–37
[10] Bak C, Zahle F, Bitsche R, Kim T, Yde A, Henriksen L C, Natarajan A and Hansen M 2013 *DTU Wind Energy Report-I-0092* **5**
[11] Kroll N, Rossow C C, Becker K and Thiele F 2000 *Aerospace Science and Technology* **4** 223–237
[12] Weihing P, Meister K, Schulz C, Lutz T and Krämer E 2014 Cfd simulations on interference effects between offshore wind turbines *Journal of Physics: Conference Series* vol 524 (IOP Publishing) p 012143
[13] Schulz C, Klein L, Weihing P, Lutz T and Krämer E 2014 Cfd studies on wind turbines in complex terrain under atmospheric inflow conditions *Journal of Physics: Conference Series* vol 524 (IOP Publishing) p 012134
[14] Sayed M, Lutz T and Krämer E 2015 Aerodynamic investigation of flow over a multi-megawatt slender bladed horizontal-axis wind turbine *Renewable Energies Offshore* (CRC Press) pp 773–780
[15] Meister K, Lutz T and Krämer E 2009 *EUROMECH [508]-Wind turbine wakes, Madrid*
[16] Wilcox D C 1988 *AIAA journal* **26** 1311–1320
[17] Burton T, Sharpe D, Jenkins N and Bossanyi E 2011 *Wind energy handbook* 2nd ed (John Wiley & Sons) ISBN 9780470699751

Journal of Physics: Conference Series **749** (2016) 012009 doi:10.1088/1742-6596/749/1/012009

The impact of geometric non-linearities on the fatigue analysis of trailing edge bond lines in wind turbine rotor blades

Pablo Noever Castelos, Claudio Balzani

Leibniz Universität Hannover, Institute for Wind Energy Systems, Hannover, 30167, Germany

E-mail: pablo.noever@iwes.uni-hannover.de

Abstract.
The accurate prediction of stress histories for the fatigue analysis is of utmost importance for the design process of wind turbine rotor blades. As detailed, transient, and geometrically non-linear three-dimensional finite element analyses are computationally weigh too expensive, it is commonly regarded sufficient to calculate the stresses with a geometrically linear analysis and superimpose different stress states in order to obtain the complete stress histories.

In order to quantify the error from geometrically linear simulations for the calculation of stress histories and to verify the practical applicability of the superposition principal in fatigue analyses, this paper studies the influence of geometric non-linearity in the example of a trailing edge bond line, as this subcomponent suffers from high strains in span-wise direction. The blade under consideration is that of the IWES IWT-7.5-164 reference wind turbine. From turbine simulations the highest edgewise loading scenario from the fatigue load cases is used as the reference. A 3D finite element model of the blade is created and the bond line fatigue assessment is performed according to the GL certification guidelines in its 2010 edition, and in comparison to the latest DNV GL standard from end of 2015.

The results show a significant difference between the geometrically linear and non-linear stress analyses when the bending moments are approximated via a corresponding external loading, especially in case of the 2010 GL certification guidelines. This finding emphasizes the demand to reconsider the application of the superposition principal in fatigue analyses of modern flexible rotor blades, where geometrical nonlinearities become significant. In addition, a new load application methodology is introduced that reduces the geometrically non-linear behaviour of the blade in the finite element analysis.

1. Introduction

In order to capture enough wind energy modern multi-MW wind turbines are equipped with large rotor blades with lengths of 80 m and more. Such blades are slender and flexible structures, and they experience large deformations in flap- and edgewise direction. These deformations provoke geometrically non-linear effects when it comes to ultimate strength analysis in the design stage [1]. The computational costs for a single geometrically non-linear static ultimate strength analysis is relatively low. However, geometrically non-linear effects within a fatigue load history will result in huge computational efforts in the framework of transient stress analyses for the calculation of accurate stress results.

This paper is concerned with geometrically non-linear effects of 3D finite-element-based stress analyses in the context of fatigue lifetime predictions of trailing edge bond lines in large rotor blades. The simulations are based on the reference wind turbine model IWT-7.5-164 [2], which is a 7.5 MW direct drive wind turbine with blades of 80 m length. The analysis is further focused on the shear stress proof according to the 2010 edition of the Germanische Lloyd (GL) certification guidelines [3] and a multiaxial stress proof as demanded by the latest DNV GL standard [4]. Both standards disregard any potential geometric non-linearity of the blade arising during fatigue analyses.

Geometric non-linearities may appear when the strains in a material exceed more than a few percent (large strain), or when large deflections result in additional offsets of loadings, and thus in additional internal moments and stresses [5]. Modern wind turbine rotor blades are subjected to large structural deformations [6], not only in ultimate load situations but also in fatigue time histories. Since the accuracy in stress response predictions is crucial for a correct estimate of the fatigue lifetime, it is of utmost importance to take into consideration geometric non-linearities in the design phase of blades, as will be pointed out in the framework of this paper. Otherwise, a structural blade design suffers from underestimations of the structural stresses, and consequentially from a nonconservative overestimation of the fatigue lifetime. The consequences will be severe damages up to failure of blades and probably of entire turbines in real life.

2. Aero-servo-elastic turbine simulations and finite element model of the blade

The loads for the fatigue and strength analysis are calculated through an aero-servo-elastic simulation that couples a multi-body system of several non-linear beam models [7] for the entire structure of the turbine with an aerodynamic model of the rotor. In this work, we utilized the turbine simulation software HAWC2 [8] to compute the loads. The relevant output for the fatigue analysis of the blade are internal forces and/or deflections in different cross sections along the blade, taking into account large deflections of the blade and the tower.

For the stress analysis a finite element (FE) model of the IWT-7.5-164 reference blade [2] is utilized. The FE model is generated with a Model Creator and Analysis tool called MoCA that has been developed at the Institute for Wind Energy Systems at Leibniz Universität Hannover. MoCA is encoded in MATLAB [9] and creates fully parameterized three-dimensional finite element models of rotor blades for ANSYS [5]. The stress post processing in this paper is limited to a span-wise region between 43 m and 47 m. A mesh convergence study has been performed to guarantee reliable results. The resulting global FE mesh and a local view of the adhesive discretization are shown in Fig. 1.

The boundary conditions have to be applied to the FE model in a reasonable way. The application of internal moments to discrete points along the blade is impossible without cutting the blade into sections, and would result in a stepwise increase of the moments along the blade. Therefore, it is common practice to apply external lateral single substitute forces $Q_{sub,j}$ at different positions along the blade that are denoted by r_j, see Fig. 2 (b). The substitute forces

(a) (b)

Figure 1. Finite element model of the IWT-7.5-164 reference wind turbine blade: Global mesh (a) and local mesh of the adhesive (b).

$Q_{sub,j}$ result in a stepwise linear distribution of bending moments and are adjusted to obtain a good approximation of the real bending moment distribution, see Fig. 2 (c). This methodology is similar to the loading strategy in a full-scale blade test, which is depicted in Fig. 2 (a).

Note that with this load application the applied transverse force distribution is not well approximated, since single forces yield stepwise jumps, which are not realistic in a real model. The normal forces in span-wise direction and the torsional moment are not included in the load application. However, since the bending moments are the major design drivers, this load application is considered sufficiently accurate to reproduce the stresses and strains in the blade.

In order to distribute the single forces $Q_{sub,j}$ across the cross-sections at the respective positions r_j along the blade, we introduce master nodes located in the shear centers in order not to create erroneous torsional moments. Those master nodes are connected to slave nodes in the cross-sections via rigid multi point constraints (MPC) in ANSYS. These MPCs represent the load clampings as seen in a full-scale test, see Fig. 2 (a).

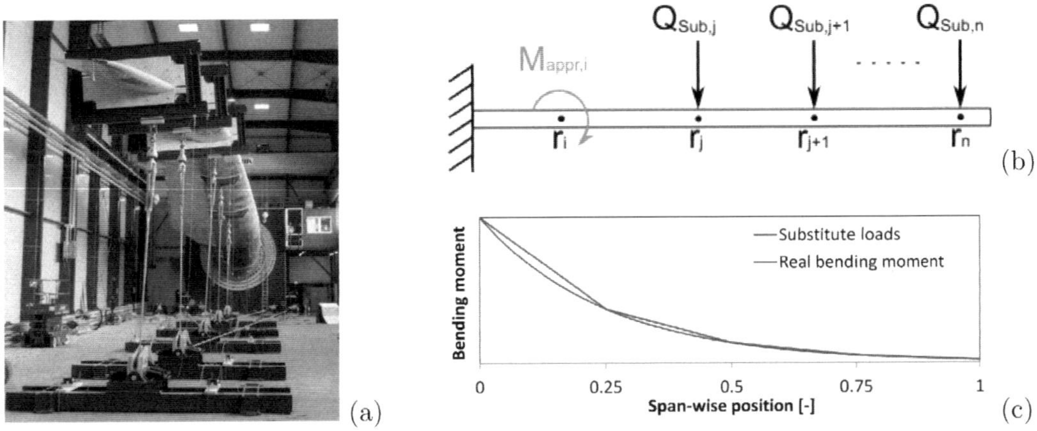

Figure 2. Load application on wind turbine blades: Loading fixtures in a full-scale test [10] (a), bending moment approximation with substitute loads on a beam (b), and comparison of approximated and real bending moments plotted against the span-wise position along the blade (c).

3. Fatigue analysis with substitute forces

After computing the load time series for specific wind speeds and operation modes via aero-servo-elastic turbine simulations with HAWC2, the stress histories have to be derived for all nodes in the finite element model. For this purpose the linear superposition principal is normally utilized, which states that the total response of a linear system is the sum of the responses of all contributing individual forces [11].

The fatigue analysis methodology based on the superposition principal works as follows: First, we define the number of single substitute loads n to be applied for the approximation of the bending moments. Then, we apply those loads, at this stage as unit loads, individually to the 3D finite element model. For each of the unit loads we carry out a finite element analysis and save the nodal stresses in an output file. Then we calculate the magnitudes of the substitute loads for each time step of the load time series that are required to have a sufficiently accurate approximation of the bending moments. The next step is to scale the stresses from the n unit load FE simulations by the real magnitudes of the substitute loads, and to superimpose all n scaled stress values for the entire time series. In this way we transfer the loads analysis to the 3D FE model. In other words, we transfer the internal forces and moments from beam theory in the framework of turbine simulations to stresses from 3D FE simulations. These stresses then enter a fatigue life prediction in a postprocessing calculus according to Palmgren & Miner theory, including rainflow counting, Markov matrix generation, Goodman diagram and S-N-curve evaluation, and linear damage accumulation.

Note that for this methodology n detailed finite element simulations have to be carried out. Usually, $5 \leq n \leq 8$, i.e. the number of substitute loads that is required to have a good approximation of the bending moments is quite low. Consequentially, the computational costs for the FE simulations is also limited to an acceptable degree. However, we would like to emphasize that geometrical and material linearity is required for the validity of the superposition principal. It is seen in the next section that this is a rather rough assumption, which is generally not valid for large and flexible rotor blades.

4. Linear versus non-linear finite element simulations

To evaluate the validity of the aforementioned methodology, a comparison of linear and non-linear FE simulations is carried out. In this context we use the highest shear stress τ_{yz} (the y-axis is pointing in chord direction of the cross-section, and the z-axis in span-wise direction) for the proof according to the 2010 edition of the GL guidelines [3] and the Drucker-Prager equivalent stress [12] for the multi-axial stress proof demanded by the latest DNV GL guideline [4]. The Drucker-Prager equivalent stress is calculated according to

$$\sigma_v = \frac{m-1}{2}(\sigma_1+\sigma_2+\sigma_3) + \frac{m+1}{2}\sqrt{\frac{1}{2}[(\sigma_1-\sigma_2)^2+(\sigma_2-\sigma_3)^2+(\sigma_3-\sigma_1)^2]}. \qquad (1)$$

Herein, $m = R_t/R_c$ is the ratio between the tensile and compressive strength of the adhesive, R_t and R_c, and σ_1, σ_2, and σ_3 are the principal stresses, respectively, following the convention $\sigma_1 \geq \sigma_2 \geq \sigma_3$. The highest deviations between linear and non-linear stresses emerge at large deflections, thus high loads. For the example of the trailing edge bond lines the maximum edgewise load is assumed to be the critical load case, as the strains in the bond lines reach their maxima due to the large distance to the principal bending axis. Table 1 shows the load set for maximum edge wise bending during the simulation of the design load case (DLC) 1.2 according to IEC standard 61400-1 [13], representing the normal power production load case for fatigue life prediction.

To determine the influence of geometric non-linearities, the stresses in the bond line were computed performing both, geometric linear (lin) and non-linear (nlin) calculations using the load application methodology explained in section 3. In the following, the stresses of both calculation methods (linear and nonlinear) are compared. The fatigue proof of the trailing edge according to the GL certification guidelines uses the shear stresses in the adhesive to determine

Table 1. Load set for maximum edgewise bending for DLC 1.2.

Span-wise position [m]	Edgewise moment [kNm]	Flapwise moment [kNm]
29	5,632	-2,663
39	3,737	-3,664
51	1,943	-2,948
59	1,011	-1,818

the material effort. Fig. 3 shows the mean deviation of the shear stresses from linear to non-linear analysis and the range of the three-fold standard deviation for all finite elements in the corresponding bond line cross section. The relative deviation is related to the linear calculations, as these are considered as the state of the art. Disregarding the outer regions of the analyzed bond line, where free edge effects could affect the stress intensity, the mean value of the deviation from non-linear to linear FE-solution of the shear stress is approximately 48 %. Respecting the range of the threefolded standard deviation the value can even reach up to 80 %.

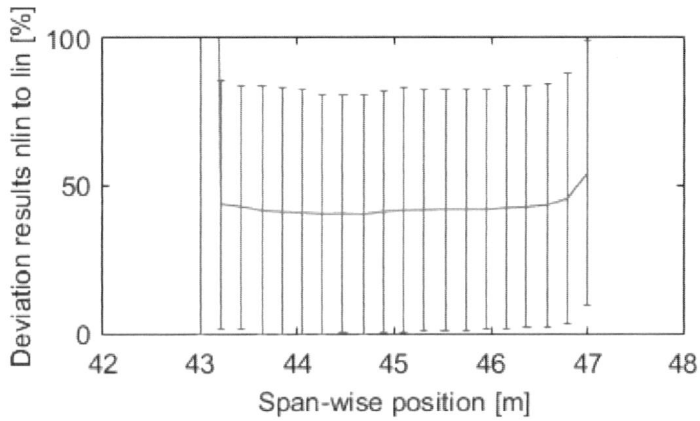

Figure 3. Relative deviation of the non-linear to linear calculated shear stresses in the bond line cross sections (mean value and three-fold standard deviation).

For the estimation of the error of the fatigue life prediction according to the new DNV-GL guideline the Drucker-Prager failure hypothesis is used to evaluate the multi-axial stress state. This criterion is developed to model materials with different tensile and compressive strengths [14], which in fact is also the nature of commonly used adhesives (EP and PU) showing a significantly higher compressive than tensile strength [15]. Fig. 4 illustrates the deviation of the computed Drucker-Prager equivalent mean stresses in the bond line cross section with their corresponding three-fold standard deviation range. The deviations for the multi-axial stress proof differ from the shear stress proof. Mean values for the relative difference of geometric linear to non-linear calculation are only about 3 to 4 %. According to the graph in Fig. 4 the maximum expected values reach up to 9 %, disregarding the outer regions of the bond line.

The large deflections of the blade and the large distance between the evaluation region and the load application positions lead to a significant geometrically non-linear effect in the stress space. The shear stresses tend to suffer from geometric non-linearities to a high degree, see Fig. 3. The effect is smaller for the Drucker-Prager euqivalent stress (Fig. 4), which is dominated by normal span-wise stresses, see Fig. 10. Nevertheless, the non-linearity is remarkably high, and could even increase if the torsional moments and normal span-wise forces are applied in addition to the bending moments. We can thus conclude that the linearity assumption is a strong weakness in the fatigue life prediction methodology presented in section 3.

5. Direct application of internal forces and moments

In order to reduce the influence of geometric non-linearities from the load application, a new method is proposed to introduce the loads which is based on a sectionwise analysis of the blade. As we receive non-linear internal forces and moments from the aero-servo-elastic simulations these can be applied directly to the model for the purpose of modelling the boundary conditions as accurate as possible. Here the internal forces and moments are applied behind the analysis

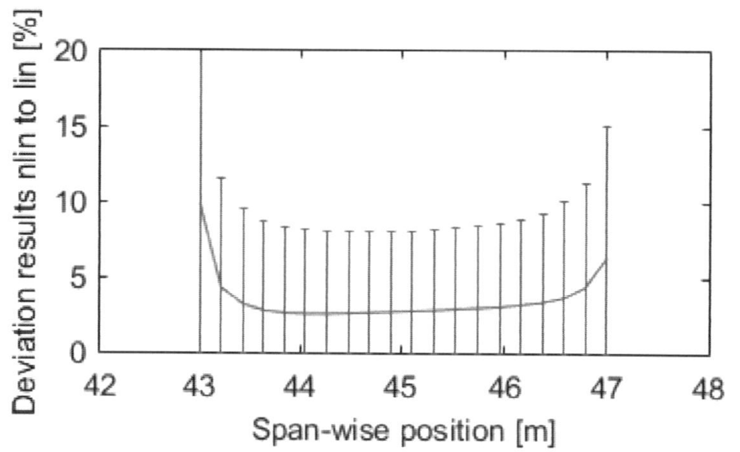

Figure 4. Relative deviation of the non-linear to linear calculated Drucker-Prager equivalent stresses in the bond line cross sections (mean value and three-fold standard deviation).

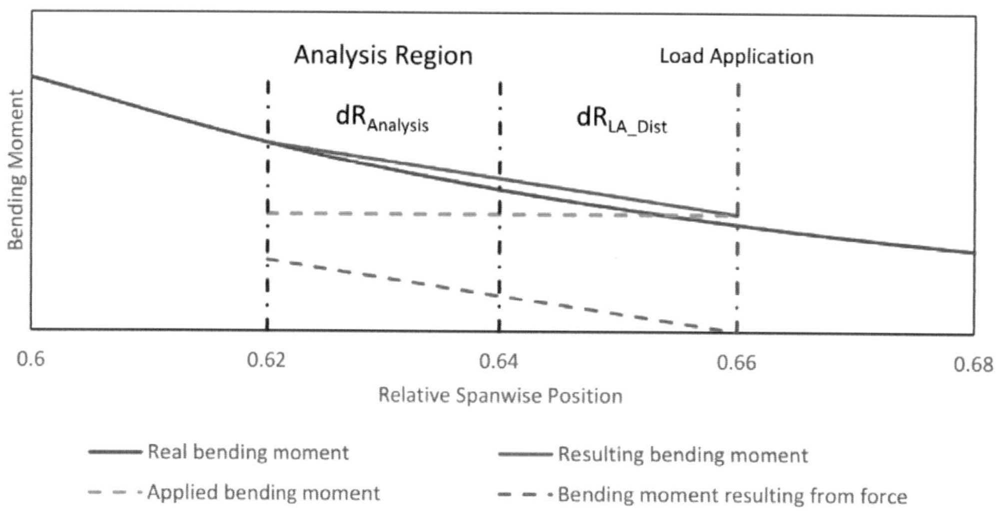

Figure 5. Composition of the applied internal bending moments in the proposed load application.

region in span-wise direction to the tip. The analysis region is reduced to a short span-wise portion of the blade $dR_{Analysis}$ to guarantee a good approximation of the loads in this region. In order to avoid influences from the load application modelled with rigid MPCs a distance dR_{LA_Dist} is kept to the analysis region. A big advantage of this method is that all forces and moments can be applied as they were calculated in the loads analysis. The applied internal forces with the distance dR_{LA_Dist} to the analysis region would introduce additional bending moments in the structure, therefore the applied internal moments are reduced by the amount of the bending moment produced by the internal forces as shown in Fig. 5. The resulting bending moment of the superposition of the applied internal moment and the moment provoked by the internal force is triggered to the most conservative point at the edge of the analysis region that is closer to the blade root.

Using this method the finite element analysis is only valid for the analysis region. Only there the internal forces and moments are accurately approximated. For the superposition principle in the fatigue analysis the necessary stress response can be derived from the results of the elements inside the analysis region. To perform the fatigue analysis for the whole blade the analysis region and load application has to be shifted repeatedly in order to obtain information for all blade regions. The larger the analysis region is the less computational effort is necessary to examine the whole blade. However, the larger the analysis region is the poorer the approximations of the internal forces and moments becomes. Hence, the user has to find a compromise between computational costs and accuracy.

In order to compare both load application methods we have chosen the same span-wise region from 43 m to 47 m for the analysis. All internal loads except the torsional moment and the force in span-wise direction were applied to obtain the same scenario as for the analysis described in section 4. Fig. 6 shows the deviation of the shear stress from geometric non-linear to linear analysis. The mean deviation increases in span-wise direction from 33 % to 43 %, but respecting the standard deviation the maximum value can reach from 45 % to 65 %. The deviation is very high and indicates a significant impact of geometrical non-linearity on the stress response. However, in comparison to the former load application method described in section 4 the new method reduces the mean deviation value by 5-15 % and the maximum by 15-35 % for the shear stress analysis of the highest edgewise loads in fatigue load histories.

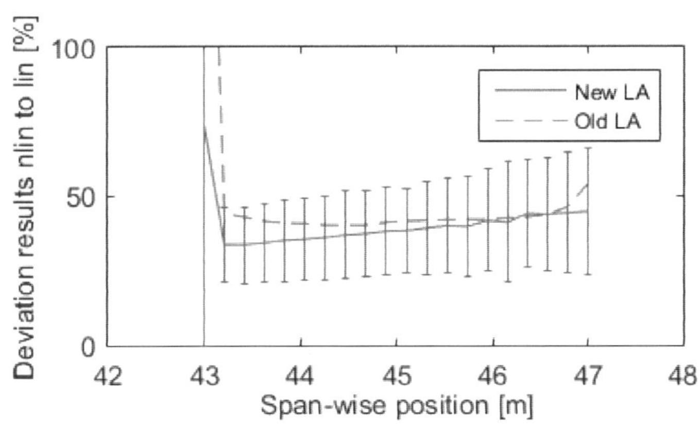

Figure 6. Relative deviation of the shear stresses in the bond line cross sections with the new load application method (New LA) calculated with the non-linear or the linear solver (mean value and three-fold standard deviation) and additional mean results for old load application method (Old LA).

Regarding the Drucker-Prager equivalent stress for the multi-axial stress proof shown in Fig. 7 according to the new DNV GL standard [4] the mean deviation between the geometrically non-linear and linear analysis is only about 2 % and the maximum expected deviation is up to 3.5 %, neglecting the outer regions of the bond line. These results show that the new application method can halve the deviation between geometrically non-linear and linear results in comparison to the former load application.

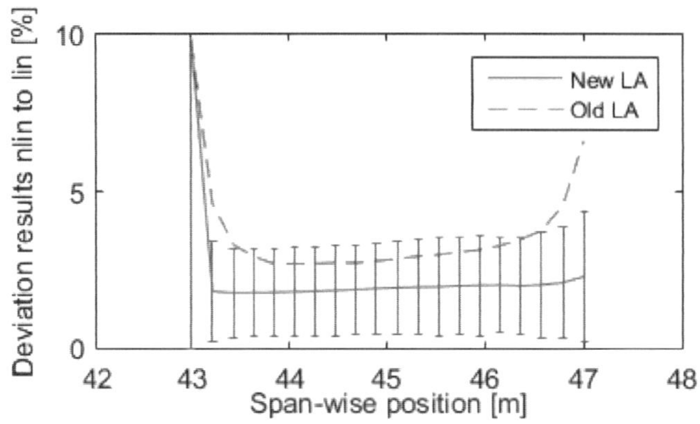

Figure 7. Relative deviation of the Drucker-Prager equivalent stresses in the bond line cross sections with the new load application method (New LA) calculated with the non-linear or the linear solver (mean value and three-fold standard deviation) and additional mean results for old load application method (Old LA).

The new load application method allows the application of torsional moment and span-wise normal forces. The inclusion of these additional loads intensifies the non-linear effect for the shear stress dramatically as shown in Fig. 8. The mean value of the deviation increases to 150 %. While the shear stress is very sensitive to geometric non-linearities introduced by torsional moments the Drucker-Prager equivalent stress for the multi-axial stress proof increases only slightly in its maximum mean deviation up to 3 %, see Fig. 9. This might be due to normal stresses in span-wise direction, which dominate the Drucker-Prager equivalent stress, being relatively insensitive to geometric non-linearity. However, more likely a low impact of geometrical non-linearity on the absolute values of all stresses, see Fig. 10, results in a dramatic relative change in shear stresses (SXY,SXZ,SYZ) (as the absolute values are very small), but a moderate relative change in Drucker-Prager equivalent stress (as that is governed by the normal stress in span-wise direction (SZ), of which the absolute value is orders of magnitude larger than the shear stresses).

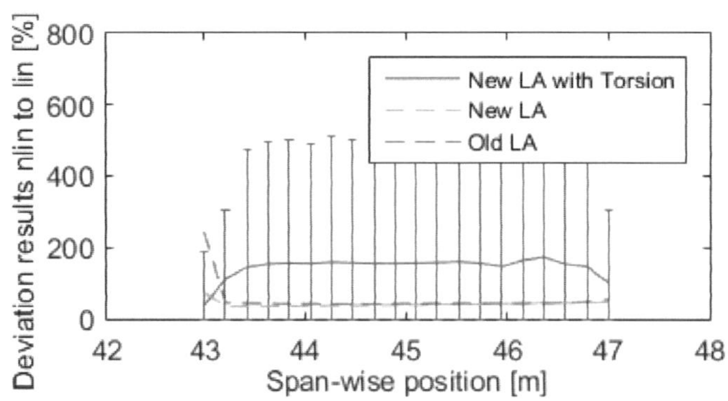

Figure 8. Relative deviation of the shear stresses in the bond line cross sections with the new load application method respecting torsional moments and normal span-wise force (New LA with Torsion) calculated with the non-linear or the linear solver (mean value and three-fold standard deviation) and additional mean results for new load application without torsional moments (New LA) as well as old load application method (Old LA).

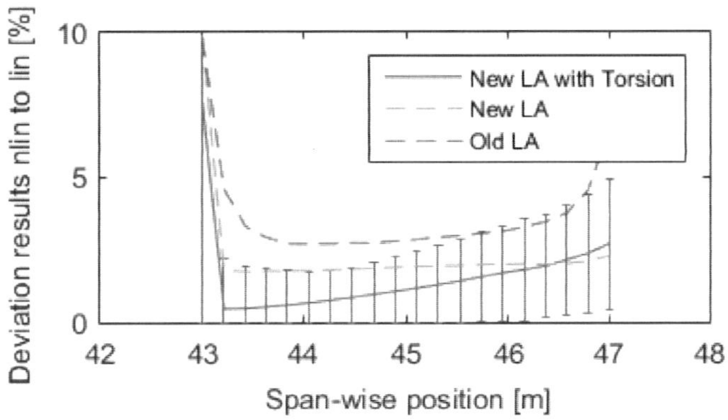

Figure 9. Relative deviation of the Drucker-Prager equivalent stresses in the bond line cross sections with the new load application method respecting torsional moments and normal span-wise force (New LA with Torsion) calculated with the non-linear or the linear solver (mean value and three-fold standard deviation) and additional mean results for new load application without torsional moments (New LA) as well as old load application method (Old LA).

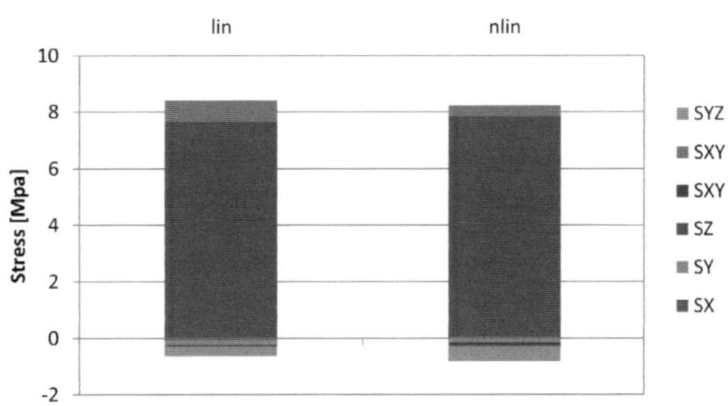

Figure 10. Absolute Stresses for one element in the center of the trailing edge bond line cross section at a span-wise Position of 44 m, calculated with the linear solver (left) and the non-linear solver (right). The normal stresses SX points in thrust direction, SY in rotor tangential direction and SZ in blade span-wise direction.

6. Conclusion

This paper has demonstrated that large wind turbine rotor blades behave geometrically non-linear in finite element analyses, not only in extreme load cases but also in fatigue load histories. This is a fundamental information that every blade designer has to be aware of, because the standard methods for fatigue life estimations are no longer valid. Neglecting the nonlinearities can result in nonconservative designs, and thus to severe damages in real rotor blades that may probably lead to the failure of blades or even entire turbines.

Depending on the type of stress proof for the bond lines in the rotor blade, the non-linear effects can have a high influence on the stress response, e.g. up to 80 % with a shear stress proof as shown in the paper. The utilization of a multi-axial stress proof, in this case based on the Drucker-Prager equivalent stress, reduces the impact of geometric non-linearities to a maximum of 9 %, at least in the analyzed trailing edge bonding region. A side effect is that the major failure mode, namely cracks due to normal stresses in span-wise direction, are captured – in contrast to former design guidelines that only asked for a simplified shear strength proof.

It has further been shown that load introduction methodologies have a significant impact on the stress response. Although computationally very attractive, a methodology based on scaled and superimposed unit loads, as described in section 3, is principally not applicable due to the requirement of linearity for its mathematical and physical validity. A novel load introduction technique has been proposed where all internal loads can be considered. The novel concept significantly reduces nonlinearity effects in the stress space. The disadvantage is that the computational costs are quite high, because the evaluation regions are strongly limited in span-wise direction, and thus a sectionwise analysis has to be performed.

The loading state chosen for the investigations in this paper was taken from a normal operation simulation (design load case 1.2 according to the IEC standard), where an instance of time was chosen where the bending moments are maximum. For other time instances, the loads may be much lower, and the nonlinearity effect may be less significant. An evaluation of the impact of geometrical non-linearity on the fatigue lifetime prediction, and not just one instance of time, is ongoing. That study will include the scatter in loading over time, and will be published in the near future.

Principally, the results of this paper indicate that further investigations are needed in order to develop an accurate and computationally not too expensive load application methodology for finite element analyses in the context of fatigue life assessments for wind turbine rotor blades.

References

[1] Rosemeier M, Berring P and Branner K 2015 *Wind Energy*
[2] Sevinc A, Rosemeier M, Bätge M, Braun R, Meng F, Shan M, Horte D, Balzani C and Reuter A 2016 *Reference Wind Turbine IWT-7.5-164 Specification*
[3] Germanischer Lloyd Industrial Services GmbH 2010 *Richtlinie für die Zertifizierung von Windenergieanlagen* (Hamburg: Germanischer Lloyd)
[4] DNV GL AS 2015 *DNVGL-ST-0376 - Rotor blades for wind turbines*
[5] ANSYS Inc 2013 *ANSYS Mechanical APDL Theory Reference*
[6] Rezaei M M, Behzad M, Moradi H and Haddadpour H 2016 *Renewable Energy* **94**
[7] Pavese C, Wang Q, Kim T, Jonkman J M and Sprague M A Hawc2 and beamdyn: Comparison between beam structural models for aero-servo-elastic frameworks *EWEA Annual Event 2015*
[8] Larsen T J and Hansen A M 2015 *How 2 HAWC2, the user's manual* (*Risø R, Report* vol 1597) (Roskilde: Risø National Laboratory)
[9] The MathWorks I 2016 *MATLAB Primer*
[10] Fraunhofer IWES 2015 *IWES statischer Test eines Rotorblatts*
[11] Gross D, Hauger W, Schröder J and Wall W A 2012 *Technische Mechanik 2* (Berlin, Heidelberg: Springer Berlin Heidelberg)
[12] Alejano L R and Bobet A 2012 *Rock Mechanics and Rock Engineering* **45**
[13] IEC 61400-1 2007 *International Standard: Wind turbines - Part 1: Design requirements* 3rd ed (*International standard Norme internationale* vol CEI 61400-1) (Geneva: International Electrotechnical Commission)
[14] Stark R 2006 *Festigkeitslehre: Aufgaben und Lösungen* (Vienna: Springer-Verlag/Wien)
[15] Zarouchas D and van Hemelrijck D 2012 *Journal of Adhesion Science and Technology* **28**

Power performance optimization and loads alleviation with active flaps using individual flap control

Vasilis Pettas, Thanasis Barlas, Drew Gertz and Helge A Madsen

Technical University of Denmark Department of Wind Energy, Aerodynamic Design
Risø Campus, Frederiksborgvej 399, 4000 Roskilde Denmark

Corresponding author: vas.pettas@gmail.com

Abstract. The present article investigates the potential of Active Trailing Edge Flaps (ATEF) in terms of increase in annual energy production (AEP) as well as reduction of fatigue loads. The basis for this study is the DTU 10 MW Reference Wind Turbine (RWT) simulated using the aeroelastic code HAWC2. In an industrial-oriented manner the baseline rotor is upscaled by 5% and the ATEFs are implemented in the outer 30% of the blades. The flap system is kept simple and robust with a single flap section and control with wind speed, rotor azimuth, root bending moments and angle of attack in flap's mid-section being the sensor inputs. The AEP is increased due to the upscaling but also further due to the flap system while the fatigue loads in components of interest (blade, tower, nacelle and main bearing) are reduced close to the level of the original turbine. The aim of this study is to demonstrate a simple and applicable method that can be a technology enabler for rotor upscaling and lowering cost of energy.

1. Introduction

The design trend in the wind turbine industry leads towards larger multi MW rotors where the reduction of design loads and aerodynamic optimization becomes more significant. The application of ATEFs is an important part of these concepts as has been shown in previous works [**1, 2**]. In the present work all simulation activities are based on HAWC2 aeroelastic code [**3**] using the verified flap model implementation [**4**] with the unsteady aerodynamics using the ATEF dynamic stall model implementation [**5**]. In order to calculate the loads according to the industrial standard for load certification, the calculations are based on DLC 1.2 NTM (as stated on IEC standard 61400-1, third edition) an overview of which is presented in [**6**] along with the post processing tools in [**7**]. The assumption that the total fatigue is derived from this DLC is valid, since in the channels of interest more than 98% of the fatigue is attributed to DLC 1.2 NTM. In the same manner the AEP calculations are performed with different turbulence intensity values without yaw misalignment for all speeds with six turbulence seeds for each calculation point, since there is not a standardized method for numerical derivation of power curves. The approach in this work focuses on increasing AEP and reducing design fatigue loads to the original baseline load envelope using the active flap system in order to enable efficient rotor upscaling. This approach, along with a robust and simplified control implementation, establishes a close connection to a possible industrial application of such a system.

2. Wind turbine upscaling

The DTU 10 MW RWT **[8]** is considered as the baseline turbine. There have been investigations on the topic of upscaling multi-MW turbines and its limitations, an overview of which can be found at **[9, 10]**. In the present case in order to formulate a business case and get a clear picture of the flap implementation potential, all the components except from the rotor are kept the same with the baseline. The already highly aerodynamically optimized rotor, has the same hub in the upscaled case while the blades are elongated to a 5% total increase. In order to keep the same characteristics the solidity was kept the same as well as the airfoil profiles at the same non-dimensional positions. Hence, chord and thickness are increased to the same level while the structural properties are kept exactly the same (same mass per length, cross sectional stiffness etc.) leading to a same (performance wise) but 'softer' blade. The maximum power is limited to 10MW in order to lower the loads and utilize the same generator/electronics setup. Moreover, in order to be consistent with the upscaling procedures **[10]** and have a 'fair' case for the flaps the maximum steady state thrust of the upscaled rotor was decreased to the same level as the baseline by introducing an earlier pitch scheduling which in turn decreased slightly the rated wind speed. Finally, the gains of the DTU WE controller were tuned using the DTU HwacStab2 tool **[11, 12]**. The resulting key operational characteristics and dimensions of both original and upscaled rotors are shown in **Table 1.**

Initially, the baseline turbines (original and 5% upscaled) were simulated according to IEC Design Load Basis (DLB). The results show an increase of 2-10% for the fatigue loads in the channels investigated with the highest identified at the blade root bending moment channels where torsional moment MzBR is increased the most by 18%. Previous results in load alleviation potential using ATEFs **[13, 14, 16]** agree that this reduction margin can be feasible. In terms of AEP the difference for the class IA wind climate ($V_{mean} = 10\ m/s$) and 10% turbulence for all speeds is 3.41% while for the other turbine classes this increase becomes higher.

3. Implementation and control of ATEF

The flap configuration applied to the upscaled rotor involves one flap section at the last 30% of the blade (29.7 m) close to the tip with individual actuators for each blade. At this section of the blade the airfoil used is the FFA-W3-241 and the flaps are occupying the 10% of the chord. The variation of aerodynamic characteristics by the flap deflection is based on 2D CFD results from the in-house code EllipSys 2D **[17]** for a Reynolds number between $6 \cdot 10^6 - 12 \cdot 10^6$, for free transition with turbulence intensity 0.1, 3D corrected using Bak's model **[18]**. The deflection limit of the flap is +/- 15 degrees and the one-per-blade actuator dynamics are taken into account using a linear servo model in Hawc2 with a first order system **[19]**, using a time constant whose value is an investigated parameter.

Table 1. Comparison of key parameters of the baseline and upscaled turbines.

Parameter	DTU 10 MW RWT	Upscaled turbine
Wind Regime	IEC Class IA	Same
Control	Variable Speed/ Collective Pitch	Same
Cut in speed [ms^{-1}]	4	Same
Cut out speed [ms^{-1}]	25	Same
Rated Wind Speed [ms^{-1}]	11.4	11
Rated Power [MW]	10	Same
Rotor Diameter [m]	178.3	187.2
Hub Diameter [m]	5.6	Same
Hub Height [m]	119.0	Same
Drivetrain	Medium Speed, Multiple stage Gearbox	Same
Minimum Rotor Speed [rpm]	6.0	5.0
Maximum Rotor Speed [rpm]	9.6	9.2
Maximum Generator Speed [rpm]	480.0	459.3
Gearbox Ratio	50	Same
Maximum Tip Speed [ms^{-1}]	90	Same
Shaft Tilt Angle [deg]	5.0	Same
Rotor Precone Angle [deg]	-2.5	Same
Blade Prebend [m]	3.33	3.50
Rotor Mass [kg]	227,962	239,360
Nacelle Mass [kg]	446,036	Same
Tower Mass [kg]	628,442	Same

The main controller of the turbine (pitch and torque regulation) is the same as the baseline **[12]** with retuned gains to fit the upscaled rotor and earlier pitching scheduling to limit maximum thrust to the baseline level. The controller features both partial and full load operation as well as switching mechanisms between modes of operation, utilizing measurements of rotor speed, tower accelerations and pitch angles as inputs and the generator torque and collective pitch angle as outputs. The proposed flap controller is completely decoupled from the main controller and is divided into two parts.

3.1 AEP increase objective

The first involves the partial load operation of the turbine. The flap angle command for each blade is based on measurement of the angle of attack in the mid-flap section. This requires the derivation of the local angle of attack from an inflow or pressure sensor. Alternatively, a model-based estimation of the angle of attack from load measurements could be utilized. The control algorithm uses predefined parameters which are based purely on the known $\mathbf{C_L}$-$\boldsymbol{\alpha}$ polar and a tunable flap angle offset parameter, as seen in **Equation 1**.

$$\boldsymbol{\beta} = \frac{C_{L\alpha}\cdot(\alpha-\alpha_{set})}{C_{L\beta}} - C_{L\alpha}\cdot\alpha_0 + \boldsymbol{\beta}_{offset} \quad (1)$$

$\mathbf{C_{L\alpha}}$ and $\mathbf{C_{L\beta}}$ are the lift slopes for variations in angle of attack and flap angle respectively and α_0 is the angle of attack for zero lift. The $\boldsymbol{\alpha}_{set}$ is chosen as the angle for optimal gliding ratio $\mathbf{C_L/C_D}$. Then $\boldsymbol{\alpha}$ is the measured (or estimated) angle of attack at the mid-flap section which is the only controller input. The

Journal of Physics: Conference Series **749** (2016) 012010 doi:10.1088/1742-6596/749/1/012010

$\boldsymbol{\beta}_{offset}$ can be chosen based on prior calculations. The flap system dynamics time constant $\boldsymbol{\tau}$ is also chosen as a tunable parameter in this case in order to optimize the required (slow) response of the flap.

The choice of optimal parameters for the flap controller ($\boldsymbol{\beta}_{offset}$ and $\boldsymbol{\tau}$) is based on derivation of the optimal flap angle (beta) variation from simulations using HAWC2. A full sweep parametrical study is carried out for each wind speed and a range of the two basic flap parameters. The comparison of achieved average power gain for the parametric sweep is shown in **Figure 1**. Maximum power extraction is achieved with a flap angle offset of -7.5 degrees for all cases. Increasing the flap system time constant is shown to improve performance, as then the flap actuator compensates to the system's phase delay observed between flap actions and actual change in forces applied as well as the delay between measured angle of attack and control inputs. The sensitivity analysis on the chosen time constant indicates that there is a clear optimum setting for $\boldsymbol{\tau}$ =10s in the whole range of below rated operating points. Moreover, angle of attack input signal is low pass filtered so that the flap does not react to high frequency variations.

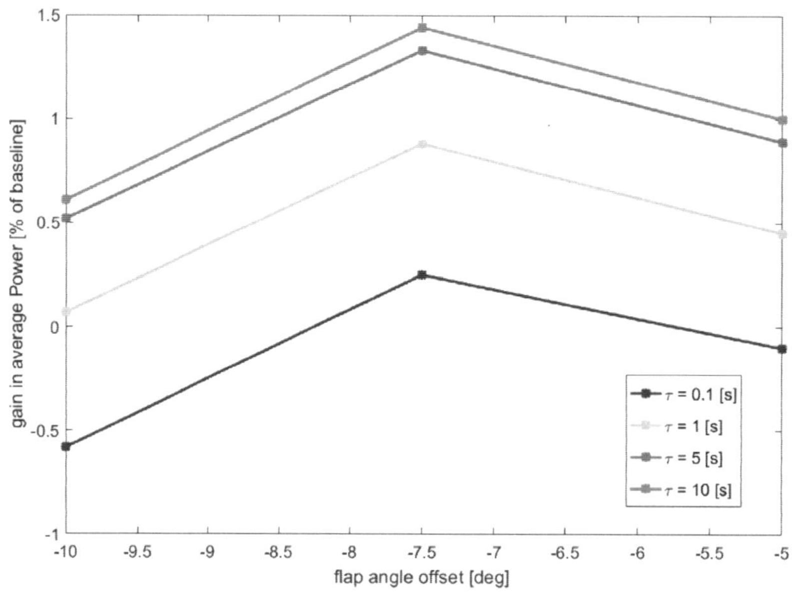

Figure 1. Average power gain for a range of flap offset angles and flap time constants.

3.2 Fatigue reduction objective

The second part of the controller is a classical Proportional Derivative (PD) individual flap controller. It is based on high-pass (HP) filtered flapwise blade root moment (MxBR) with a gain scheduling for proportional and derivative terms based on low-pass (LP) filtered wind speed, operating at the full load regime which was observed to contribute the most to the total fatigue damage equivalent load for the majority of the channels.

The MxBR signals pass through a HP filter in order to avoid flap reacting to steady and low frequency variations. Furthermore, a signal from main controller indicating full power is used in order to limit controller's actions to full load regime. This signal is a LP filtered indicator taking values between 0 and 1 (full load reached). Finally, the flap controller utilizes the main controller's grid status signal in order to assign a static angle when the turbine is not producing since it was decided not to operate the controller in parked cases.

The tuning method used in the present work consists of two parts. Firstly, a rough tuning is done using Ziegler-Nichols method **[20]** based on a high fidelity state space representation of the turbine (in open and closed loop) per speed obtained from HawcStab2. These gains along with their linear scheduling over wind speeds are shown in **Figure 2**. Subsequently, these values were manually tuned approximately for saturation limits based on a Matlab script including no-feedback and finally representative example cases were checked with results from applications of the controller in HAWC2 simulations. Finally, having the values from the previous step as a starting point parametrical studies were executed by running DLC 1.2 simulations in order to find the optimal gains for fatigue reduction. This control approach for load alleviation in full load regime has also been demonstrated and validated in previous works **[13, 14]**. Finally, the transition region is treated with a scaling factor added based on the full load indicator which makes sure that the flap controller is switching progressively from one region to another when operating close to rated speed.

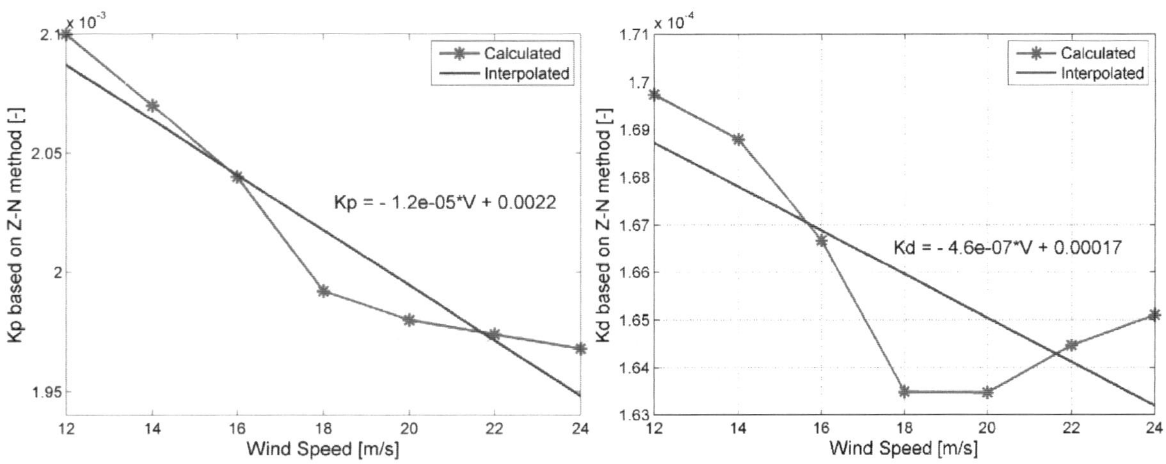

Figure 2. Proportional and Derivative term gains derived with Ziegler-Nichols method scheduled over wind speeds.

3.3 Integrated controller

The two parts of the controller are integrated in a holistic approach in order to realistically evaluate the load impact in the turbine lifetime, as well as, the influence in the individual effectiveness of each controller. The switching between the two controllers is based on wind speed and the rated power signal from the controller. When wind speed and rated indicator are below defined thresholds the AEP part is activated, while in all other cases (turbine producing power) the PD controller is active. These values were found with parametrical sweeps based on DLC 1.2 HAWC2 simulations in order to fulfill the objectives of smooth activity of the controller around rated speed and decoupling the control objectives so that the AEP increase and fatigue reduction is same as the individual applications. The block diagram of the integrated controller is shown in **Figure 3**.

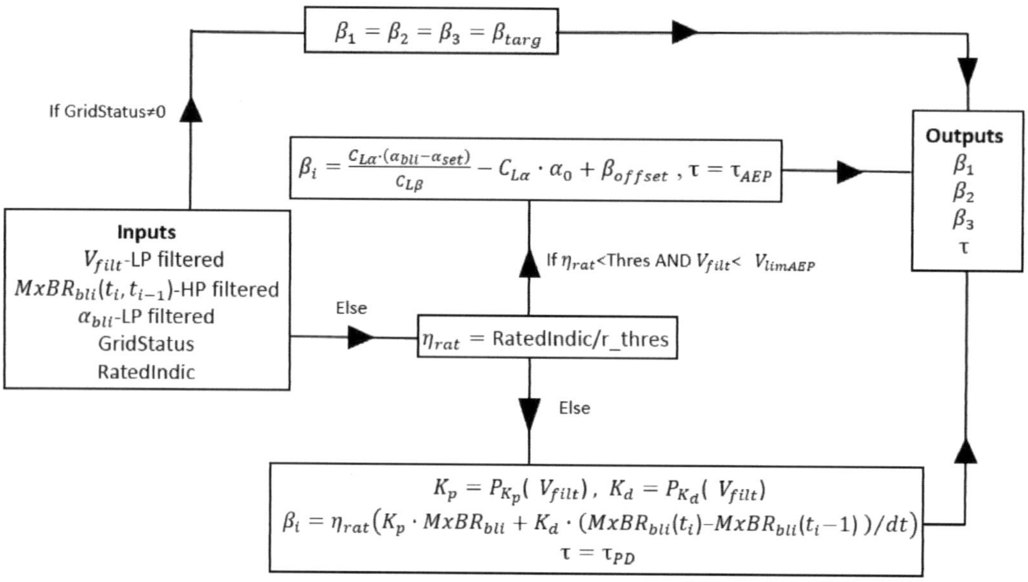

Figure 3. Flap controller's block diagram.

4. Results

The simulations were carried out at the baseline turbine as well as the upscaled with and without flaps while the controllers where applied both individually and integrated. The results indicate that the flap implementation can improve AEP up to 0.41% compared to the upscaled rotor. The AEP potential increase is evaluated with 10% turbulence intensity, as it is commonly performed in industry, where the flap contribution is found to be 0.19% while the total increase is 3.62%. All simulations were carried out with a wind shear exponent of 0.2, as stated in IEC standard, and tower shadowing effect included.

The comparison of power performance for every wind speed between all cases is shown in **Figure 4**. The variation of the achieved AEP for different turbulence intensity values is shown in **Table 2.** It is observed that the maximum AEP gain of 0.41% is achieved at the most ideal no turbulence conditions and decreases down to 0.11% when full IEC turbulence occurs. This is expected since the controller parameters are derived based on steady state simulations and as turbulence increases the non-linear effects become more dominant resulting to assignment of suboptimal flap angle values. The power increase trend is evident over the whole range of partial load region except for the transition regions at cut-in (where the turbine operates in an off-design point due to the constraints imposed by main controller for the rotational speed) and switching to full power.

Table 2. AEP results for compared cases.

case	flap angle offset [deg]	flap time constant [s]	turbulence intensity [%]	AEP gain [% 5% up baseline]	AEP gain [% original baseline]
1	-7.5	10	0	0.41	-
2	-7.5	10	5	0.25	-
3	-7.5	10	10	0.19	3.62
4	-7.5	10	IEC	0.11	3.51

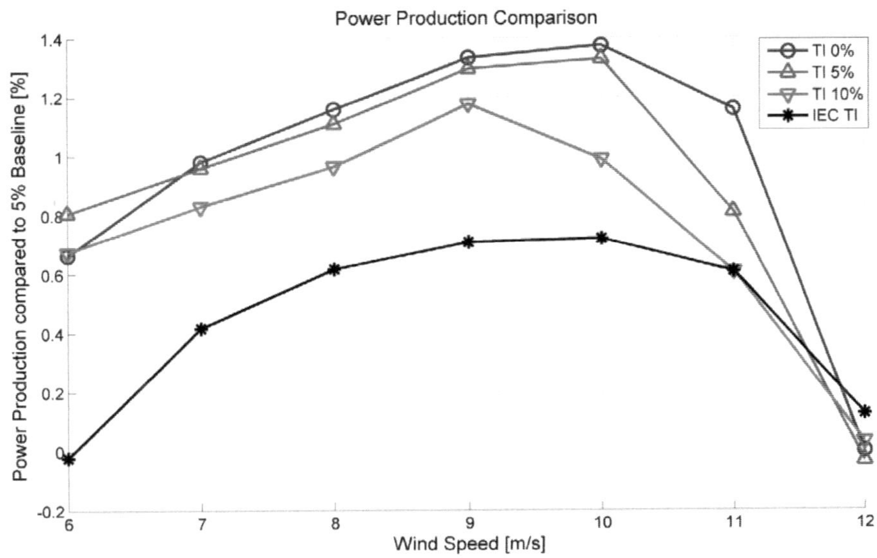

Figure 4. Power production per wind speed compared to 5% baseline with no flaps.

The previous results showed that the integrated controller reaches the same increase in AEP as the individual controller. Regarding the fatigue load reduction objective, results show that the trade-offs of this approach is mainly the small increase in the Damage Equivalent Loads (DEL) contributed from speeds below rated which is not possible to be compensated for by the PD part. This is in the level of 1-2% for most channels and only tower bottom channels are affected more in the level of 5-6%. The short term equivalent fatigue loads per wind speed shown in **Figure 5** for the main load channels can verify this observation where the fatigue loads over rated are decreased and below are increased due to the AEP part.

The PD controller alone (not shown), presents a decrease of 2%-15% to the lifetime fatigue loads (except for tower fore-aft, blade root edgewise and blade root torsion channels) close to the level of the baseline turbine and no influence to power generation. With the integrated controller applied, blade root bending moment and main bearing channels reach the baseline level while the tower loads are increased in the order of 2% to 4% due to the controller's AEP targeting operation. In above rated operation the reduction of the target blade root fatigue load is in the order of 10%, which brings the load level down to the baseline without upscaling. The blade root torsion is considerably increased, as expected, as a result of the increased aerodynamic pitching moment due to the flap operation. In general, the fatigue load targeting implementation suggested is found to be able to influence significantly MxBR and main bearing while the nacelle (tower-top) channels are less reduced when control is applied integrating both fatigue and AEP targeting modules. Tower bottom related channels are marginally decreased while blade and main bearing torsion channels seem to be only negatively affected by the flap implementation.

The lifetime fatigue equivalent loads comparison is shown in **Figure 6**. An overview of the abbreviations used in this figure and the corresponding load channels can be seen in **Table A1**. In general the integrated controller is a trade-off between power increase and load alleviation resulting in a 3.7% total increase in AEP while the fatigue loads in most channels are on the level of the baseline.

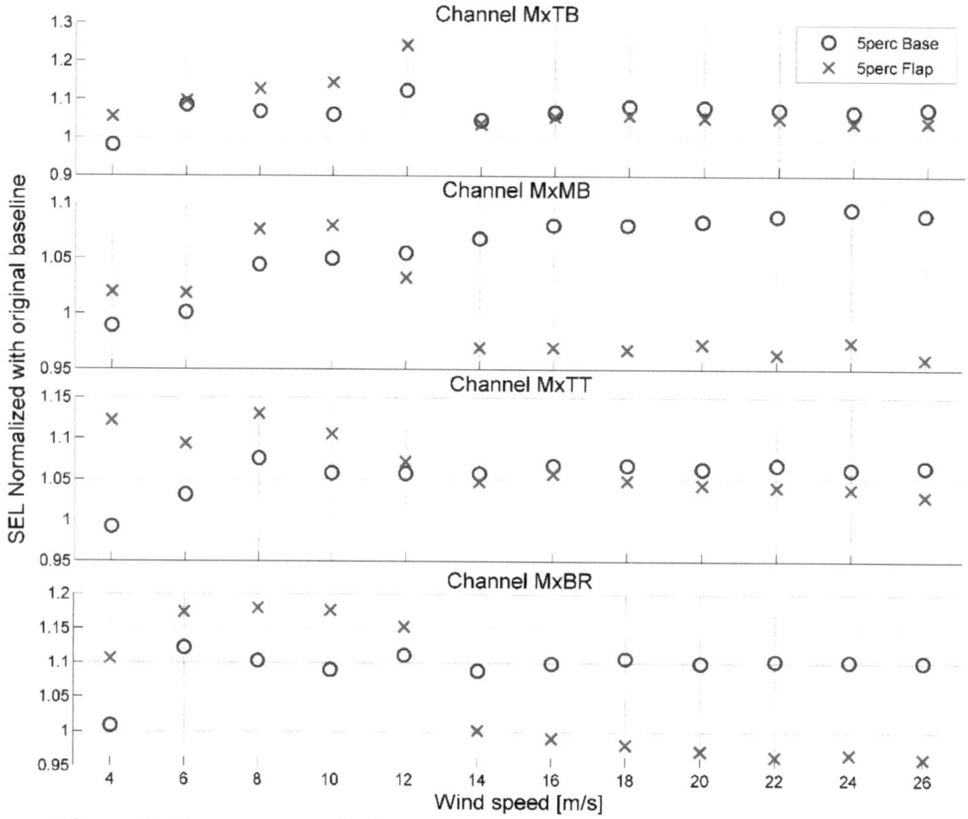

Figure 5. Short-term Equivalent Load (SEL) comparison to original baseline.

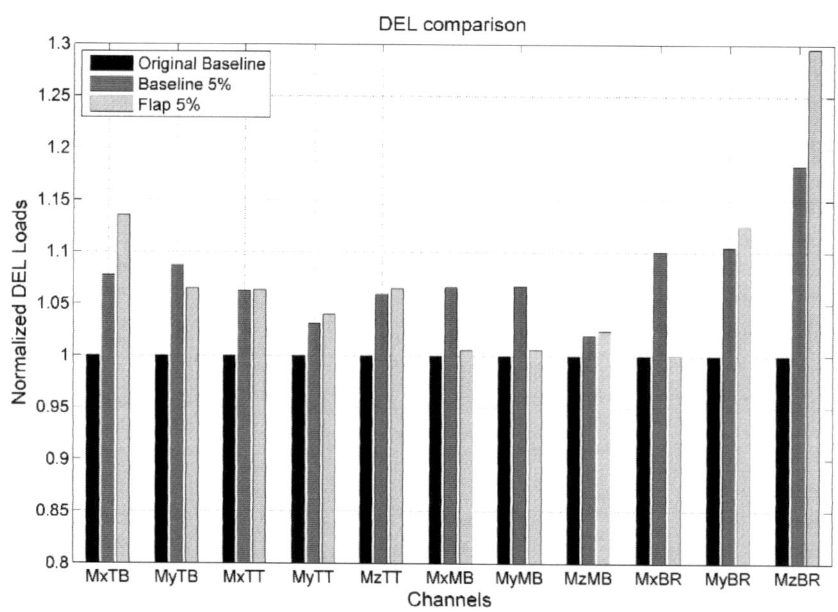

Figure 6. Comparison of lifetime equivalent fatigue loads on main channels for all cases.

5. Conclusions

The main conclusions are summarized below:

- In terms of AEP the individual flap control can increase AEP depending on the level of aerodynamic optimization of the rotor and the site-specific wind shear and turbulence intensity.
- The controller delivers significant fatigue load reduction in load channels of interest. The fatigue margin of the baseline components' design should be taken into account in order to evaluate whether the achieved reduction is sufficient to keep the same turbine platform.
- This control method is considered more robust and easy to apply than previous approaches since no significant increase to pitch activity is observed and the turbine–specific prescribed values are less prone to errors than other methods like model-based control algorithms.
- The presented flap controller approach establishes AEP increase and load reduction integration capability of the system in enabling further rotor upscaling.

Suggested future work should also consider the extreme load alleviation. A controller approach like a threshold cut-off could also be integrated in order to alleviate extreme loads which are another significant design driver. This fully integrated controller could reveal the full potential of a practical flap implementation in large rotor wind turbines.

Appendix

Table A1. Abbreviations of the load channels considered

Abbreviation	Load Channel
MxTB	Tower bottom fore-aft moment
MyTB	Tower bottom side-side moment
MxTT	Tower top tilt moment
MyTT	Tower top roll moment
MzTT	Tower top yaw moment
MxMB	Main bearing tilt moment
MyMB	Main bearing yaw moment
MzMB	Main bearing torsion
MxBR	Blade root flapwise moment
MyBR	Blade root edgewise moment
MzBR	Blade root torsion

Acknowledgments

The presented work is partly based on activities within the INDUFLAP2 project, funded by The Danish Ministry of Climate, Energy and Building, EUDP-2015-I, J.nr. 64015-0069 and with participation by Rehau, Siemens Wind Power and DTU.

References

[1] Barlas T K and van Kuik G A M Review of state of the art in smart rotor control research for wind turbines *Progress in Aerospace Sciences - 2010* **46** 1 pp 1-27 2010

[2] Madsen H A. et al. Towards an industrial manufactured morphing trailing edge flap system for wind turbines *Proceedings of EWEC 2014*, Barcelona, Spain 2014

[3] Larsen T J et al. How 2 Hawc2, the user's manual *Technical report, Risø-R-1597(ver. 4-4)(EN)* 2013

[4] Bergami L. Adaptive Trailing Edge Flaps for Active Load Alleviation in a Smart Rotor Configuration *PhD thesis*, DTU Wind Energy 2013

[5] Bergami L and Gaunaa M ATEFlap Aerodynamic Model, a dynamic stall model including the effects of trailing edge flap deflection *Technical report, Risø-R-1792(EN)* 2012

[6] Hansen M H et al. Design Load Basis for onshore turbines *Technical report, DTU Vindenergi-E-0174(EN)* 2015

[7] Pedersen, M. M., Post processing of Design Load Cases using Pdap *Technical report,, DTU Vindenergi-I-0371(EN)*, 2014.

[8] Christian Bak H et al. Description of the DTU 10 MW Reference Wind Turbine, *DTU Wind Energy Report-I-0092* 2013

[9] Sieros G et al. Upscaling wind turbines: theoretical and practical aspects and their impact on the cost of energy *Wind Energy,* ***15*** pp 3–17 2012

[10] UpWind Design limits and solutions for very large wind turbines, *UpWind EU project report* 2012

[11] Henricsen L C et al. HAWCStab2 User Manual *Technical report,, DTU Vindenergi (EN)*, 2015.

[12] Hansen M H et al. Basic DTU Wind Energy Controller, *DTU Wind Energy Report-E-0018* 2013

[13] Barlas et al. Load alleviation potential of active flaps and individual pitch control in a full design load basis, *Proceedings of the EWEA Annual Event and Exhibition* 2015

[14] Validation of new control concepts by advanced fluid-structure interaction tools, *Innwwind EU Project Deliverable 2.3.2* 2015

[15] Smit et al. Sizing and control of trailing edge flaps on a smart rotor for maximum power generation in low fatigue wind regimes. *Wind Energy, 19: 607–624* 2015

[16] Chen, Z. J. et al. System identification and controller design for individual pitch and trailing edge flap control on upscaled wind turbines. *Wind Energy, doi: 10.1002/we.1885*1 2015

[17] Sørensen, N. N. (2008). The EllipSys2D/3D code and its application within wind turbine aerodynamics [Sound/Visual production (digital)]. *Workshop on future aero-elastic tools for wind turbines*, Risø (DK) 2008

[18] C. Bak, J. Johansen, and P. B. Andersen. Three-dimensional corrections of airfoil characteristics based on pressure distributions. *Proc. European Wind Energy Conference and Exhibition (EWEC), 27. Feb. - 2. Mar. 2006, Athens, Greece*, 2006.

[19] T. Barlas and H. A. Madsen, Influence of actuator dynamics on the load reduction potential of wind turbines with distributed controllable rubber trailing edge flaps (CRTEF) *Proceedings of ICAST2011, pp. 1–12, 2011.*

[20] Z. JG and N. Nichols, Optimum settings for automatic controllers *Trans. ASME, no.* ***64****, pp. 759–768, 1942.*

Journal of Physics: Conference Series **749** (2016) 012011 doi:10.1088/1742-6596/749/1/012011

Application of two passive strategies on the load mitigation of large offshore wind turbines

Rasoul Shirzadeh and Martin Kühn

ForWind-Center for Wind Energy Research, Institute of Physics, Carl von Ossietzky University of Oldenburg, Küpkersweg 70, 26129 Oldenburg, Germany

E-mail: `rasoul.shirzadeh@forwind.de`

Abstract. This study presents the numerical results of two passive strategies to reduce the support structure loads of a large offshore wind turbine. In the first approach, an omnidirectional tuned mass damper is designed and implemented in the tower top to alleviate the structural vibrations. In the second approach, a viscous fluid damper model which is diagonally attached to the tower at two points is developed. Aeroelastic simulations are performed for the offshore 10MW INNWIND.EU reference wind turbine mounted on a jacket structure. Lifetime damage equivalent loads are evaluated at the tower base and compared with those for the reference wind turbine. The results show that the integrated design can extend the lifetime of the support structure.

1. Introduction

In recent years, the size of offshore wind turbines is further increased beyond the 5 MW class to reduce the cost of energy. The cost effective design of support structures for such large machines is still a challenging task as the rotor diameter and tower height are exceeding 100 m. The support structure eigenfrequencies are analysed in the early stage of the design procedure to prevent significant resonances between the structural eigenfrequencies and excitations from waves, rotor frequency and its multiples. This can be acquired via the Campbell diagram which plots the eigenfrequency of the entire wind turbine system against the rotor speed including the harmonic excitations 1P, 3P, etc. Figure 1 illustrates qualitatively the trends for the rotor rotational frequency (1P) and blade passing frequency (3P) ranges along with the first eigenfrequency of the support structure as a function of the wind turbine size. In the design procedure of large offshore wind turbines, i.e. 7.5+ MW, with jacket structures, a strong and severe blade passing (3P) resonance is expected at low rotor speeds [1].

Figure 2 shows the Campbell diagram for the 10MW INNWIND.EU reference turbine [2]. It can be seen that at rotor speed of 5.7 rpm, the blade passing frequency (3P) coincides with the first natural frequency of the system. The rotor frequency (1P) is however not problematic as it is found far outside of the operational region. In order to reduce the dynamic excitation, a rotor speed exclusion zone between 5.2 rpm and 6.3 rpm is considered. Compared to the initial exclusion zone chosen in the preliminary stage [3], the exclusion zone is shifted downward by a reduction of the cut-in speed from 5 rpm to 4.5 rpm.

The larger the blades and the support structure of the wind turbine, the higher the bending loads. Consequently, strongly increased fatigue loads are experienced by critical components

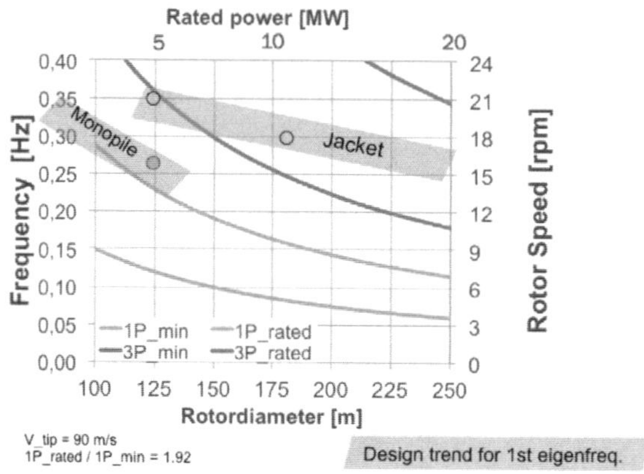

Figure 1. Design trends on the first design eigenfrequency and both the rotor and the blade resonance ranges for three-bladed offshore turbines in the 5 to 20MW class [1].

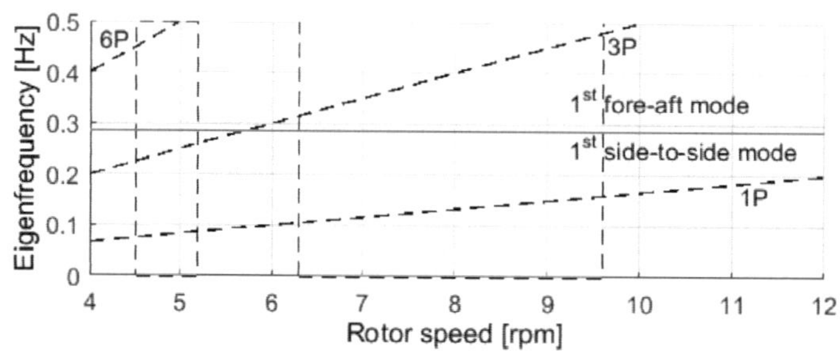

Figure 2. Campbell diagram of the INNWIND.EU 10MW reference turbine. The yellow marked areas represent the operational ranges of the turbine. The cut-in speed is lowered to avoid overlapping with the exclusion zone. The first fore-aft and side-side modes are closely spaced.

e.g. the tower base. Innovations are required in lowering the loads experienced by the support structure. This goal can be met through a variety of load mitigation strategies e.g. implementation of passive or (semi)-active damping devices and different control and regulation concepts. Nowadays, the application of a tuned mass damper (TMD) is becoming increasingly practical for the load mitigation of offshore wind turbines. Previous works have shown the potential for the integration of TMDs in the tower top location [4]. In addition, different solutions have been proposed for the application of semi-active or active dampers, e.g. toggle-brace viscous fluid damper (VFD), tuned liquid column damper (TLCD), magnetorheological (MR) damper and hybrid mass dampers, see [5–7].

To control the dynamic excitation of the support structure, two innovative concepts are designed and integrated in the INNWIND.EU 10MW reference wind turbine (RWT). In the first study, the results of the integration of a passive structural damper to reduce the support structure loads are presented. A passive tuned mass damper (TMD) is designed to realize the best configuration according to the calculated tower base lifetime and damage equivalent loads (DELs) for the reference number of cycles of $N_{ref} = 10^7$ and the S-N curve slope of $m = 4$. During the second analysis, the numerical model of a toggle-brace VFD is developed. Aeroelastic simulations are performed in DNV GL Bladed software [8] for the offshore 10MW INNWIND.EU

RWT mounted on a jacket structure. The loads obtained from two concepts are then compared with those for the reference turbine. We show that the applied loads can be effectively mitigated by the use of damping strategies which would result in an extension in the lifetime of the support structure.

2. Implementation of Wind Turbine Innovations

2.1. Description of the performed studies

The performed studies are based on the offshore 10MW INNWIND.EU RWT [9]. A schematic picture of the jacket and the reference wind turbine (RWT) structures as well as the considered coordinate system are displayed in Figure 3. The design water depth is applied according to the North Sea site with a water depth of 50 m. The support structure consists of a 4-leged, x-braced jacket structure. The 3D turbulent wind field with the Kaimal spectrum is generated for six random seeds while the wave kinematics is modelled based on the irregular wave model with the JONSWAP spectrum. No yaw misalignment angle is assumed in this analysis. The design load case (DLC) 1.2 based on wind class IA according to IEC61400-1 standard [10] is considered and fatigue loads are calculated during the full operating wind speed range. The wind and wave characteristics are listed in Table 1.

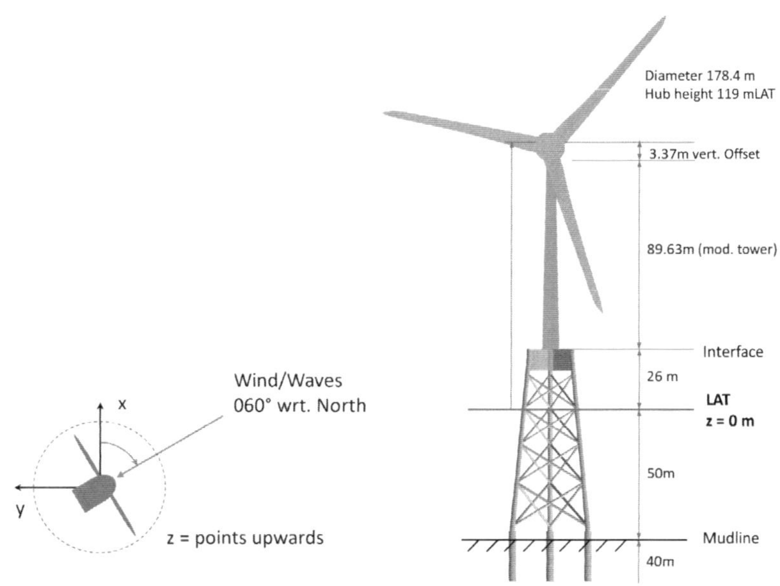

Figure 3. The global coordinate system (left) and a schematic drawing of the RWT and jacket structures (right).

2.2. Description of the load mitigation concepts

Two load mitigation concepts namely the tuned mass damper and the viscous fluid damper which have already been introduced in [11] are considered. These concepts are integrated into the final design of the support structure to ensure an optimised and affordable design. Figure 4 illustrates schematic models of these two concepts. Time series of combined wind and hydrodynamic loads are extracted for both concepts to calculate the Fatigue Limit State (FLS) of the offshore wind

Table 1. Environmental conditions for the wind and wave parameters.

Wind speed [m/s]	Long. Turb., I_u [%]	Lat. Turb., I_v [%]	Vert. Turb., I_w [%]	H_s [m]	T_p [s]	Occurrence [hours/year]
4	20.4	16.30	10.20	1.10	5.88	874.7
6	17.5	14.00	8.75	1.18	5.67	992.8
8	16.0	12.80	8.00	1.31	5.67	1181.8
10	15.2	12.16	7.60	1.48	5.74	1076.3
12	14.6	11.68	7.30	1.70	5.88	1137.2
14	14.2	11.36	7.10	1.91	6.07	875.6
16	13.9	11.12	6.95	2.19	6.37	764.7
18	13.6	10.88	6.80	2.47	6.71	501.3
20	13.4	10.72	6.70	2.76	6.99	336.0
22	13.3	10.64	6.65	3.09	7.40	289.4
24	13.1	10.48	6.55	3.42	7.80	130.4

turbine in operational conditions. In the current study, no wind-wave misalignment is considered. Fatigue loads are evaluated at the tower base and compared with the reference design.

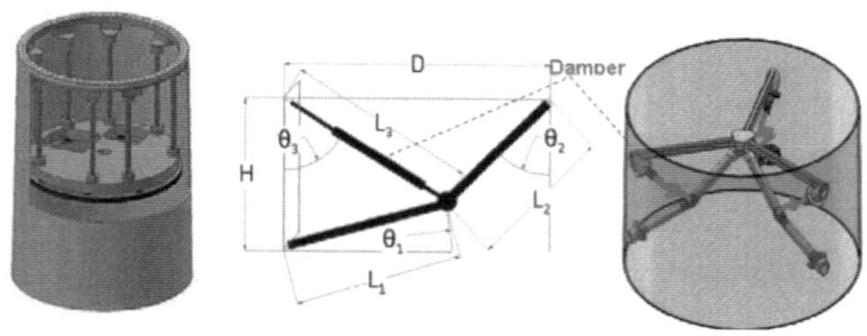

Figure 4. Two passive strategies integrated in the tower top of the 10MW INWWIND.EU RWT, the TMD [12] (left) and the toggle brace viscous fluid damper [13] (right).

During the first concept, the numerical modeling of a passive tuned mass damper mounted on the tower top is performed (see Figure 4). This consists of a mass-spring-damper system which is characterised by its mass, the resonance frequency and the damping factor. The damper is omnidirectional meaning that it can vibrate at all direction. Tuned mass dampers are widely implemented and tested in recent commercial wind turbines, e.g. Vestas V90-3MW. However, the initial INNWIND.EU 10MW reference jacket is designed without considering any damper in the tower. As a consequence, the initial jacket design is not optimised and needs further improvements. The integration of a TMD has the potential to improve the jacket design by reducing interface loads. The interface spot, as shown in Figure 3, is the connection point between the tower and the jacket foundation, i.e. the tower base.

According to [4], the TMD mass is approximately either 8% of the nacelle mass or 6% of the tower top mass. Other studies propose to use a value between 2-4% of the modal mass associated with the fundamental lowest eigenfrequency which is commonly used for civil engineering structures [4]. In this analysis, simulations are performed with a TMD with mass ratios of 1% and 2% with respect to the modal mass. The resonance frequency of the TMD is tuned to the first fundamental mode of the support structure (the first fore-aft mode) and is able to reduce the amplitude of vibrations around this frequency.

The second concept describes the mathematical representation of a VFD. VFDs have already been used in civil engineering structures to dissipate seismic energy. Up to now, the application of VFDs in wind turbines did not come into practice and further investigations are required to accommodate a damper network system of this type in the tower. The optimal location of the damper is a challenging task. In this study, we assume that the VFD is installed at the tower top, whereas maximum vibrations occur and hence a higher damping efficiency is desired.

The design formulas for a VFD system are based on the study described in [14]. For a system represented in Figure 4, the following relation exists:

$$u_D = fu \tag{1}$$

where u_D is the relative displacement along the axis of the damper, u is the relative displacement of the attachment points and f is the magnification factor. The magnification factor depends on the configuration of the bracing system and is usually larger than 1. For a diagonal toggle-brace-damper system, it equals to $\cos\theta$ where θ is the inclined angle of the damper [14]. The horizontal component of the force exerted by the damper on the structure, F, is obtained from:

$$F = fF_D \tag{2}$$

with F_D, the damper force. For a VFD, the damper force along its axis is proportional to the velocity and can be written as below:

$$F_D = C\dot{u}_D^{\alpha} \tag{3}$$

where C is the damping coefficient of the damper, $\dot{u}_D$ represents the relative velocity between the attachment points of the damper and α is the damper nonlinearity. In this paper, we assume $\alpha = 1$ which models a linear damper. Combining Eqs. 1 and 3 and knowing that any real number can be expressed as the product of its absolute value and its sign function, the magnitude of the damping force can be expressed as:

$$F_D = Cf^{\alpha}|\dot{u}|^{\alpha}sgn(\dot{u}) = C_0|\dot{u}|^{\alpha}sgn(\dot{u}) \tag{4}$$

The effective damping coefficient, $C_0 = Cf^{\alpha}$, has an important impact on the damper force. The damper force is applied to the structure and mitigates the external forces exerted on the structure. Several strategies can be used to calculate the viscous damper force. One of the most practically available parameters is the acceleration which can be easily measured using accelerometers.

Figure 5 demonstrates the mechanism to calculate the viscous damper force using the tower top simulated accelerations. A Simulink model of a VFD is developed and connected to Bladed via an external Dynamic-Link Library (DLL). The simulated time series of accelerations at the ending points of the damper where it is attached to the tower are recorded. Velocity signals can be attained by integrating the simulated accelerations. The damper force, F_D, is then obtained in Simulink using both velocities and the effective damping coefficient. This force is divided in two and returned to the Bladed whereas it will be applied as reaction forces at the damper ending points (see Figure 5). For multiple dampers, the damper force is split up into several components based on the orientation of the viscous dampers. The damper model performs as a passive device if a constant damping coefficient is assumed. Alternatively, for a semi-active device, the effective damping coefficient can be provided as a lookup table which is calculated and scheduled based on external or operating conditions. For such a damper, it acts semi-actively as the damper force is adapted based on the environmental conditions. In this study, however, a constant effective damping coefficient is chosen at all wind speeds.

3. Design Load Verification

In this section, results of the load verification with the integration of two concepts are discussed. Figure 6 demonstrates the DELs of the tower base moment in the fore-aft (M_y) and side-to-side

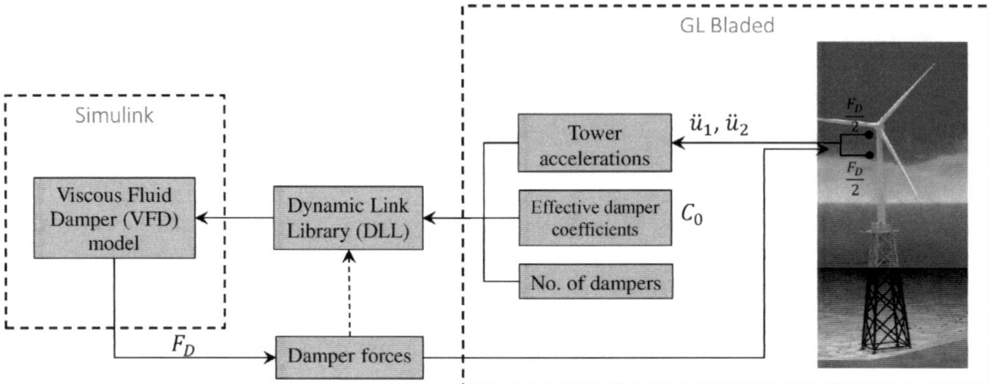

Figure 5. Numerical modeling of the VFD shows the mechanism to calculate the damper forces using the tower accelerations.

(M_x) directions, respectively. M_x and M_y are represented in the global coordinate system with X and Y axes pointing towards the north and west, respectively, when nacelle is faced to the wind inflow. The results are only for one set of load setup where the wind-wave misalignment angle is zero. The simulations are carried out with six random seeds for the turbulent wind and DELs are averaged at each wind speed. For all results shown here, the exclusion zone is activated. The influence of the TMD is not significant in the fore-aft direction, especially near the rated wind speed, while the TMD can effectively mitigate the fatigue loads in the sideways direction. It should be noted that despite the rotor speed exclusion zone is active, a resonance can be seen for the reference design which is further mitigated by the TMD even in the fore-aft direction.

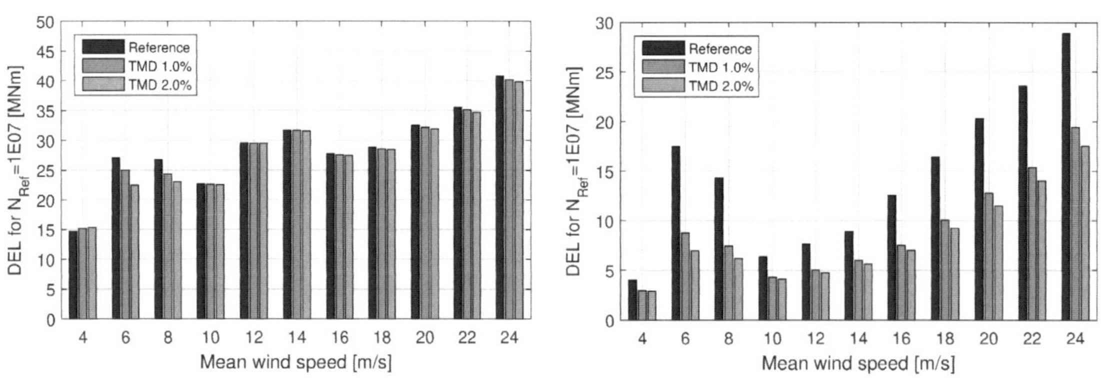

Figure 6. DELs of the tower base moment in the fore-aft, M_y, (left) and sideways, M_x, (right) directions for different mass ratios of the TMD.

This is in agreement with findings of Kuhnle [11] where it has been shown that the DELs can be improved by the application of a tuned mass damper at the tower top location. In addition, the TMD is more effective for the sideways direction which is due to the low aerodynamic damping in this direction. The aerodynamic damping is the dominant phenomenon in the fore-aft direction and therefore the performance of the TMD is marginal in this direction. Except for DELs in the fore-aft direction in a region around the rated wind speed, the TMD effectively dissipates side-to-side loads in the whole operational range.

The maximum reduction of DELs occurs in the sideways direction with a TMD with the mass ratio of 2% at the wind speed of 6 m/s which corresponds to a 60% reduction with respect to the reference turbine configuration. For a TMD with the mass ratio of 2%, the lifetime weighted DELs are reduced by 4% and 24.3% in the fore-aft and side-to-side directions respectively.

As explained in the previous section, the second concept is a viscous fluid damper installed at the tower top location. Three toggle brace viscous fluid dampers with 120 phase shifting i.e. 0° towards north, 120° and 240°, are considered. The VFD model is created in Simulink and called through a DLL at each time step. Since the damper force is proportional to the effective damping coefficient, C_0, a larger damper force is achieved with increasing the damping coefficient. Therefore, the effective damping coefficient is increased step by step up to a value for which the simulations became unstable. This corresponds the damping coefficient which causes maximum load reduction.

A typical time interval of the nacelle displacements in the fore-aft and side-to-side directions are demonstrated in Figure 7. It is obvious that the tower top vibration, particularly in the sideways direction, is considerably dissipated compared to the reference configuration.

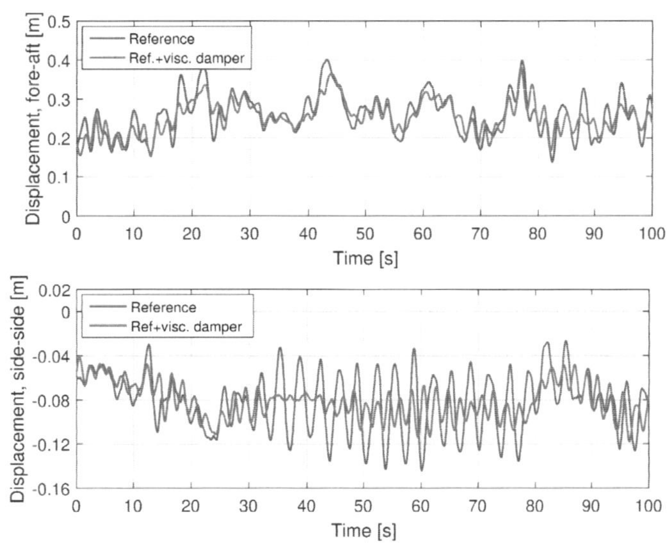

Figure 7. Nacelle displacement with and without VFD in the fore-aft (top) and side-to-side (bottom) directions at wind speed of 12 m/s.

The tower base DELs of the bending moments in the fore-aft and sideways directions are plotted in Figure 8. It should be noted that these plots are for a loading setup where the wind-wave misalignment is zero. It can be seen that the DELs are remarkably decreased in both directions. The maximum load reduction, i.e. around 69%, occurs in the sideways direction at the wind speed of 6 m/s. In addition, the lowest loads reduction occurs near the rated wind speed where the highest variations of aerodynamic forces take place. Although these results are calculated without the wind-wave misalignment angle, the interface loads and moments calculated for the full load setup are likely to be resulted in optimized jacket geometry.

The lifetime weighted equivalent loads of the tower base moments for the reference turbine and both strategies are compared in Figure 9. The values are normalised with respect to the DEL of the reference turbine in the fore-aft direction. It can be concluded that the DELs are

Journal of Physics: Conference Series **749** (2016) 012011 doi:10.1088/1742-6596/749/1/012011

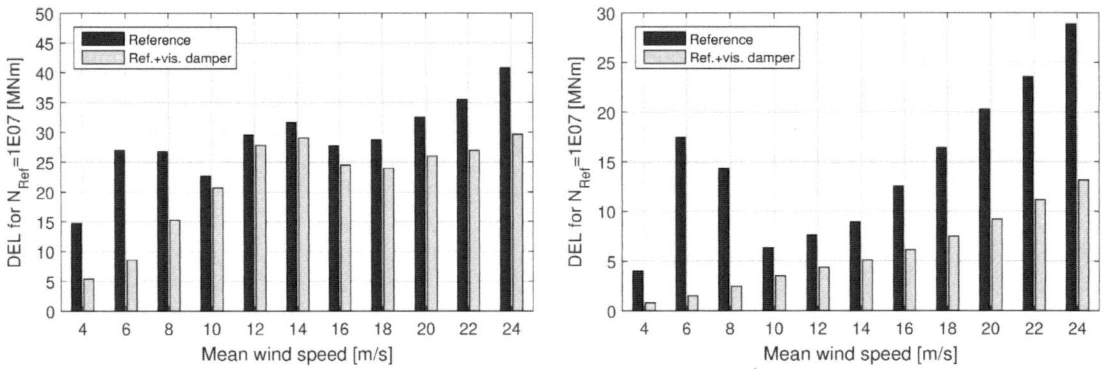

Figure 8. Tower base moment DELs in the fore-aft (left) and side-to-side (right) directions with integration of a VFD.

strictly improved with passive viscous dampers. The application of the VFD is however more effective than TMD.

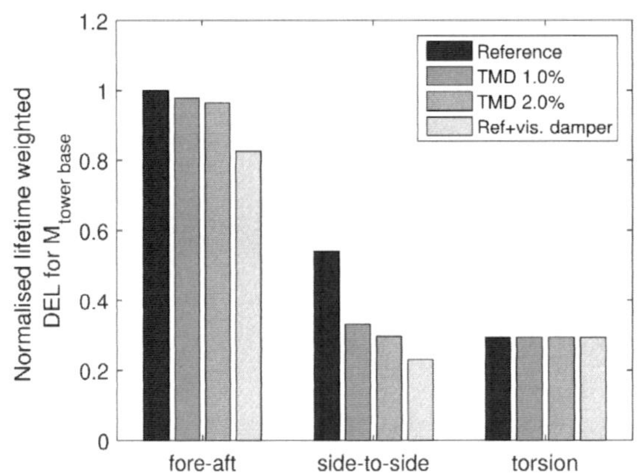

Figure 9. Normalised lifetime weighted DELs of the tower base moments calculated for different damping strategies.

4. Conclusions

This paper presents the results of the integration of two damping concepts in the 10MW INNWIND.EU reference wind turbine. These two concepts are: a passive tuned mass damper (TMD) and a toggle brace viscous fluid damper. The improved tower base loads are obtained and compared with the reference design.

It can be concluded that for the load setup where inflow wind is aligned with the wave direction, the DELs in the sideways direction, where no aerodynamic damping is active, can be lowered up to 29% and 69% compared to the reference case for respectively a TMD and viscous fluid damper. In this condition, the impact of the TMD in the fore-aft direction is however not significant while the viscous damper gives remarkably lowered loads in both directions.

The integration of the damping concepts could have a positive impact on the lifetime of the system. The integration of the developed strategies could be considered for an optimized jacket design with reduced entire mass and applied loads for the INNWIND.EU 10MW RWT.

Acknowledgments
The presented work was funded by the European Community's Seventh Framework Programme FP7-ENERGY-2012-1-2STAGE under grant agreement No.308974 (INNWIND.EU).

References
[1] Berger F, Dalsgaard Sörensen J, Gintautas T, Kühn M and Kraft M 2015 INNWIND.EU Deliverable D1.35 System design assessment for innovative support structures, www.innwind.eu, accessed 2016-04-10 Tech. rep.
[2] Bak C, Zahle F, Bitsche R, Kim T, Yde A, Henriksen L C, Natarajan A and Hansen M 2013 INNWIND.EU Deliverable D1.21 Description of the DTU 10 MW Reference Wind Turbine, www.innwind.eu, accessed 2016-04-10 Tech. rep.
[3] Kuhnle B and Kühn M 2015 Unfavourable trends of rotor speed and systems dynamics for very large offshore wind turbines analysis of the 10mw innwind.eu reference turbine *EWEA offshore 2015 conference*
[4] Lackner M A and Rotea M A 2011 *Wind energy* **14** 373–388
[5] Carcangiu C, Pineda I, Fischer T, Kuhnle B, Scheu M, Martin M *et al.* 2011 Wind turbine structural damping control for tower load reduction *Civil Engineering Topics, Volume 4* (Springer) pp 141–153
[6] Karimi H R, Zapateiro M and Luo N 2010 Semiactive vibration control of offshore wind turbine towers with tuned liquid column dampers using H_∞ output feedback control *Control applications (CCA), 2010 IEEE international conference on* (IEEE) pp 2245–2249
[7] Caterino N 2015 *Journal of Sound and Vibration* **345** 1–17
[8] Bossanyi E 2010 *Garrad Hassan and Partners Ltd., Bristol, UK*
[9] von Borstel T 2015 INNWIND.EU Deliverable D4.31 Design Report Reference Jacket, www.innwind.eu, accessed 2016-04-10 Tech. rep.
[10] IEC. International Standard IEC-61400-1 Wind Turbines – Part 1: Design requirements, International Electrotechnical Commission (IEC): Geneva, Switzerland, 2005
[11] Scholle N, Radulovi L, Nijssen R, Ibsen L B, Kohlmeier M, Foglia A, Natarajan A, Thiel J, Kuhnle B, Kraft M, Kühn M, Brosche P and Kaufer D 2015 INNWIND.EU Deliverable D4.13 Innovations on component level (final report), www.innwind.eu, accessed 2016-04-10 Tech. rep.
[12] URL `ESM GmBh, www.esm-gmbh.de/EN/Products/Tuned-mass-dampers, accessed 2016-07-10`
[13] Soong T 2005 *Active, Hybrid, and Semi-active Structural Control: A Design and Implementa* (John Wiley & Sons)
[14] Hwang J S, Huang Y N and Hung Y H 2005 *Journal of structural engineering* **131** 1035–1043

Journal of Physics: Conference Series **749** (2016) 012012 doi:10.1088/1742-6596/749/1/012012

Aerodynamic Simulation of the MARINTEK Braceless Semisubmersible Wave Tank Tests

Gordon Stewart and Michael Muskulus

Department of Civil and Transport Engineering, NTNU, Høgskoleringen 7A, 7491 Trondheim, Norway

E-mail: gordon.stewart@ntnu.no

Abstract. Model scale experiments of floating offshore wind turbines are important for both platform design for the industry as well as numerical model validation for the research community. An important consideration in the wave tank testing of offshore wind turbines are scaling effects, especially the tension between accurate scaling of both hydrodynamic and aerodynamic forces. The recent MARINTEK braceless semisubmersible wave tank experiment utilizes a novel aerodynamic force actuator to decouple the scaling of the aerodynamic forces. This actuator consists of an array of motors that pull on cables to provide aerodynamic forces that are calculated by a blade-element momentum code in real time as the experiment is conducted. This type of system has the advantage of supplying realistically scaled aerodynamic forces that include dynamic forces from platform motion, but does not provide the insights into the accuracy of the aerodynamic models that an actual model-scale rotor could provide. The modeling of this system presents an interesting challenge, as there are two ways to simulate the aerodynamics; either by using the turbulent wind fields as inputs to the aerodynamic model of the design code, or by surpassing the aerodynamic model and using the forces applied to the experimental turbine as direct inputs to the simulation. This paper investigates the best practices of modeling this type of novel aerodynamic actuator using a modified wind turbine simulation tool, and demonstrates that bypassing the dynamic aerodynamics solver of design codes can lead to erroneous results.

1. Introduction

The offshore wind energy resource worldwide is one of the largest renewable energy sources in terms of total potential energy. However, much of this resource is located over deep water, so floating offshore wind turbines are being developed that have the potential to economically capture this energy. Floating wind turbines have added benefits, as the ability of being towed out to the energy production site allows for assembly in port, and the potential to be located farther from shore to reduce visibility impacts. However, for useful design work to be possible, accurate computer modeling tools are essential. As full-scale prototypes are scarce, expensive, and often proprietary, many research institutions are conducting small-scale experiments in wave tanks to provide data both for platform design and modeling tool validation. One difficulty of small-scale experiments of floating wind turbines is the scaling of forces between the hydrodynamics and the aerodynamics. A variety of methods have been used to modify the rotor to scale the aerodynamic forces correctly, including using a simple actuator disk [1], or modifying the chord and other blade properties of the rotor [2, 3, 4]. Recently, there have been experiments conducted that utilize an actuator in place of a rotor to provide realistic forces calculated in real time by

Journal of Physics: Conference Series **749** (2016) 012012 doi:10.1088/1742-6596/749/1/012012

Figure 1. Image of the MARINTEK braceless semisubmersible (photo courtesy of MARINTEK.

aerodynamic codes during an experiment. These devices may take the form of a ducted fan that can be controlled to provide thrust forces [5, 4], or what is used in the recent series of experiments at MARINTEK, a frame with wires tensioned by small motors that provide the aerodynamic forces [6, 7]. A picture of the experimental setup can be seen in figure 1. The advantage of this type of actuator is that unlike a ducted fan, which can only provide force in the thrust direction, the tensioned frame can provide forces in five directions; all three rotational directions, surge, and sway; the MARINTEK actuator does not provide forces in the vertical heave direction, as aerodynamic forces in this direction were found to be insignificant.

The calculation of the aerodynamic forces is done in real time during the experiment, using both a given turbulent or steady simulated wind input and the location and velocity of the platform. This creates both correctly scaled wind-induced aerodynamic forces as well as forces due to platform motion and rotor interaction, including aerodynamic damping forces [7]. The MARINTEK experiment uses the aerodynamic solver from the National Renewable Energy Laboratory's (NREL) design code suite, called AeroDyn, with some modifications for the integration into the experiment. In this way, all of the forces that are applied to the nacelle are known. To model this system in simulation, there are two options for the aerodynamics. Since the actual aerodynamic forces are known, the force time series could be directly applied to the simulation, bypassing the built-in aerodynamic solver of the code, or the wind speed time series from the experiment could be used as the simulation input, and the internal aerodynamic solver would calculate forces based on this wind input and the dynamics of the simulation. The first method has the potential to be inaccurate because the dynamic feedback forces from platform motion on the aerodynamic are uncoupled from the actual platform motion in the simulation, as those forces are calculated from the motion of the experiment. This paper will investigate the consequences of bypassing the aerodynamic calculation in this manner.

2. Methods

As the data from the MARINTEK experiment has not been released at the writing of this paper, simulation will be used to replicate the experiment and actuator for this work. The National Renewable Energy Laboratorys wind turbine design software FAST will be used with the spar buoy model from the Offshore Code Comparison Collaboration project (OC3) [8]. FAST is a medium-fidelity simulation software that uses blade-element momentum theory for the aerodynamics, and includes hydrodynamic, structural, and control modules. The spar buoy model features a controller with modified pitch control gains to reduce the effect of surge instabilities.

To replicate the novel aerodynamic actuator used in the experiment, an addition to the FAST software had to be written that allows the aerodynamics calculation to be bypassed, by applying the sum of all aerodynamic and inertial forces from the blades directly onto the hub. In this way, a reference simulation could be run with the standard version of FAST, and the aerodynamics loads from this baseline simulation could be applied to a second simulation. Parameters and inputs to this second simulation could be modified to replicate experimental uncertainty. In this way, the baseline simulation represents the experiment, and the simulation that uses the predefined aerodynamic loads represents the simulation of the experiment. To model the uncertainty in the platform modeling of an experiment, the hydrodynamic drag coefficient was modified. The drag coefficient of the platform is chosen as an example of a hydrodynamic parameter that could be different between the simulation model and the experiment, any of the potential flow parameters could have been modified with a similar conclusion. A series of simulations are then run, using identical wind and wave inputs, but bypassing the aerodynamic calculation for one set of simulations.

3. Results and Discussion

A range of wind speed and wave heights are simulated, similar to the operational design load cases in the IEC standard. The wind and wave inputs used in the simulations can be seen in table 1. These significant wave height and peak spectral period values were taken from a study of the NOAA floating data buoys, and correspond to the expected value for the specified wind speed at a representative site from the east coast of the United States [9].

Table 1. Wind and wave parameters used in the simulations.

Mean Wind Speed (m/s)	4	6	8	10	12	14	16	18	20	22	24
Significant Wave Height (m)	1.1	1.2	1.3	1.5	1.8	2.2	2.6	3.1	3.6	4.0	4.5
Peak Spectral Period (s)	8.5	8.3	8.0	7.7	7.4	7.5	7.6	8.0	8.5	9.0	9.4

A baseline simulation of the OC3 spar buoy was run for each of these wind speeds, using the corresponding wave parameters. Next, these 11 simulations were repeated with the same wind and wave inputs (keeping the random seeds, and thus the turbulence and random waves the same), but with the platform hydrodynamic drag reduced from 0.6 to 0.5. Finally, a third set of simulations was run using the lower C_d value, but bypassing the aerodynamic calculation and using the aerodynamic forces directly from the baseline simulation with the higher C_d value. Figures 2 through 3 compare timeseries from various outputs for one windspeed (10 m/s). First, in figures 2 and 3, the platform surge displacement (translational movement of the platform in the direction of the wind) is plotted.

In figure 2, the plots show the surge output for the two simulations that used the baseline aerodynamic calculation but two different C_d values. These outputs are very close to each other,

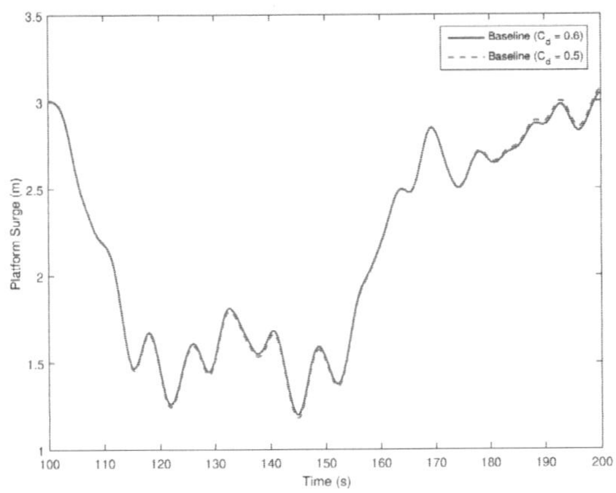

Figure 2. Comparison of surge displacements for baseline simulations with two C_d values

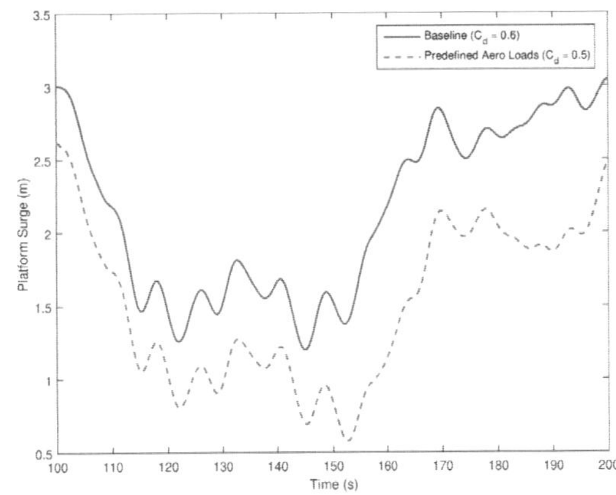

Figure 3. Comparison of surge displacements for baseline simulation and predefined aerodynamics

but slightly more variance can be seen in the output from the lower C_d simulation due to the lower damping. Figure 3, which compares the baseline simulation with the simulation with both the predefined aerodynamic forces and the lower C_d, shows much more deviation of the two outputs. As the platform surges, the apparant wind speed seen by the rotor is increased or decreased by the surge velocity, which changes the rotor thrust magnitude, which in turn feeds back to the surge velocity. By predefining the aerodynamic load, this feedback is not preserved, and the small difference in the surge displacement early in the simulation lead to large differences later in the simulation.

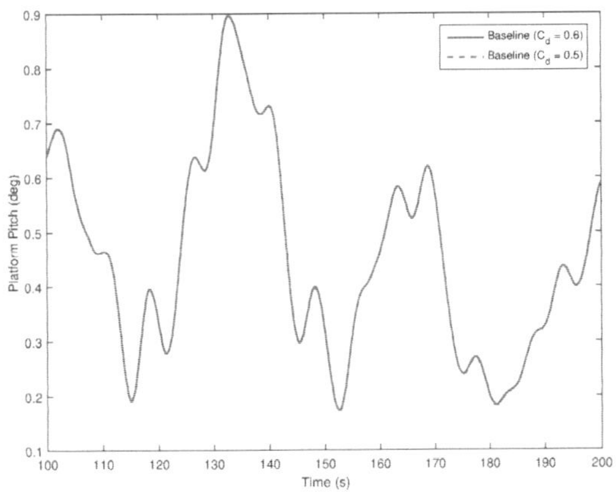

Figure 4. Comparison of platform pitch displacements for baseline simulations with two C_d values

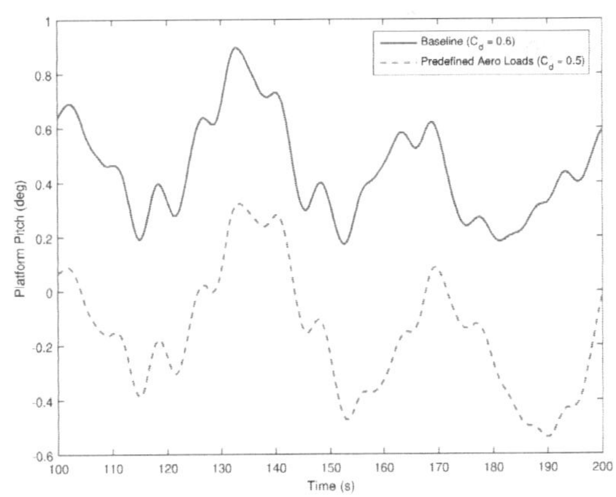

Figure 5. Comparison of platform pitch displacements for baseline simulation and predefined aerodynamics

Figures 4 and 5 repeat the comparisons made in figures 2 and 3, but look at the platform pitch angle (rotational displacement in the wind direction) instead. Similar trends to the surge displacement figures can be seen in the platform pitch, with a large divergence for the simulation

Journal of Physics: Conference Series **749** (2016) 012012 doi:10.1088/1742-6596/749/1/012012

with the predefined aerodynamic loads. Interestingly, the phasing of the predefined rotor thrust has resulted in a platform pitch that has a mean pitch angle that actually leans into the wind (which can be seen as a negative mean pitch angle in figure 5), which is a good indicator of the potential instability and erroneous results that using this method of aerodynamic simulation can produce.

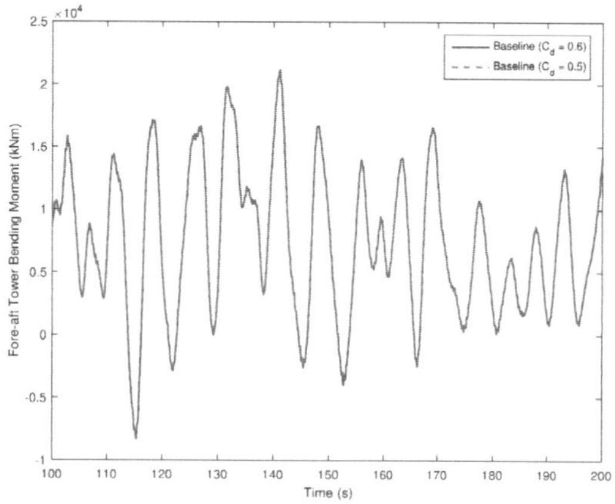

Figure 6. Comparison of tower base fore-aft bending moments for baseline simulations with two C_d values

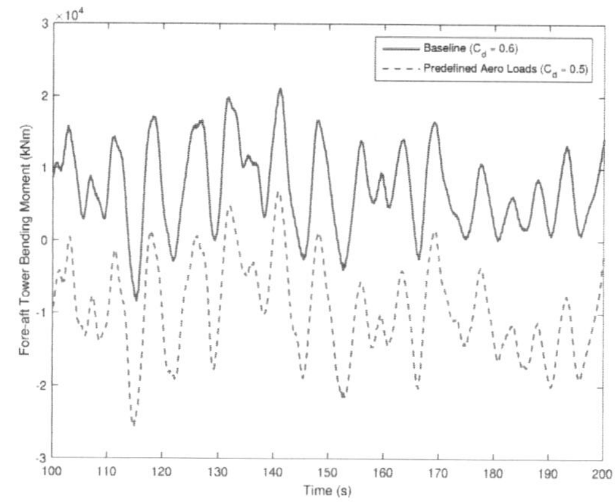

Figure 7. Comparison of tower base fore-aft bending moments for baseline simulation and predefined aerodynamics

Figures 6 and 7 show comparisons of the tower base bending moment. As this output is correlated with platform pitch (due to inertial loading), the trends in bending moment resemble the platform pitch results. The mean bending moment in figure 7 has also become negative for the predefined aerodynamics simulation, similar to figure 5.

To look at trends for the entire operational wind envelope for this platform, figures 8 through 11 show the standard deviation of various outputs plotted as a function of wind speed. Note that the wave height and peak spectral period also vary as wind speed varies (see table 1). In figures 8 and 9 the standard deviation of the surge displacement is compared for the baseline aerodynamic calculation for the two C_d values (figure 8), and for the predefined aerodynamics (figure 9). The difference from the baseline in the standard deviation of surge is larger for the predefined loads, especially for the higher wind speeds (and thus higher wave hights).

In figures 10 and 11, the standard deviation of fore-aft tower base bending moment as a function of wind speed is compared. Here there is a negligible difference for the baseline case when varying the C_d (figure 10), but the standard deviation of the bending moment increases by as much as 25% when using predefined aerodynamic loads (figure 11).

4. Conclusions and Future Work

The simulations presented in this paper show that using a predefined aerodynamic force time series can lead to large differences in both loads and platform motions when comparing simulations to experiments. These errors are due to the lack of realistic feedback between the platform motion and the aerodynamics captured in the simulations with predefined aerodynamic loads. However, as the data from the experiment are not available at the time of this writing, more investigation is needed using the data to confirm this conclusion. It is likely that this method is reasonable to use for simpler cases, such as regular waves with steady wind, as the phasing of the aerodynamic forces can be carefully matched with the wave loading. An upcoming

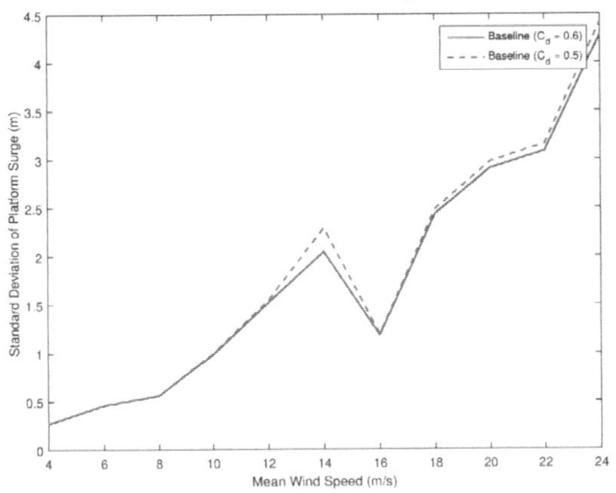

Figure 8. Platform surge standard deviation as a function of wind speed for baseline simulations with two C_d values

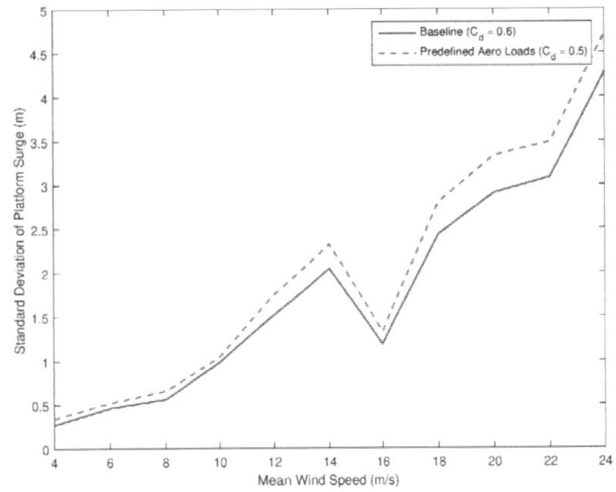

Figure 9. Platform surge standard deviation for baseline simulation and predefined aerodynamics

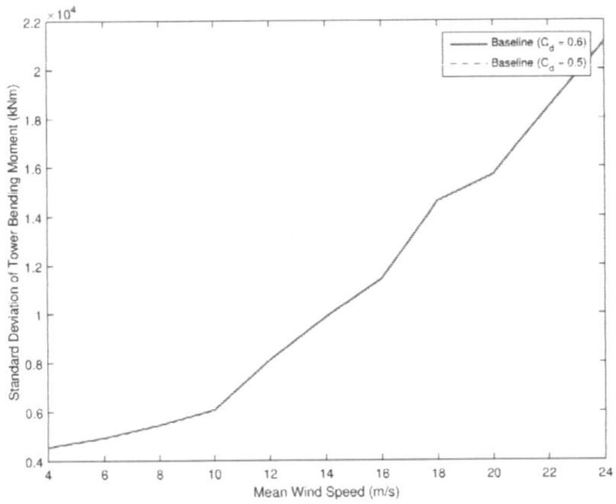

Figure 10. Fore-aft tower bending moment standard deviation as a function of wind speed for baseline simulations.

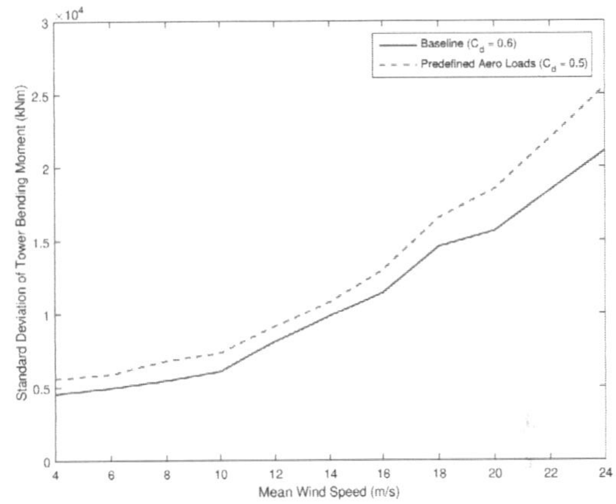

Figure 11. Fore-aft tower bending moment standard deviation for baseline simulation and predefined aerodynamics

paper will use the experimental data to test the conclusion of the simulation work presented in this paper, as well as allowing the method of aerodynamic modeling that uses only the wind time series as an input to be tested.

Acknowledgements
The research leading to these results has received funding from the IRPWind program, part of the European Unions Seventh Framework Program (FP7/2007-2013) under grant agreement no. 609795.

References
[1] Roddier D, Cermelli C, Aubault A and Weinstein A 2010 *Journal of Renewable and Sustainable Energy* **2**

[2] Martin H R 2011 *Development of a scale model wind turbine for testing of offshore floating wind turbine systems* Ph.D. thesis Maine Maritime Academy
[3] Coulling A J, Goupee A J, Robertson A N, Jonkman J M and Dagher H J 2013 *Journal of Renewable and Sustainable Energy* **5**
[4] Sandner F, Amann F, Azcona J, Munduate X, Bottasso C L, Campagnolo F and Robertson A 2015 *EERA Deepwind*
[5] Azcona J, Bouchotrouch F, González M, Garciandiá J, Munduate X, Kelberlau F and Nygaard T A 2014 *Journal of Physics: Conference Series* **524**
[6] Kvittem M I, Bachynski E E and Moan T 2012 *Energy Procedia* **24** 351–362
[7] Tande J O G, Kvamsdal T, Muskulus M, Bachynski E E, Chabaud V and Sauder T 2015 *Energy Procedia* **80** 2 – 12
[8] Jonkman J and Musial W 2010 Offshore Code Comparison Collaboration (OC3) for IEA Task 23 Offshore Wind Technology and Deployment Tech. Rep. December National Renewable Energy Laboratory
[9] Stewart G M, Robertson A, Jonkman J and Lackner M A 2016 *Wind Energy* **19** 1151–1159

CFD simulations of a wind turbine for analysis of tip vortex breakdown

K Kimura[1], Y Tanabe[2], T Aoyama[2], Y Matsuo[2], C Arakawa[1] and M Iida[1]

[1] The University of Tokyo, 7-3-1 Hongo, Bunkyo-ku, Tokyo, 113-8656, Japan
[2] Japan Aerospace Exploration Agency, 7-44-1 Jindaijihigashimachi, Chofu-shi, Tokyo, 182-8522, Japan

E-mail: k.kimura@gg.cfdl.t.u-tokyo.ac.jp

Abstract. This paper discusses about the wake structure of wind turbine via the use of URANS and Quasi-DNS, focussing on the tip vortex breakdown. The moving overlapped structured grids CFD Solver based on a fourth-order reconstruction and an all-speed scheme, rFlow3D is used for capturing the characteristics of tip vortices. The results from the Model Experiments in Controlled Conditions project (MEXICO) was accordingly selected for executing wake simulations through the variation of tip speed ratio (TSR); in an operational wind turbine, TSR often changes in value. Therefore, it is important to assess the potential effects of TSR on wake characteristics. The results obtained by changing TSR show the variations of the position of wake breakdown and wake expansion. The correspondence between vortices and radial/rotational flow is also confirmed.

1. Introduction

For economical use of wind turbine, it is important to install the wind turbines collectively at one site (wind farm). The collective layout leads to good effects such as the reductions of the total length of power transmission lines, the load for maintenance etc... However, the wake behind the wind turbines is known to cause major problems when the turbines are in a collective layout. A wind turbine converts the kinetic energy of wind into mechanical energy; consequently, the speed of the wind in the wake region is greatly reduced. Moreover, the existence of tip vortices impedes momentum exchange between the wake and the main stream [1]. Therefore, it needs a long distance to recover the wake velocity. In addition, when these vortices directly flow in a subsequent wind turbine, its fatigue damage gets accumulated faster. For the above reasons, each wind turbine in a wind farm is placed away from the others. From past researches and some experiences, distance of 10D (D is the diameter of the wind turbines) between two rotors is recommended [2]. However, this criterion is determined empirically, and the optimum distance is supposed to vary depending on the operational conditions. Therefore, the wake structure must be investigated by using scientific approaches for the optimum layout of the wind farms.

Experimental or/and numerical approaches have been considered to study the wake structure. The experimental approaches' results typically tend to be reliable because it can deal with the actual phenomena. However, future commercial wind turbine becomes increasingly large, and experimentally performing the grasping of the flow will become difficult in view of the scaling law in the flow dynamics such as the Reynolds number problems. At present, the rotor diameters of the commercial wind turbines are approximately 100 m and will increase to approximately 200 m in the future, making it very expensive to perform experiments. Alternatively, numerical studies do not require such construction costs. In addition, the entire field data can be obtained via computational fluid dynamics (CFD). At present, CFD can predict the average flow fields around the wind turbines and become a tool for understanding the wake structure. Vermeer *et al.* [3] summarized several methods of CFD for the wake simulation. They reported that Reynolds-averaged Navier-Stokes (RANS) simulation was able to give fair results in attached aerofoil flow conditions. On the other hand, Large Eddy Simulation (LES) or Detached Eddy Simulation (DES) were better to simulate separated aerofoil flow condition. Up to the present time, many researchers have developed the CFD methods for wake simulation and studied about various conditions to reproduce the detailed structure of wake. Troldborg *et al.* [4] investigated the effect of CFD models for wind turbine on the wake such as actuator disc (AD) and actuator line (AL) models. The distribution of vortices is greatly influenced by the type of rotor modelling and it is significant to resolve the rotor geometry for capturing the wake vortices. AbdelSalam *et al.* [5] also reported the advantages of fully resolving the rotor geometry in terms of prediction of the wake velocity. Modelling of the natural wind is also important to simulate actual operations. Li *et al.* [6] have implemented the velocity fluctuations into the inflow condition to model the natural wind. Considering the disturbance of the natural wind, it is understood that wake recovery gets faster than uniform flow.Nilay *et al.* [7] have studied the effect of wind shear on wake flow structure. The existence of a wind sheer causes vortices become close each other, which increases the mutual interaction between vortices. These CFD results can be used to develop wake models oriented to predict velocity deficits. Göçmen *et al.* [8] reviewed the wake models developed at DTU. Compared with the field test and CFD results, they show the availability of the wake models for considering the wind farm layout in terms of velocity deficits. Although wake models are getting reliable to use in industry, it has a difficulty for reproducing the values of velocity deficit when inflows fluctuate and the turbulence intensity get increased. To be more precise, understanding and modelling of vortex breakdown is important because it generates a large velocity gradient. Discussion on the behaviour of the tip vortices is ongoing and it needs to be solved by using CFD.

In this study, the CFD simulations of the wind turbine were conducted to study the vortex structure and the breakdown process by using a higher order scheme and dynamic calculations. To consider various operational conditions, the tip speed ratio λ (TSR) is changed and the simulations are executed. It is carried out to investigate the relation between TSR and the tip vortex structures.

2. Approach

2.1. Case study

The model wind turbine that was previously used in the Model Experiments in Controlled Conditions (MEXICO) project, was selected as case study [9-12]. One of the special features of this test is that it includes particle image velocimetry (PIV) measurements of near-wake regions. Therefore, such data can be used to validate CFD results of velocity distributions and/or vorticity distributions. The specifications for the MEXICO experiment are shown in table 1. The Reynolds number is based on the characteristic chord length (at 82% spanwise position) and the rotational speed of blade tip. Several researchers have previously endeavoured to simulate the MEXICO rotor by CFD. For example, Micallef *et al.* [13] studied the effects of yawed conditions on wake characteristics and Carrion *et al.* [14] executed TSR cases of MEXICO using a compressible multi block solver and an all-Mach scheme; both of their calculations ultimately reproduced the flow around the MEXICO rotor with precision. Moreover, Réthoré *et al.* [15] investigated wind tunnel effects on wake characteristics in MEXICO experiments, concluding that the impact of wind tunnels on wakes is not significant in PIV measurement region.

Table 1. Specifications of MEXICO experiments [9].

Rotor diameter (D) [m]	4.5
Number of blades [-]	3
Speed of main flow[m/s]	10, 15, 24
Tip speed ratio [-]	10, 6.67, 4.17
Rated rotational speed [rpm]	424.5
chord length [m]	0.113 (at 82% span-wise position)
Reynolds number	8×10^5

2.2. Analysis code

The Moving Overlapped Structured Grids CFD Solver rFlow3D [16] is used for wake analysis. rFlow3D was developed by the Japanese Aerospace eXploration Agency (JAXA) for analysing rotorcrafts such as helicopters. The governing equations are the compressible Navier–Stokes equations, which are discretised using a finite volume method (FVM). rFlow3D can address moving overlapped grids; thus, it can simulate the blade rotations and deformations. By using an all-speed numerical scheme SLAU [17], dissipations in low-speed area such as root of blades can be kept small. Furthermore, rFlow3D is good at capturing the blade tip vortices by adapting a fourth order reconstruction method FCMT [16]

2.3. Computational grid

The Computational grid is composed of two background Cartesian grids and three rotating blade grids for analysis. Furthermore, an inner background grid, which has high grid resolution was established to capture the small structure of the tip vortices. The grid covers from 0.5 rotor diameters upwind of the rotor plane to 10 rotor diameters downwind. Details of the computational grids are presented in table 2 and figure 1. x, y and z denote the axial, horizontal and vertical direction each other. The origin of the coordinate is the rotational canter of the wind turbine.

Table 2. Computational grid information.

	Blade Grid	Inner Background Grid	Outer Background Grid
Number of partition (x×y×z)	$121 \times 121 \times 61$	$1307 \times 187 \times 187$	$151 \times 101 \times 101$
Computational domain[a]	-	$10.5D \times 1.3D \times 1.3D$	$14.5D \times 5D \times 5D$
Minimum space[b]	$y^+ \approx 1$	0.32 c	0.32 c
Total number of cells	900 thousands	46 millions	1.5 millions

[a] diameter of the rotor D = 4.5 [m].
[b] chord length c = 0.113 [m] (at 82% span-wise position).

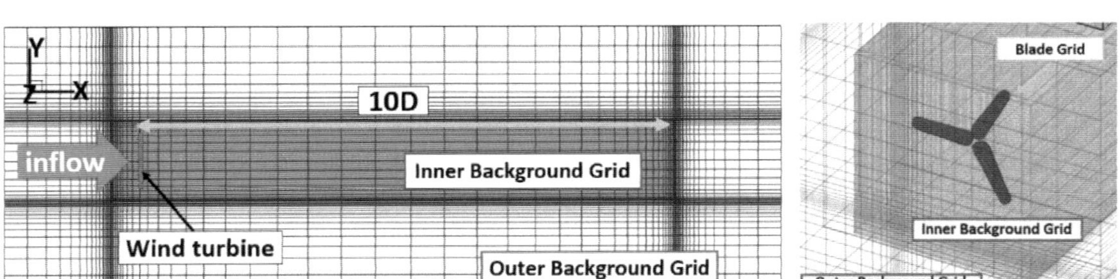

(a) (b)

Figure 1. Computational grid (a) wake region and (b) around Blade region.

2.4. Numerical analysis condition

In MEXICO experiments, tripping tape is used to transient from laminar to turbulent on blades [9]. Therefore, attached flows are expected in high TSR conditions and uses of turbulence model is considerable. However, such RANS simulations have difficulty for prediction of separated aerofoil flows. When TSR decreases, attached flow on the blades becomes separated flow because of increase the angle of attack. According to Tanabe *et al.* [18], some turbulence models of RANS tend to predict attached flows till a higher angle of attack. Therefore, In high TSR condition, RANS simulation should fail to reproduce the actual phenomena. To simulate separated flows, Quasi Direct Numerical Simulation (QDNS) is effective [18]. The QDNS does not use any turbulence model and it is utilized for capturing vortices of wake region [19]. In this study, Unsteady RANS (URANS) simulation with the Spalart-Allmaras (SA) turbulence model [20] and QDNS of the wind turbines were executed. The numerical analysis conditions are shown in table 3. A fourth-order compact MUSCL TVD [16] interpolation scheme was adopted as the reconstruction method. Thus, the resolution in space is fourth-order, enabling us to precisely capture even small vortices. The operational conditions are summarized in table 4. Please note that the inflow is uniform and steady.

Table 3. Numerical analysis condition.

Governing equations	Unsteady Reynolds-averaged Navier-Stokes equation / Navier-Stokes equation
Space discretization	Cell-vertex FVM (Background)/Cell-centered FVM (Blade Grid)
Inviscid flux	SLAU [17]
Reconstruction	4th order Compact MUSCL TVD interpolation [16]
Turbulence model	Spalart-Allmaras [18] / without turbulence model (Quasi-DNS)
Rotation of wind turbine	Moving Overlapped Structured Grids
Time integration	4th order Runge-Kutta (Background) LU-SGS/LU-DUR implicit method (Blade Grid)

Table 4. summary of computation in this study

case	Inflow speed [m/s]	Tip speed ratio [-]	Simulation model
1	10	10	URANS (SA)
2	15	6.67	URANS (SA)
3	24	4.17	URANS (SA)
4	24	4.17	QDNS

3. Comparison of macro characteristics

First, results must be validated by comparing them with results of the first MEXICO project experiments. Hence, three cases were simulated with λ = 4.17, 6.67 and 10, which are the conditions of MEXICO. The axial velocity along the span wise position (r = 1.8) at an azimuth angle of 270° is shown in figure 2. The results of New MEXICO [12], which was carried out to validate the previous MEXICO measurement, are plotted together with the MEXICO findings. Therefore, the specifications such as blade length are same as MEXICO experiment. As for $\lambda = 6.67$, which is the design point of the MEXICO rotor, although the URANS result slightly over predicts the velocity deficit around the rotor position (x/D = 0), it shows good agreement with both MEXICO and New MEXICO in the wake region. The simulation for the case of $\lambda = 10$ also predicted the velocity deficit well even around the rotor position (x/D = 0). However, the URANS simulation of the case of $\lambda = 4.17$ tends to over predict the velocity deficit. As mentioned in subsection 2.4, this tendency can be caused by the attached flow predicted by SA model. As a result of attached flow, velocity deficit and force acting on the blade are overestimated. On the other hand, the results of QDNS shows good agreement with MEXICO results.

Next, the axial force distributions along the blade are shown in figure 3. Axial force is directly related to velocity deficit. Spanwise positions and axial force distributions are normalised by the rotor radius and dynamic pressure of inflows respectively. The magnitude of axial force is proportional to rotational speed; thus, axial force is found to increase as the radial position moves towards the tip. In contrast, axial force is found to sharply decrease around the blade tip because of tip loss. Focusing on each TSRs, the URANS simulations of $\lambda = 6.67$ and $\lambda = 10$ show good agreement with the MEXICO results qualitatively. However, these simulations slightly overestimate MEXICO results. This tendency could be caused by fully turbulent assumption of URANS, while tripping tape was used for laminar-turbulent transition in MEXICO rotor [21]. Same as the axial velocity, the URANS simulations of $\lambda = 4.17$ is found to over predict the axial force and QDNS shows consistency between MEXICO results.

From these two points of view (velocity deficit, axial force), it was determined that, in this study, the URANS and QDNS simulations successfully captured the basic characteristics of wind turbine wake. However, it should be necessary for the low-TSR case to adjust the numerical conditions.

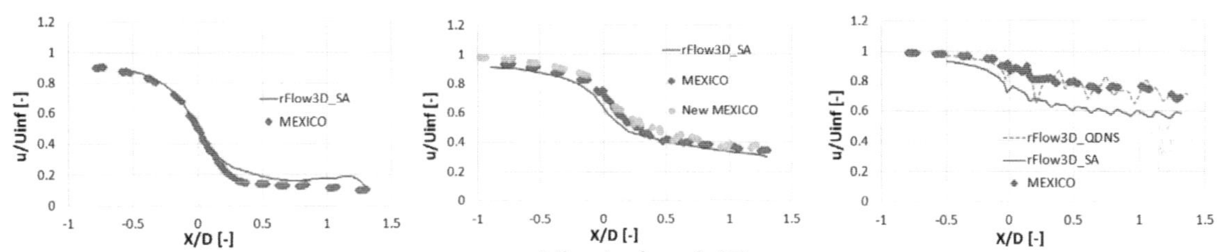

$u = 10$ m/s, $\lambda = 10$ $u = 15$ m/s, $\lambda = 6.67$ $u = 24$ m/s, $\lambda = 4.17$

Figure 2. Axial traverse of axial velocity[9,12]

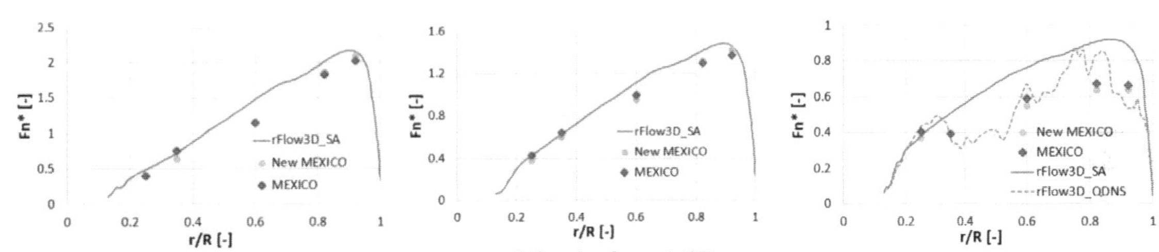

$u = 10$ m/s, $\lambda = 10$ $u = 15$ m/s, $\lambda = 6.67$ $u = 24$ m/s, $\lambda = 4.17$

Figure 3. Radial distribution of axial force along with blade [9,12]

4. Analysis of wake structure

4.1. Wake expansion and vortex breakdown

Further, analysis of vortex was executed to focus on the relationship between the vortices dynamics and TSR. As for $\lambda = 4.17$, please note that the result of QDNS simulation is used for analysis from this section because QDNS shows better agreement with MEXICO. The positions of the vortex core are shown in figure 4. The results of rFlow3D found to reproduce the MEXICO experiments with precision. When TSR increases, the radial expansion of tip vortices becomes greater. This behaviour is supposedly caused by changing the ratio of centrifugal force to wake velocity. Moreover, the radial expansion of wake should involve the vortex instability. The iso-surface of vorticity and the vorticity contour are shown in figures 6 and 7, respectively. Vorticity is normalised by the chord length in table 1 and the axial velocity of the main stream in each case. In general, the tip vortices are stable and minute structures do not exist at the initial stage of the wake. Subsequently, the vortices begin to interact with one another

Journal of Physics: Conference Series **749** (2016) 012013 doi:10.1088/1742-6596/749/1/012013

and short wavelength instabilities appear. Finally, the helical structures completely break down. Similarly, root vortices are combined with each other and resultantly diffuse into the entire wake region. Following this phenomenon, tip and root vortices merge and ultimately shape the complex structure. Such actions correspond to mixing behaviours with the main stream.

The space between each vortex decreases as TSR increases. Consequently, the interaction of the vortices as a whole ultimately increases. For the case of λ = 4.17, the wake disturbance begins at the rotor position (x = 0) and the vortex core is maintained at x ≅ 2D. In this case, mid-span vortex is found as well as tip and root vortices. Mid-span vortex should be caused by separation flow. For λ = 6.67, however, the disturbance begins at x ≅ 1D and the vortices start to combine with each other. Around x ≅ 3D, the vortices become minute structure and vorticities become weak. and for λ = 10, the vortex combination begins at about x ≅ 0.25D, with the helical structure ultimately broken at x ≅ 1D. Subsequently, wake diffusion occurs and is greatly diffused prior to x = 2D. Each of these cases depicts a 'similar' process with regard to the vortex breakdown phenomena. In terms of attached flow conditions, However, the vortex breakdown is accelerated by an increasing TSR.

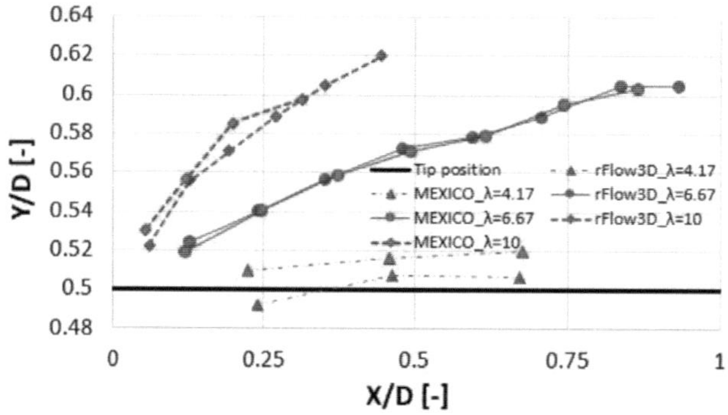

Figure 4. Positions of tip vortex core.[9]

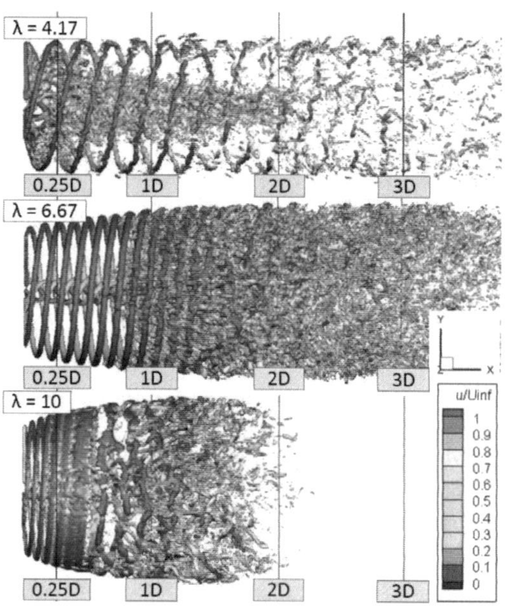

Figure 5. Iso-surface of normalized vorticity = 0.5 (top: λ = 4.17, middle: λ = 6.67, bottom: λ = 10).

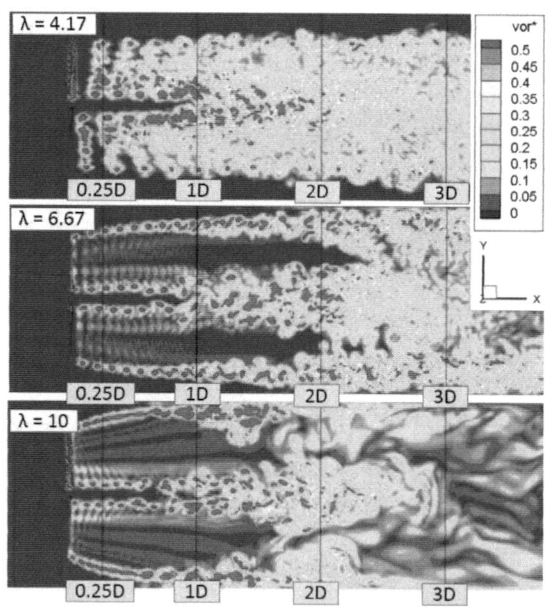

Figure 6. Vorticity contour of wake region(top: λ = 4.17, middle: λ = 6.67, bottom: λ = 10).

4.2. Velocity distribution

4.2.1. Frame of reference.
Next, analysis about velocity fields of wake region is investigated. Besides axial velocities, the radial flows and rotational flows behind the rotor plane should be investigated because they are supposed to represent the dynamics of rotational blade. Radial flows must correspond to radial expansion of the tip vortices shown in figure 4. And rotational flows of wake can be generated by the reaction force of blades. To visualize them, the Cartesian velocities (v, w) was converted into cylindrical velocities, as shown in figure 7; subsequently, the radial velocity V_r (1) and azimuthal velocity V_ψ (2) are derived. V_r and V_ψ can represent radial expansion and rotational flow of wake region respectively.

$$V_r = v\sin\psi + w\cos\psi \tag{1}$$
$$V_\psi = v\cos\psi - w\sin\psi \tag{2}$$

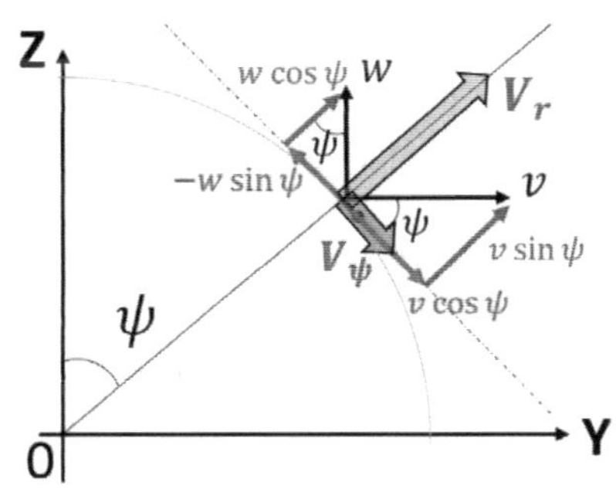

Figure 7. Rotating reference frame

4.2.2. Wake recovery process.
Characteristics of wake flow can be visualised by using a rotational reference frame. At First, the axial velocity contours of wake region are shown in figure 8. It is seen that when TSR increases, the velocity deficit also becomes quite significant. This phenomenon is caused by an increase in thrust. Associated cross-sectional contours of axial velocity (at x = 0.25D, 1D and 2D) are shown in figure 9. In the case of $\lambda = 4.17$, the region of velocity deficit is not expanded toward the radial direction, with associated distributions remaining essentially constant through each position. In contrast, for the cases of $\lambda = 6.67$ and $\lambda = 10$, the region of velocity deficit is notably expanded, with the axial velocity consequently decreasing in a manner similar to a tube-flow scenario. Furthermore, main stream mixing occurs by an exchange of momentum with radial and azimuthal velocities.

The distribution of V_r is shown in figure 10. The left column illustrates the plane at x = 0.25D as an example of the 'stable region' for the tip vortices in the case of $\lambda = 6.67$ and 10. These cases display a red circular area (i.e. a relatively strong radial flow toward the radial direction). Because of these radial velocities, wake expansion occurs. Micallef et al. [13] also pointed out that strong radial velocities exist near and around tip regions. Note that the strong red and blue regions represent cross sections of the tip vortices. In the case of $\lambda = 4.17$, the plane at x = 0.25D which can be affected by dynamic separation flow shows relatively weak radial flow. The planes at x = 2D represent the regions of vortex breakdown for each TSR. Because of the occurrence of tip vortex breakdown, small vortices are generated and diffused into the whole wake region, leading to the disappearance of radial flow. Alternatively, a wake expansion may cease, and in such a case, the tip vortex becomes unstable. As mentioned in the previous section, when TSR increases, the tip vortex breakdown process starts at an early stage; however, it should be noted that V_r correspondingly strengthens during the same period of time. Hence, the existence of V_r can act as a barrier that impedes recovery of the wake velocity because the existence of radial flow should prevent a wake from taking a main stream.

The azimuthal velocity (V_ψ), which stands for rotational movement of wake region is shown below in figure 11. As for $\lambda = 4.17$, which is the separated condition, the rotational flow is relatively weak and disappear quickly because of dynamic mixing by mid-span vortex, with the same tendency presenting itself for V_r. In terms of attached flow conditions, rotational flow is maintained until x = 1~2D which is much longer than the case of $\lambda = 4.17$. In comparison between $\lambda = 6.67$ and $\lambda = 10$, the case of $\lambda = 10$ displayed a rapid vortex breakdown and azimuthal velocity V_ψ disappears more rapidly. At x = 2D, the area depicting strong rotational flow essentially disappears. This indicates that tip vortex breakdown also corresponds to disappearance of azimuthal velocity. When comparing the position of disappearance for V_r and V_ψ, V_r tends to disappear earlier than V_ψ. If it is assumed that their disappearance has a direct correlation to mixing with the main stream, this tendency can be deemed conclusively valid because V_r distributes to the outside area, whereas V_ψ distributes inside of the wake. Therefore, V_r distributions that are close to the main stream tend to disappear more rapidly.

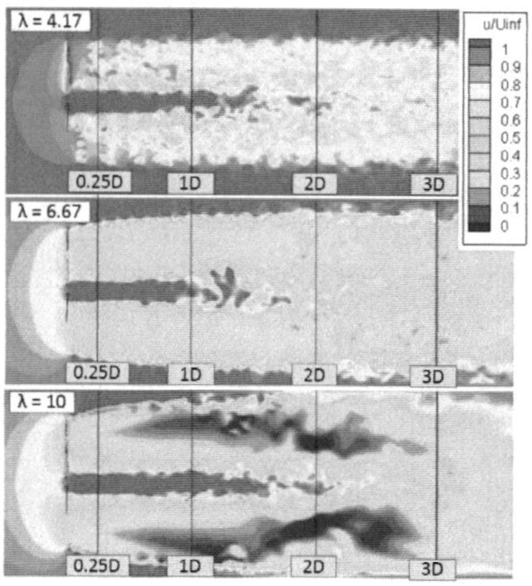

Figure 8. Axial velocity contour of wake region.

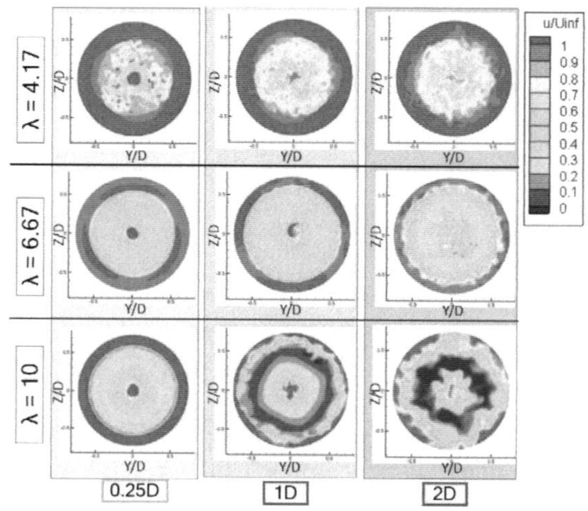

Figure 9. Axial velocity contours of wake (top: $\lambda = 4.17$, middle: $\lambda = 6.67$, bottom: $\lambda = 10$).

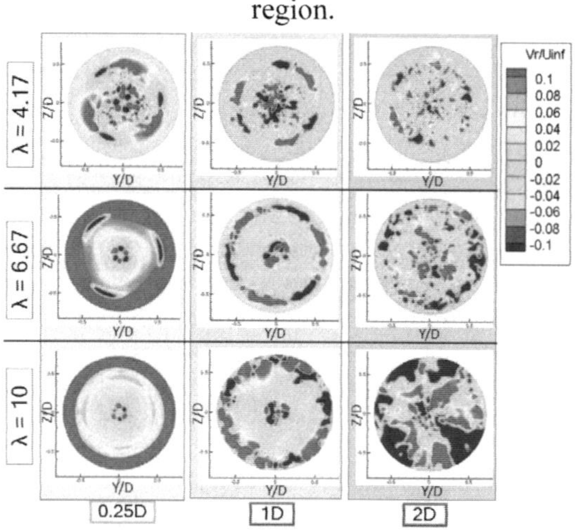

Figure 10. V_r contours of wake (top: $\lambda = 4.17$, middle: $\lambda = 6.67$, bottom: $\lambda = 10$).

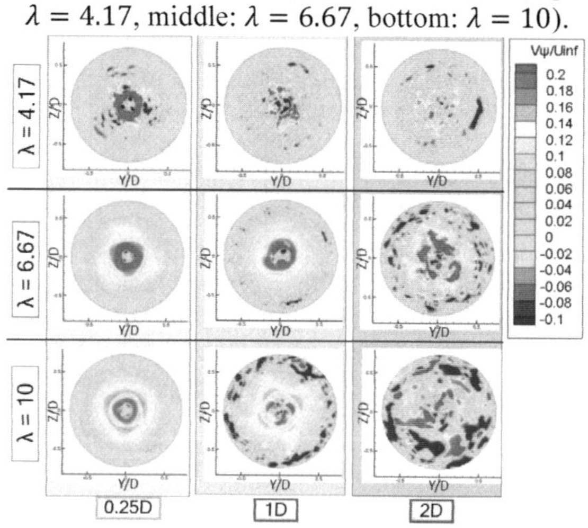

Figure 11. V_ψ contours of wake (top: $\lambda = 4.17$, middle: $\lambda = 6.67$, bottom: $\lambda = 10$).

5. Conclusion

URANS simulations and QDNS of the MEXICO rotor were executed by using rFlow3D and validations of the near-wake characteristics and axial force were implemented by comparing the results of the MEXICO projects. Furthermore, the effect of TSR on wake was investigated by focussing on the vorticity and velocities. From these results, the conclusions are following:

- The basic characteristics such as velocity distribution and axial force obtained by rFlow3D are generally consistent with the results of the MEXICO experiments.
- In low TSR, numerical conditions such as turbulence model need to be adjusted to predict the wake characteristics quantitatively.
- When TSR increases, the space between each vortex decreases. Therefore, the interaction between vortices increases and the tip vortex breakdown occurs more rapidly.
- During the beginning of vortex instabilities, wake expansion ceases because the radial flow produced by centrifugal forces disappears as a result of the breakdown of the tip vortices.
- Rapid breakdown of the tip vortices does not always correspond to rapid mixing of the wake with the main stream because this condition is often accompanied with a strong radial flow, which can be a barrier to wake mixing.
- Rotational flows by V_ψ tend to remain within the wake region longer than radial flows because the former primarily distribute inside of the wake, whereas radial flows typically exist in the outer area (proximal to the main stream).

6. Future works

To attain practical data, analytical conditions should approach actual field conditions. In particular, turbulence intensity is vital for accurately reproducing the actual phenomena; however, such simulations are normally difficult to duplicate and validate. Hence, via future studies, attempts will be set forth to conduct similar evaluations by comparing the attained results with light detection and ranging measurements.

References

[1] Uchida T, Ohya Y and Sugitani K 2006 Comparison between the wake behind wind turbine generator under optimal tip speed ratio and the wake behind stationary plate *Proc. of the 19th national symp. on wind engineering* pp187–92

[2] New Energy and Industrial Technology Development Organization 2008 *The guidebook for Installation of wind power energy*. http://www.nedo.go.jp/content/100079735 (2016/04/13 accessed)

[3] Vermeer L, Sørensen J and Crespo A 2003 Wind turbine wake aerodynamics *Prog. Aerosp. Sci.* **39** 467–510

[4] N Troldborg, F Zahle, PE Réthoré, N Sørensen 2012 Comparison of the wake of different types of wind turbine CFD models *50th AIAA Aerospace Sciences Meeting including the New Horisons Forum and Aerospace Exposition* 2012-0237

[5] AM AbbelSalam, V Ramalingam, 2013 Wake prediction of horizontal-axis wind turbine using full-rotor *Journal of Wind Engineering and industrial Aerodynamics* **124** 7-19

[6] Y Li, AM Castro, T Sinokrot, W Prescott, PM Carrica 2015 Coupled multi-body dynamics and CFD for wind turbine simulation includeing explicit wind turbulence *Renewable Energy* **76**: 338-361

[7] N Sezer-Uzol and O Uzol 2013 Effect of steady and transient wind shear on the wake structure and performance of a horizontal axis wind turbine rotor *Wind Energy* **16** pp 1-17

[8] T Göçmen, *et al.* 2016 Wind turbine wake models developed at the technical university of Denmark: A review *Renewable and Sustainable Energy Reviews* **60** pp 752-769

[9] Schepers JG, *et al.* 2012 *Final report of IEA Wind Task 29: Mexnext* (Phase 1) ECN-E 12-004

[10] Schepers JG, *et al.* 2014 *Final report of IEA Wind Task 29: Mexnext* (Phase 2) ECN-E 14-060

[11] Schepers JG and Snel H 2008 *Model Experiments in Controlled Conditions Final Report* ECN-E 07-042
[12] Schepers JG and Boorsma K 2014 *New MEXICO experiment* ECN-E 14-048
[13] Micallef D, Bussel G, Ferreira CS and Sant T 2012 An investigation of radial velocities for a horizontal axis wind turbine in axial and yawed flows. *Wind Energy* **16** 529–44
[14] Carrión M, Steijl R, Woodgate M, Barakos G, Manduate X and Gomez–Iradi S 2015 Computational fluid dynamics analysis of the wake behind the MEXICO rotor in axial flow conditions *Wind Energy* **18** 1023–45
[15] PE Réthoré, *et al.* 2011 MEXICO wind tunnel and wind turbine *29th AIAA Applied Aerodynamics Conf.* pp 2011–3373
[16] Tanabe Y and Saito S 2009 Significance of all-speed scheme in application to rotorcraft cfd simulations *The 3rd Int. Basic Research Conf. on Rotorcraft Technology*
[17] Shima E and Kitamura K 2009 On new simple low-dissipation scheme of AUSM-family for all speeds *AIAA paper* **136** pp 1–15
[18] Tanabe Y, Aoyama T, Oe H, Uemura Y and Sugawara H 2015 Simulations of horizontal Axis Wind Turbines in Complex Operational Conditions *The AHS 71st Annual Forum*
[19] JP Boris, FF Grinstein, ES Oran and RL Kolbe 1992 New insights into large eddy simulation *Fluid Dynamics Research* **10** pp 199–229
[20] Spalart PR and Allmaras SR 1992 A one-equation turbulence model for aerodynamic flows *AIAA J.* **92** 0439
[21] A Bechmann and NN Sørensen and F Zahle 2011 CFD simulations of the MEXICO rotor *Wind Energy* **14** pp 677–689

Journal of Physics: Conference Series **749** (2016) 012014 doi:10.1088/1742-6596/749/1/012014

Analysis of the high Reynolds number 2D tests on a wind turbine airfoil performed at two different wind tunnels

O Pires[1], X Munduate[1], O Ceyhan[2], M Jacobs[3], J Madsen[4] and J G Schepers[2]

[1] National Renewable Energy Centre (CENER), 31621 Sarriguren, Navarra, Spain
[2] Energy Research Center of the Netherlands (ECN), 1755LE, Petten, The Netherlands
[3] DNW, Bunsenstrasse 10, 37073 Göttingen, Germany
[4] LM Wind Power, Lunderskov, Denmark

E-mail: oepires@cener.com

Abstract. 2D wind tunnel tests at high Reynolds numbers have been done within the EU FP7 AVATAR project (AdVanced Aerodynamic Tools of lArge Rotors) on the DU00-W-212 airfoil and at two different test facilities: the DNW High Pressure Wind Tunnel in Gottingen (HDG) and the LM Wind Power in-house wind tunnel. Two conditions of Reynolds numbers have been performed in both tests: 3 and 6 million. The Mach number and turbulence intensity values are similar in both wind tunnels at the 3 million Reynolds number test, while they are significantly different at 6 million Reynolds number. The paper presents a comparison of the data obtained from the two wind tunnels, showing good repeatability at 3 million Reynolds number and differences at 6 million Reynolds number that are consistent with the different Mach number and turbulence intensity values.

1. Introduction

Within the EU FP7 AVATAR project (AdVanced Aerodynamic Tools of lArge Rotors) [1], 2D tests have been performed at high Reynolds numbers in order to evaluate airfoil performance under the expected conditions of the future multi-MW wind turbine blades. The DU00-W-212, a 21% relative thickness airfoil from the TU-Delft DU airfoil family, has been tested at the DNW High Pressure Wind Tunnel in Göttingen (HDG) at 5 different Reynolds numbers (3, 6, 9, 12 and 15 million) and low Mach numbers (below 0.1). In parallel, LM Wind Power has performed at his own wind tunnel facility a test on this same airfoil at two different Reynolds numbers, 3 and 6 million, and Mach numbers of 0.14 and 0.28 respectively.

The comparison of the results from these two experiments gives a good opportunity to check the repeatability of the results of airfoil aerodynamic performance data when obtained at different facilities. In addition, the effect of significant differences in Mach number and turbulence intensity conditions can be observed on the airfoil aerodynamic coefficients behavior.

2. Tests description

Both wind tunnel tests have been performed on a 2D model of the DU00-W-212 airfoil arranged horizontally in a closed test section. The models were instrumented with 90 pressure taps to capture the static pressure distribution around the airfoil and a wake rake downstream to measure the total pressure of the wake. Lift and pitching moment coefficients (Cl and Cm) were obtained by integration of the pressure distribution around the airfoil and Drag coefficient (Cd) was calculated from the wake loss of momentum by integrating the wake total pressure distribution.

A description of each of the wind tunnel facilities and their particular test arrangement is presented below:

2.1. DNW wind tunnel test

The DNW HDG is a closed return circuit wind tunnel with a closed test section of 0.6 x 0.6 m. (width x height) and 1 m. length, and a contraction ratio of 5.85. The wind tunnel speed range is 3.5 to 35 m/s and the maximum Mach number is 0.1. This tunnel can be pressurized up to 100 bars to achieve high Reynolds numbers.

A 150 mm. chord 2D airfoil model was horizontally installed in the middle of the test section and was equipped with 90 pressure taps at the mid span. A wake rake with 118 total and 8 static pressure probes was installed around 2 chords downstream of the trailing edge of the model.

The data presented in this paper has been corrected using classical wind tunnel wall corrections compiled in AGARD-AG-336 [2].

Tests have been performed at different Reynolds numbers: 3, 6, 9, 12 and 15 million. Analysis of the Reynolds effect on the airfoil performance has been reported in [3]. Measurements have been performed at airfoil surface clean configuration as well as with transition dots installed in the model to force laminar to turbulent boundary layer transition at a certain chord position.

Additional instrumentation as a 3-component balance for correlation of the aerodynamic coefficients, 5 fast response kulite sensors installed in the model for unsteady pressure measurements, a cylindrical hot-film probe for inflow speed fluctuation measurements and UV LED light and camera for oil flow visualization have also been used.

2.2. LM wind tunnel test

LM wind tunnel is an atmospheric closed return circuit wind tunnel with a closed test section of 1.35 x 2.7 m. (width x height) and 7 m. length, and a contraction ratio of 10. The maximum wind speed at the test section is 105 m/s.

A 900 mm. chord 2D airfoil model instrumented with 90 pressure taps was horizontally installed in the test section. A wake rake instrumented with total and static pressure probes is installed downstream of the model.

All the data presented here has been corrected for wind tunnel effects classical wind tunnel wall corrections as described in standards given in AGARD-AG-336 [2]

Tests have been performed at two Reynolds numbers conditions, 3 and 6 million. Measurements have been done at airfoil surface clean configuration as well as with transition zig-zag tape installed in the model to force the laminar to turbulent boundary layer transition. In addition, some configurations with vortex generators and gurney flaps have been tested. An analysis of the effect of these configurations has been reported within the AVATAR project.

Some additional instrumentation was also used in this test, as floor and ceiling pressure holes distributions along the centerline of the test section and a load cell system for measuring the airfoil lift, drag and pitching moment coefficients.

2.3. Main differences between both tests

Both tests have been done over the same airfoil geometry and using the same methodology for obtaining the aerodynamic coefficients. The instrumentation of the model was also done in the same way for both experiments, using the same number and chord-wise positions of the pressure taps.

However, there are some differences between both experiments that have to be taken into account when comparing the results. These main differences can be summarized, on one hand in the geometric set up of the models and test sections, and on the other in the different wind conditions reproduced in the test section.

From the geometric differences shown in Table 1, it can be seen how the model chord size in the HDG test is 6 times smaller than in the LM test. This should not have to make any difference in the results, but it has to be considered for the test arrangement as in a small model the height of forcing transition elements has to be smaller as well. In this way, to assure the laminar boundary layer behavior in the clean configuration cases, a good surface treatment of the model has to be made, and in the case of the HDG model, the surface roughness was set always below Rz 0.1 µm.

The model aspect ratio (ratio between span and chord of the model) difference between both experiments is also considerable, being 2.6 times bigger in the case of the HDG. This difference will affect the 3-dimensional effects that appear in the tests. These effects are mostly important when separation phenomena occur.

The geometric blockage parameter is presented as it gives an indication of the tunnel test section wall effects and therefore the order of wall interference tunnel corrections applied. For both tests, the parameter is similar and low enough to derive reliable wind tunnel wall corrections up to angles of attack (AoA) of 20°.

Table 1. Geometric set up differences between the two experiments

	HDG	LM
Test section (W x H)	0.6 m x 0.6 m	1.35 m x 2.7 m
Model span (S)	0.6 m	1.35 m
Model chord (c)	0.15 m	0.9 m
Model aspect ratio (S/c)	4	1.5
Geometric blockage (c/H)	25%	33%

The other important differences between the tests are the airflow conditions. Table 2 shows the Mach number and the Turbulence intensity for the two Reynolds number conditions that are being analysed in this paper, 3 and 6 million. The turbulence intensity (TI) is lower at LM. This could affect in principle the laminar behavior of the airfoil.

The Mach number shows important differences between the tests. The reason is that in the DNW test, the Reynolds number variation is obtained modifying the air density by pressurizing the wind tunnel airflow. Therefore the Mach number is always maintained low (below 0.1) and only a pure Reynolds number variation is obtained. On the other hand, in the LM test the Reynolds number variation is obtained modifying the wind speed and therefore the Mach number. We can see then, that for the Reynolds number 3 million condition the Mach number for both tests is similar (around 0.1). But for the Reynolds number 6 million, the LM test has a Mach number one order of magnitude greater than HDG and close to 0.3. So, compressible effects could be observed between those tests at these conditions.

Table 2. Wind condition differences between the two experiments

	Reynolds 3million test		Reynolds 6 million test	
	HDG	LM	HDG	LM
Mach number (M)	0.08	0.139	0.03	0.279
Turbulence intensity (TI)	0.1 %	0.05 %	0.2 %	0.1 %

3. Data comparison between both tests

A comparison of the results obtained in both tests is presented below. The lift coefficient (Cl) against angle of attack (AoA) and against drag coefficient (Cd), and the efficiency (Cl/Cd) against angle of attack have been plotted for the Reynolds number 3 million case and for the 6 million one.

3.1. Reynods 3 million comparison

Figure 1 shows the comparison between both tests of the lift curve against angle of attack. The curves match very well in the linear region and even at the maximum Cl value.

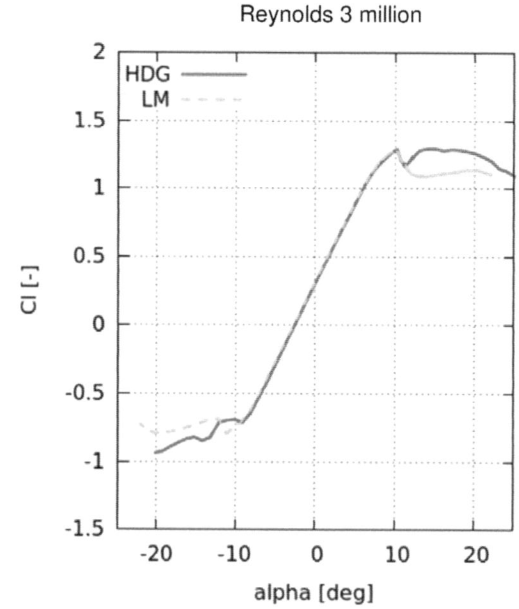

Figure 1. Lift coefficient (Cl) vs angle of attack (AoA) comparison at $Re=3\cdot 10^6$

Only different behavior is observed at the stall condition, where 3D effects can be greatly affected by formation of stall cells, whose size and number are influenced by the aspect ratio of the model [5]. The span-wise position of the pressure taps also influences the measured lift, as the level of separated flow measured by the tap depends on the typology of the stall cells created. An example of this phenomenon can be seen on Figure 2.

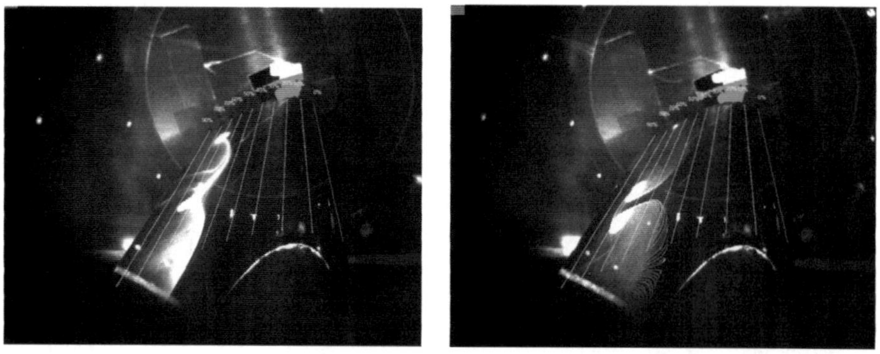

Figure 2. DNW flow visualization photos at 14° AoA (left) and 18° AoA (right)

Figure 2 shows two pictures from the flow visualization performed at the DNW test. They correspond to two different post stall angles of attack. The pictures show the model from a window on the starboard wall of the test section. Some white lines are superimposed on the pictures to show different chord-wise positions (the most right one corresponds to the leading edge of the model). As it can be seen in the pictures, the topology of the stall cells varies with the angles of attack and produces different levels of chord percentage of the model being stalled depending on the span position. Therefore, the positioning of the pressure taps influences the measured lift.

The differences in lift in the stall region between both tests can then be explained with the different aspect ratio of the model.

The measured drag matches also very well in the linear region. Figure 3 shows how the drag bucket is completely reproduced in shape and values. Since both lift and drag are well reproduced, it is expected that the airfoil efficiency (lift over drag ratio) will be also similar. We can also observe this in Figure 4.

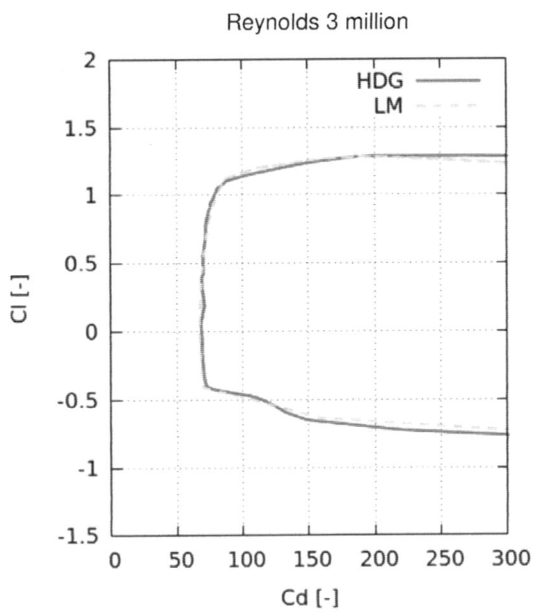

Figure 3. Cl vs Cd comparison at Re=$3\cdot 10^6$

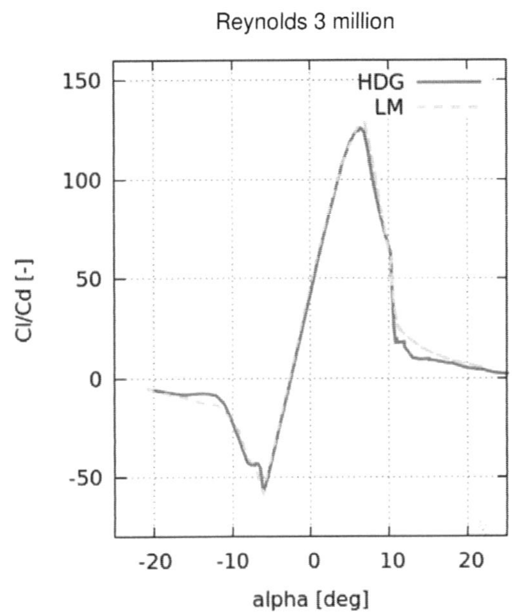

Figure 4. Cl/Cd comparison at Re=$3\cdot 10^6$

3.2. Reynolds 6 million comparison

When we look into the data at Reynolds number 6 million, we can see that in this case the results from both tests do not show such a good agreement. We can already see in Figure 5 how the lift slope is different and greater in the data from LM test. The drag values keep a very good comparison in the linear region but the bucket corners (as it can be seen in Figure 6) are different, indicating a different behavior on the boundary layer laminar to turbulent transition and separation.

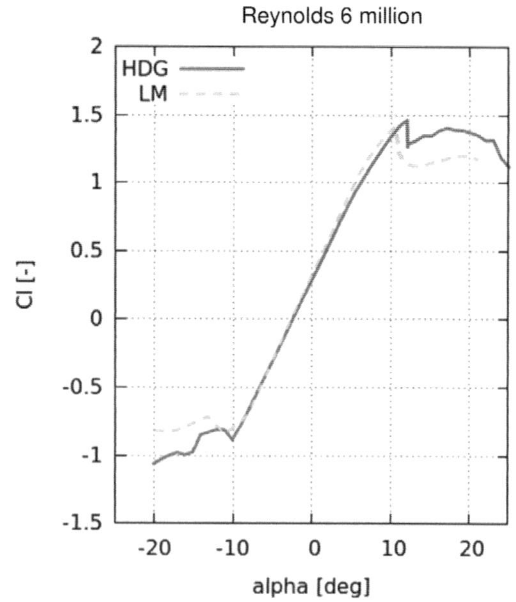

Figure 5. Cl vs AoA comparison at Re=$6\cdot 10^6$

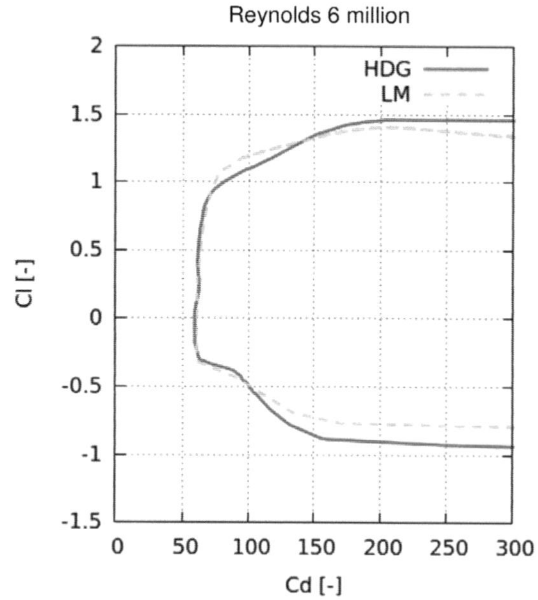

Figure 6. Cl vs Cd comparison at Re=$6\cdot 10^6$

Attending to these differences between lift and drag, the efficiency curve shows the worst agreement in the maximum values (Figure 7), as LM results have a greater lift due to its higher slope and a lower drag in the upper bucket corner.

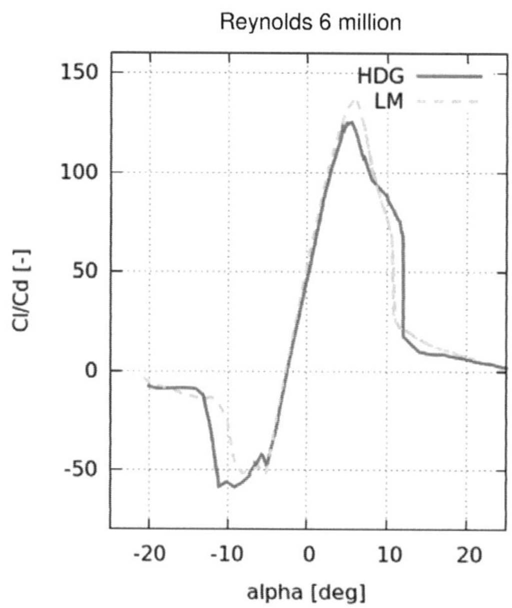

Figure 7. Cl/Cd comparison at Re=$6\cdot 10^6$

4. Analysis of the effect of the Mach number and turbulence intensity

The next step is to study the reason for obtaining different comparison results at each Reynolds number. We look therefore to tables 1 and 2 to see the differences we have at each condition. The set-up differences listed at table 1 are the same for both Reynolds. Although their effect on the results could have some variability depending on Reynolds, they affect mainly the results when separation occurs, at high angles of attack, as it has been already discussed. Table 2 shows different wind conditions at each Reynolds number. The turbulence is in both cases twice at HDG. From this different air stream turbulence, it is mainly expected to have a different measurement in drag.

The Mach number has the same order of magnitude for the Reynolds 3 million case, but for the Reynolds 6 million case is one order of magnitude higher at LM.

The higher slope of the lift coefficient is compatible with higher Mach number. If we make a simplified compressible correction using Prandtl-Glauert law [4], expressed in (1) on the HDG data to obtain the values that would be expected at a Mach number (M_∞) of 0.279, we see in Figure 8 that the linear region would match.

$$C_l = \frac{C_{l,0}}{\sqrt{1-M_\infty^2}} \quad \text{where } C_{l,0} \text{ is the lift coefficient for the incompressible flow } (M_\infty = 0) \qquad (1)$$

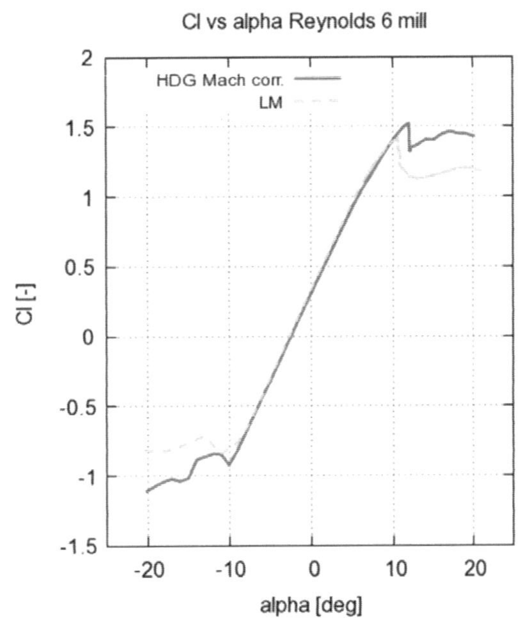

Figure 8. Lift comparison between both tests applying Prandtl-Glauert compressibility correction to HDG data

The differences in the maximum lift could be explained by the Mach effect, as the Prandtl-Glauert law is only valid for the Cl linear region, but they can also be explained by the different 3D effects created by the stall cell formations [6].

An analysis using the panel method code XFOIL version 6.96 has been performed to evaluate the differences we can expect by modifying Mach number and inflow turbulence. In Figure 9, Figure 10 and Figure 11, the experimental differences for Reynold number 6 million case (at the left) are compared to computations differences when Mach number and N-factor (from the e^N method for transition prediction) are modified.

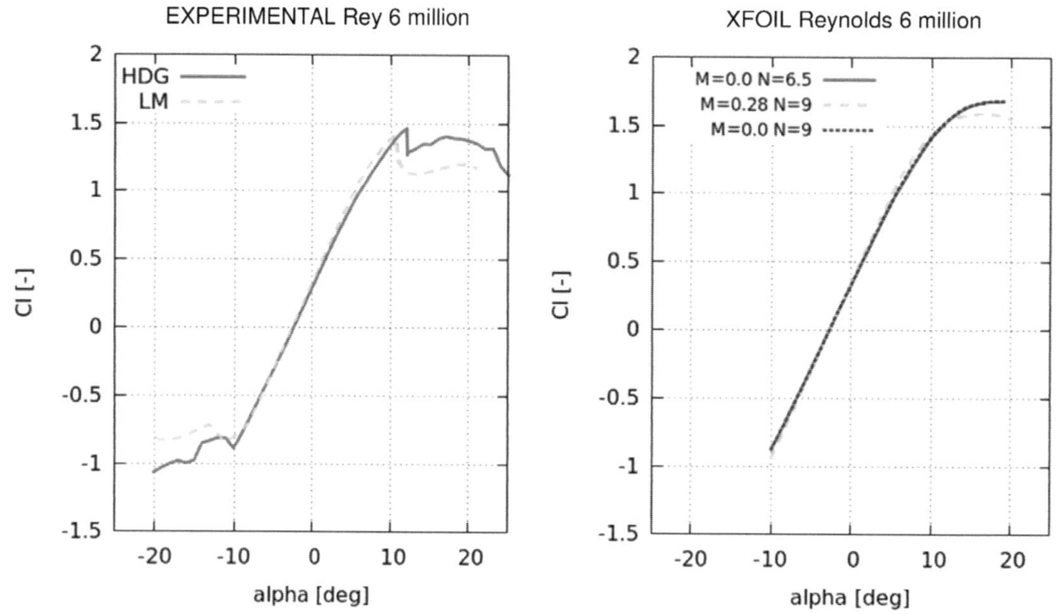

Figure 9. Experimental and XFOIL computations of Cl vs AoA at Re=$6\cdot 10^6$

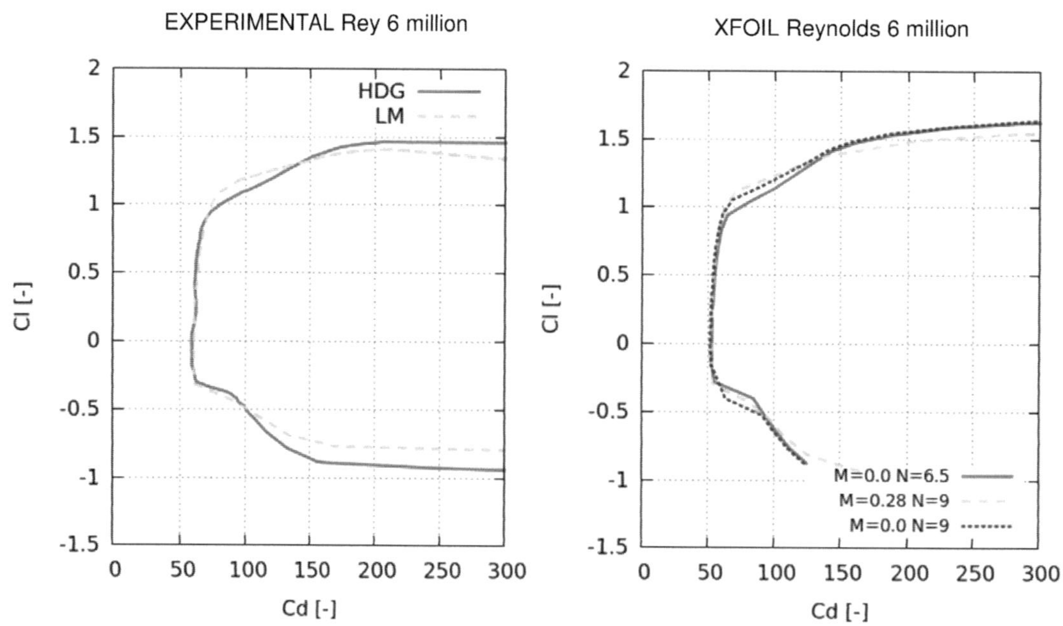

Figure 10. Experimental and XFOIL computations of Cl vs Cd at Re=$6\cdot 10^6$

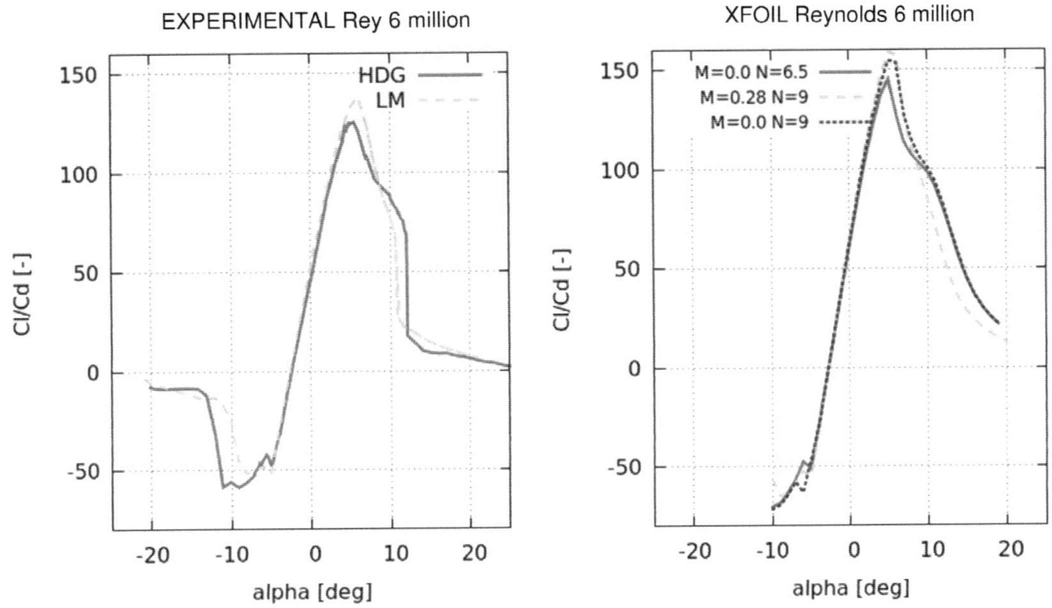

Figure 11. Experimental and XFOIL computations of Cl/Cd at $Re=6\cdot 10^6$

The XFOIL computations are coherent with the idea that Mach effect influences mainly the lift slope and the different inflow turbulence affects the drag, especially at separation. The right plot in Figure 9 shows how XFOIL predicts a negligible effect of the turbulence (different N-factor values) in lift, while the Mach effect would introduce an increase of the lift slope. On the other hand, Figure 10 shows how turbulence effect has more influence on the drag. Finally, Figure 11 shows the efficiency, which represents the combined effects of lift and drag.

In all of these last three curves, when we compare the left (experimental) plot against the right (XFOIL computation) plot, we can see that the qualitative variations of the aerodynamics coefficients between HDG and LM test are the same as the ones computed with XFOIL. Then the different values at 6 million Reynolds can be explained with the different wind conditions in each wind tunnel.

5. Conclusions

The comparison of the data obtained at 3 million Reynolds number condition at the two different wind tunnels shows a very good agreement. Only differences are shown at stall conditions, which can be explained by the 3D effects caused by the different facilities set-up.

The comparison between data at 6 million Reynolds number shows more significant differences. An analysis of these differences has been performed and the conclusion is that they are consistent with the different Mach number and turbulence intensity in both wind tunnels at this Reynolds number condition.

References

[1] Schepers G J et al 2016 Latest results from the EU project AVATAR: Aerodynamic modelling of 10 MW wind turbines. Science of Making Torque 2016
[2] Ewald B F R 1998 Wind Tunnel Wall Corrections. AGARD-AG-336
[3] Pires O et al 2016 Analysis of high Reynolds numbers effects on a wind turbine airfoil using 2D wind tunnel test data. Science of Making Torque 2016
[4] Glauert H 1928 The Effect of Compressibility on the Lift of an Aerofoil. *Proc. Roy. Soc. London*. VOL. **CXVIII**, 1928, p. 113–119
[5] Schewe G 2001 Reynolds-number effects in flow around more-or-less bluff bodies *J. Wind Eng. & Ind. Aerodyn.* **89** 1267-1289
[6] Llorente E et al 2014 Wind Tunnel Tests of Wind Turbine Airfoils at High Reynolds Numbers *J. of Phys.: Conf. Series* **524** (2014) 012012

Investigation of the validity of BEM for simulation of wind turbines in complex load cases and comparison with experiment and CFD

H. Rahimi[1], B. Dose[1], B. Stoevesandt [2], J. Peinke[1,2]

[1] ForWind, Institute of Physics, University of Oldenburg, Küpkersweg 70, 26129 Oldenburg, Germany

[2] Fraunhofer IWES, Küpkersweg 70, 26129 Oldenburg, Germany

E-mail: hamid.rahimi@uni-oldenburg.de

Abstract.
The aim of this work is to investigate the validity of simulation codes based on the Blade Element Momentum (BEM) theory for three important design load conditions. This paper includes the cases of yawed inflow, rotor tower interaction for downwind turbines and the standstill case. Computational Fluid Dynamics (CFD) and experimental data (when available) are used for the evaluation of the obtained results. For the yawed inflow, the results indicate that significant deviations between BEM and experiments & CFD can be observed. This discrepancy is caused by unsteady phenomena such as the advancing & retreating blade effect and the skewed wake effect. In the case of the rotor and tower interaction of the downwind turbine, the results show that the BEM based code overpredicts the sectional forces in terms of the normal and tangential forces by 20 %. In the case of standstill, the evaluation of the results based on tip deflections shows clear differences in the output of both numerical approaches. While the flapwise deflections show a reasonable agreement, the CFD-based coupled solver predicts much larger edgewise vibrations.

1. Introduction

One of the most common approaches to simulate the aerodynamics of wind turbines is the Blade Element Momentum (BEM) theory. Low costs of the simulations, which are a result of the two dimensional theory, make the simulation of more than thousand load cases within a short amount of time affordable. Nevertheless, complex three dimensional flow phenomena which can occur during the operation of wind turbines cannot be captured by the basic BEM theory and have to be included by correction models for the specific phenomena.

Several experimental and numerical studies showed that simulation codes based on the BEM theory are in certain cases not sufficiently accurate and reliable for predicting the complex aerodynamic behavior of wind turbine blades [1, 2]. Some of the complex cases which can arise are: The aerodynamic behavior of the rotor blades under yawed conditions, the interaction of the rotor and the tower for downwind turbines, and the interaction of the blade structure with the incoming wind during standstill.

This work investigates the validity of Blade Element Momentum (BEM) based simulation codes for the listed conditions. Computational Fluid Dynamics (CFD) and experimental data (when available) will be used for the purpose of the evaluation of the obtained results. Key aerodynamic quantities such as power, thrust and sectional forces are investigated using BEM for different turbine sizes and compared to wind tunnel measurements and CFD results. Based on that, the capabilities of BEM for the simulation of such complex conditions are discussed in this paper.

2. Approach

In this work, the wind turbine simulation code FAST [3] is used for the BEM calculations. The CFD simulations are performed using the open-source software OpenFOAM [4]. In case of the NREL Phase VI rotor, experimental results were included in the investigations. For large wind turbines, however, there is a lack of available measurement data. Therefore, only numerical results obtained by CFD are used to evaluate the validity of BEM in those cases. In the following sections, three complex scenarios are presented and the results are discussed.

3. Yawed flow

One of the common difficulties can arise during the modeling of the aerodynamic behavior of rotor blades under yawed conditions. These cases can result in different problematic situations such as: flow separation, azimuthal variation of the loads and an unbalance in the flow induced by the skewed wake [1, 5]. Thus, currently a major concern regarding BEM methods is the applicability of yaw models in the design process of wind turbines. The yaw models used are usually suffering from uncertainties such as the general validity of the model at extreme yaw angles and dynamic yaw effects[5, 2]. Therefore, it is essential to conduct more complex and realistic simulations in order to capture all the important details, which play a role in the correct modeling of wind turbines.

In this part of the work, the NREL VI wind turbine [6] is simulated using CFD and FAST. In CFD the simulations are conducted using the Spalart-Allmaras-DDES model [7] for 30 degree of yaw and a wind tunnel speed of $5m/s$. The computational grid has a spherical shape and is fully structured and based on hexahedrons. The total mesh size accounts to 22 million cells. It features 300 cells in chord wise direction around the airfoils, 250 cells in the span-wise direction for each blade, and 300 cells in the wall normal direction. The y^+ values at the surface are kept below one everywhere on the blade surface. Fig. 1 shows a representation of the grid at the mid span section for the leading edge, trailing edge, and half of the complete domain.

Different grids were created and two methods were used in order to examine the grid-independent behavior of the solution, namely the systematic grid refinement and the Grid Convergence Index (GCI). The results of the study are discussed in [8].

In order to solve the pressure-velocity coupling, the PIMPLE algorithm which is a combination of the loop structures of SIMPLE [9] and PISO [10] is used. The numerical simulations are performed using the Facility for Large-Scale cOmputations in Wind energy research (FLOW) at the University of Oldenburg. In total 264 computational cores are used for approximately 96 hours. The effects of the tower and the nacelle are neglected in the performed simulations. The physical time step is set to $\Delta t = 2 \times 10^{-4}s$. Numerical convergence is achieved after 5 rotor rotations.

For the FAST simulations, three different yaw correction models, namely the model of Pitt and Peters [11], the model from Schepers [5] and the Generalized Dynamic Wake (GDW) model [11] are used. In Fig. 2 the CFD results show a reasonable agreement with the experimental data for both normal and tangential force coefficients. It should be also mentioned that the simulation results presented in [8] and [12] show a good agreement for the case of 0 degree of yaw.

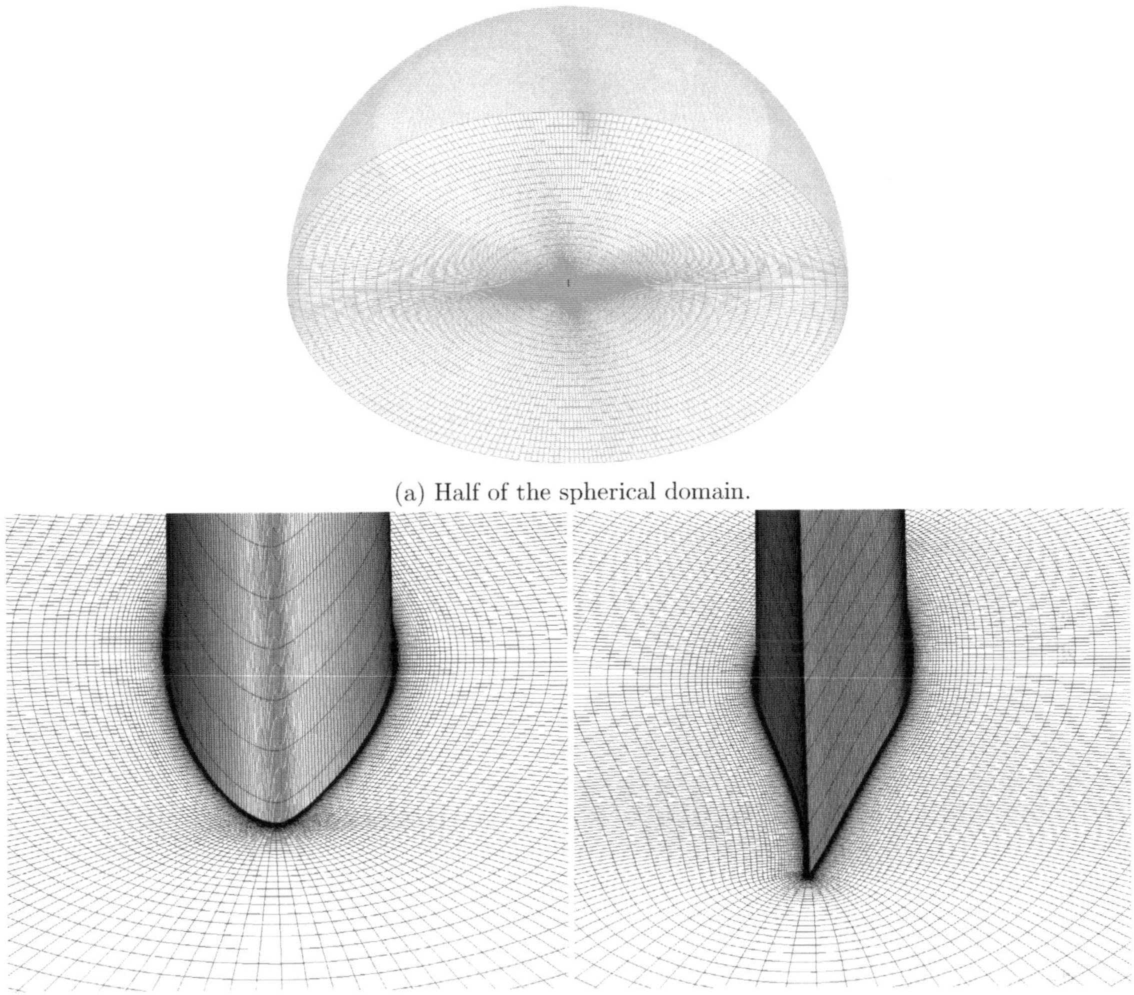

(a) Half of the spherical domain.

(b) Leading edge grid at the mid span section. (c) Trailing edge grid at the mid span section.

Figure 1: Grid

In terms of the normal force coefficient at 30% span, all three correction models are predicting the maximum forces at the same azimuthal position. The maximum is reached in the downwind side of the blade. The model by Pitt & Peters shows a smaller amplitude and predicts a pure sinusoidal behavior. This can be due to the fact that Pitt & Peters model is not taking the root free vortex effect into account. The GDW model is underestimating the forces at the upper half of the rotor blade significantly. The model from Schepers has a closer agreement to the experimental results in terms of amplitude and phase. The result for the tangential force coefficient using these three correction models shows a higher deviation from the experimental results than the CFD calculations. All the correction model results are shifted towards the downwind side of the blade. The model from Schepers has a closer agreement to the experiments since this model is considering the root free vortex effect as a function of blade radius. Moreover comparing the advancing and retreating blade effect with the skewed wake effect, it can be concluded that in this section of the blade, both effects contribute to the load variation. This can be seen from the fact that the maximum forces are shifted towards the downwind side of the blade.

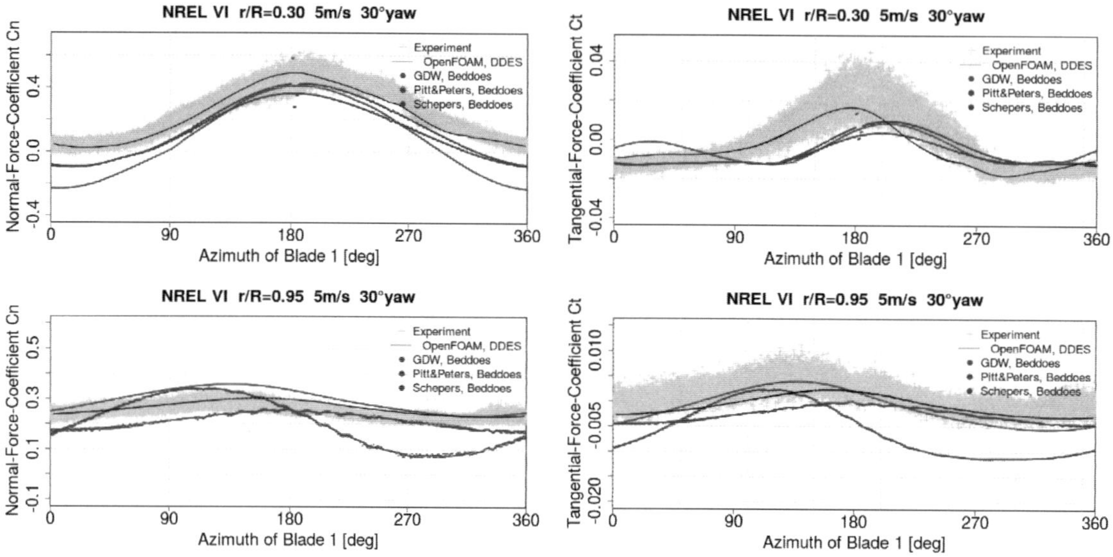

Figure 2: Normal and tangential force coefficients for 30% and 95% of span for wind speed of 5 m/s and 30 degree of yaw.

At the section of 95% of the span, both the normal and the tangential force coefficient of the experimental data and CFD results are showing a sinusoidal behavior which is induced by the tip vortex. The Pitt & Peters model is predicting a different phase and amplitude and the GDW model is predicting the correct phase but not the amplitude. The Schepers model also miss-predicts the position of maximum force and under predicts in average the forces. This could be due to the fact that these models are neglecting the effect of the tip vorticity in yawed flow in general.

Comparing the advancing and retreating blade effect with the skewed wake effect at the tip section shows that the contribution of the advancing and retreating blade on the force coefficient is smaller than the one of the skewed wake. This can be seen from the fact that the maximum forces are shifted to the upwind side of the blade, which results in a stabilizing yaw moment.

The other possible source of error for this section is the tip correction model. The capability of these correction models in the prediction of the loads in the yawed flow needs to be further investigated.

4. Rotor tower interaction

Due to new materials, control strategies and the fact, that noise is offshore not a driving factor, downwind turbines seem to go into a revival. So far Ming Yang, Hitachi and the Dutch developer 2-B Energy have been developing downwind concepts lately [13].

The effect of the tower shadow is a crucial question in the aerodynamics. In BEM based calculations mostly either a wind deficit, from potential theory, or empirical models are used [13, 14]. In this paper it is investigated, how much this assumption holds using CFD simulations.

Again, the NREL Phase VI wind turbine is used for the investigation. For the mesh generation the bladeBlockMesher, developed by ForWind and Fraunhofer IWES was used [8]. It creates a structured blockMesh around wind turbines blades. The rotor was meshed separately in a cylindrical mesh from the surrounding mesh. The latter has been meshed in a half-cylindrical way with the hub distance of 5D to the inlet and crossflow cylinder walls and 10D distance to the outflow. Overall 10 Million cells have been used to discretize the domain. Several grids were examined in order to minimize the effect of the grid errors on the solution which is discussed in [8].

Fig. 3 shows the numerical domain used in this part. The k-ω SST model has been used for unsteady RANS turbulence modeling [15]. The simulations have been done at an inflow velocity of $U_0 = 7\text{m/s}$ with the turbulence intensity of 1% .

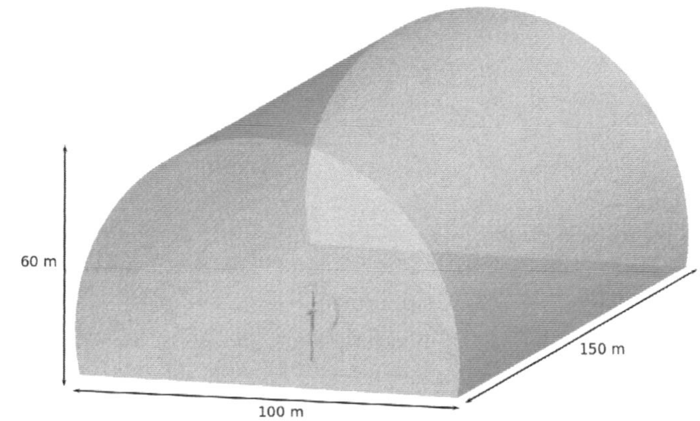

Figure 3: Computed domain with different separated rotor mesh and the overall domain in a half-cylindrical shape.

The tower influence is modeled using the FAST code based on the work of Bak et al. [16], which is based on a potential flow solution around a cylinder. The influence of the tower on the velocity is therefore applied based on the position and the drag coefficient of the tower section.

Fig. 4 shows the simulation results for the normal and tangential force coefficient at the blade sections 30% and 80% for the downwind configuration at $U_0 = 7\text{m/s}$. The CFD simulation shows a good agreement with the experimental data for both the amplitude and the location of velocity drop for both sections. However, the simulation results obtained by FAST shows 20% larger mean values. This deviating result might come from the input-polars. Nevertheless, it should be noted that the simulations using FAST are a standard certified test case provided by NREL, and no modifications were done on that.

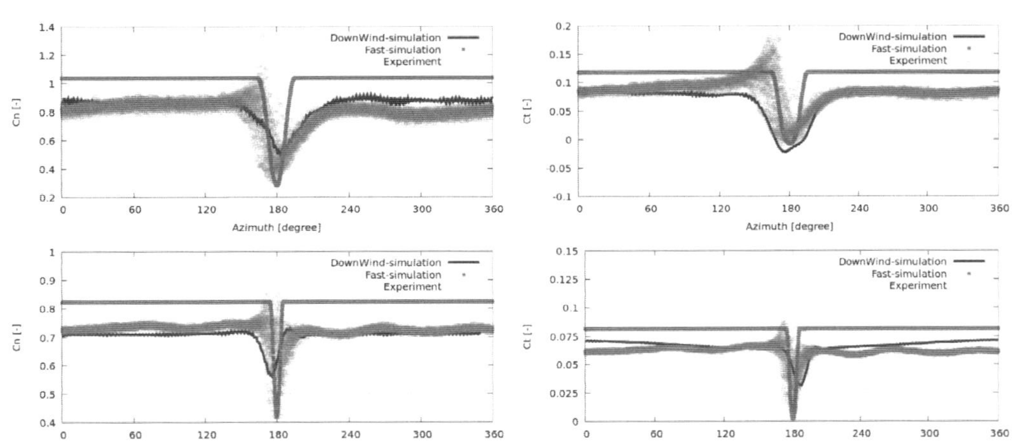

Figure 4: Normal and tangential force coefficient for 30% (upper) and 80% (lower) of span for wind speed of 7 m/s.

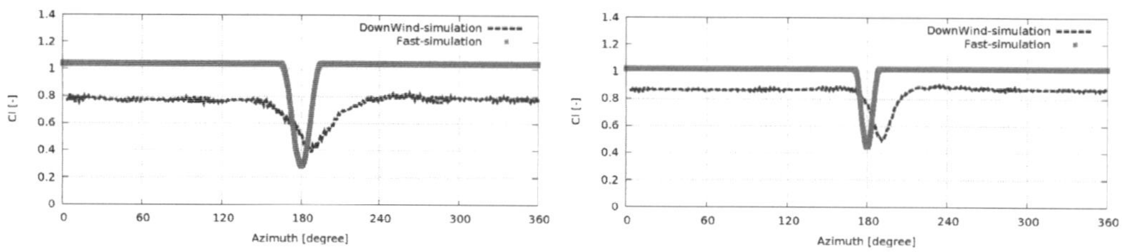

Figure 5: Lift coefficient for 30% (left) and 80% (right) of span for wind speed of 7 m/s.

Fig. 5 shows a comparison of the lift coefficient at two different sections along the blade span. For extracting the angle of attack (AOA) the technique of Shen [17] is used. This technique for extracting the AoA uses the Biot-Savart integral to determine the impact of the bound vorticity on the velocity field [17]. The approach assumes, that the longitudinal force distribution and the induced velocity caused by numerical calculations at a point of interest near the blade are numerically known. Then, the AOA is computed based on loading and velocity vectors at the point of interest. The results from FAST shows an approximately 20% higher value for the lift coefficient compared to CFD results. The CFD simulations show a drop of 50% compared to the FAST simulation which has a drop of 70%. The speed-up (increase in the force coefficients) which is observed near 170 degree for the experimental data is not captured neither by CFD nor BEM. The reason for that could be due to the fact that time stepping of the simulations were too large to capture this specific aerodynamic effect.

5. Wind turbine blades at standstill

Another complex load case relevant for wind turbine designs is the so called standstill case. This situation can occur during the erection of a wind turbine, during operation due to failure in the control system or very extreme storm cases and can result in very high loads acting on the wind turbine structure.

According to the IEC 61400-1 guideline [18], in this load case the blades have to withstand extreme wind speeds. Due to inflow angles of around 90° and therefore fully separated flows, the standstill scenario is numerically complex to simulate. As the interaction between the incoming wind and the blade structure plays a dominant role in this load case, a fluid-structure coupled solver, which combines the open source CFD code OpenFOAM with an inhouse structural beam solver, is used in this part of the work. Utilizing the Geometrically Exact Beam Theory (GEBT) [19] for the structural part and a Delayed-Detached Eddy-Simulation for the fluid part, the coupled solver allows a high fidelity analysis of complex aero-elastic problems.

In this section, the CFD based computations are based on a fully structured mesh generated with the software Pointwise [20]. The domain has the shape of a half sphere and the radius of the latter accounts to seven blade lengths. In total, 24 million cells are being used to discretize the domain in space. Three different meshes were used to check for grid convergence. Fig. 6 illustrates the used computational grid.

Fig. 7 shows a comparison of the developed high fidelity solver with the standard BEM-based engineering tool FAST [21] for the NREL 5MW reference wind turbine [22] during standstill. The inflow angle corresponds to 90°, the wind speed is set to $45m/s$. Since fully separated flows are also difficult to be accurately captured with RANS, time accurate Delayed-Detached Eddy Simulations (DDES) are performed. The simulations were initialized for 1 second to achieve a realistic field. Afterwards the flow simulation was coupled with the structural solver. In total the simulations were ran for 25 second. It can be noted, that the mean values of both edgewise and

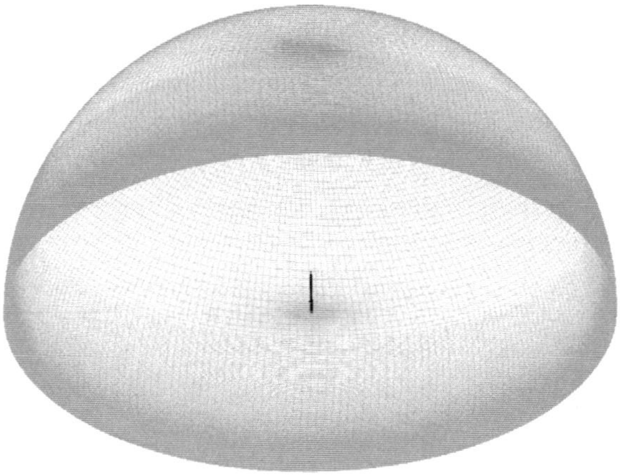

Figure 6: Half-cylindrical domain used for the fluid-structure computations of the NREL 5MW blade.

flapwise deflections show a reasonable agreement for both numerical approaches. In contrast to this, a significant difference can be observed in the dynamics of the edgewise deflection. While the FAST result shows up an almost constant edgewise deflection, the CFD coupled method predicts clear edgewise vibrations. The reason for this difference can be found in the complexity of the flow at such high angles of attack, which cannot be captured by FAST due to a missing suitable correction model. Fig. 8 gives an impression of the strongly turbulent wake behind the blade.

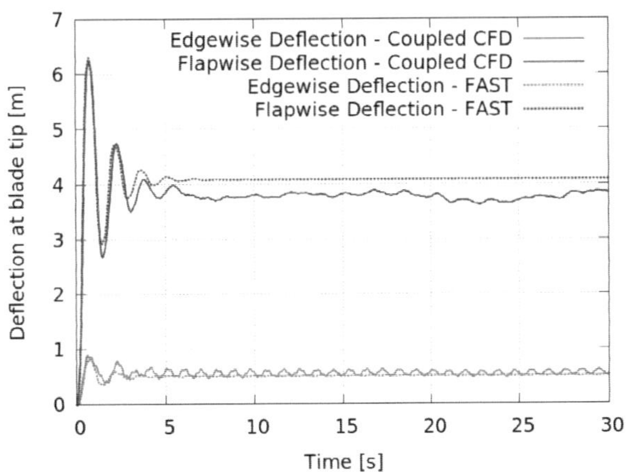

Figure 7: Deflections at blade tip for $45m/s$.

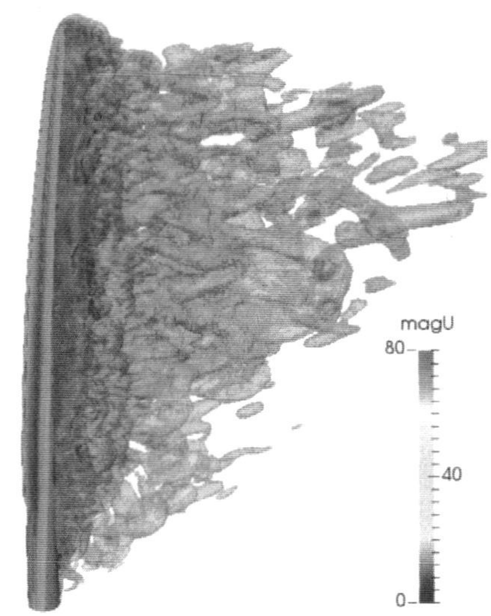

Figure 8: Instantaneous isosurface for Q coloured by velocity magnitude. Inflow speed: $45m/s$.

It can be concluded, that FAST could reproduce the mean deflections computed by the coupled CFD approach for the investigated velocity. However, blade vibrations could not be captured by the BEM based approach due to a missing suitable correction model. Potential instabilities, for example caused by overlapping vortex shedding frequencies and natural frequencies of the structure, could therefore not be captured by the BEM based approach. Following that, results obtained by FAST for this particular case should be treated with care.

6. Conclusions

While BEM can provide sufficient good results in most general load cases, care needs to be taken at cases leading to flow separation. CFD methods can help to get a more complete picture of the aerodynamics even in such complicated cases. Especially simulations of separated 3D flows with CFD give reasonable results. For load cases which are critical and difficult to grasp with BEM, CFD simulations should be used to cross check, if the critical situations are reached. In this work three different cases were simulated. For yawed flow sectional results over a rotor revolution showed a significant difference between different codes. In the root area, the skewed wake model from Schepers coincides more with the results from CFD since this model takes the root vortex into account. Close to the tip the behaviour is closer to the sinusoidal behaviour however some deviation from CFD is still observed, due the effect of advancing and retreating of the blade against the wind direction. The effect of the tower of downwind turbines was rather overestimated by the BEM. The fluid-structure coupled simulations showed that BEM and CFD predict a comparable mean deflection for the NREL 5MW blade during standstill. However, edgewise blade vibrations could be captured by the BEM based code and therefore BEM results for that specific load case should be handled with care.

References

[1] Schepers J 2012 *Engineering models in wind energy aerodynamics* november ISBN 9789461915078
[2] Micallef D 2012 *3D flows near a HAWT rotor :* (Technische Universiteit Delft) ISBN 9789995703127
[3] Jonkman J M and Buhl Jr M L 2005 *National Renewable Energy Laboratory, Golden, CO, Technical Report No. NREL/EL-500-38230*
[4] OpenCFD 2015 Openfoam®- the open source computational fluid dynamics (cfd) toolbox accessed 29 July 2015 URL `http://OpenFOAM.com`
[5] Schepers J 1998 *An engineering model for yawed conditions, developed on the basis of wind tunnel measurements*
[6] Hand M M, Simms D, Fingersh L, Jager D, Cotrell J, Schreck S and Larwood S 2001 Unsteady aerodynamics experiment phase vi: wind tunnel test configurations and available data campaigns Tech. rep. National Renewable Energy Laboratory
[7] Spalart P, Deck S, Shur M, Squires K, Strelets M and Travin A 2006 *A New Version of Detached-eddy Simulation Resistant to Ambiguous Grid Densities* vol 20 (Springer-Verlag)
[8] Rahimi H, Daniele E, Stoevesandt B and Peinke J 2016 *Wind Engineering*
[9] Patankar S and Spalding D 1972 *International Journal of Heat and Mass Transfer* **15** 1787 – 1806 ISSN 0017-9310 URL `http://www.sciencedirect.com/science/article/pii/0017931072900543`
[10] Issa R I, Gosman A D and Watkins A P 1986 *Journal of Computational Physics* **62** 66–82
[11] Moriarty P J and Hansen A C 2005 *AeroDyn theory manual* (Citeseer)
[12] Rahimi H, Hartvelt M, Peinke J and Schepers G J (IOP Publishing)
[13] Fairley P 2003 Two-bladed wind turbines make a comeback accessed 29 March 2016 URL `https://www.technologyreview.com/s/528581/two-bladed-wind-turbines-make-a-comeback/`
[14] Hagen T R 2011
[15] Menter F and Esch T 2001 *16th Brazilian Congress of Mechanical Engineering* pp 26–30
[16] Bak C, Aagaard Madsen H and Johansen J 2001 *Influence from blade-tower interaction on fatigue loads and dynamics (poster)* (WIP Renewable Energies) pp 394–397 ISBN 88-900442-9-2
[17] 2006 *etermination of angle of attack (AOA) for rotating blades* peinke, Schaumann, Barth (eds). Springer: Oldenburg, Germany,
[18] IEC 2005 *Wind Turbines - Part 1: Design Requirements, IEC 61400-1 (3rd edn.)* (Genova) ISBN 2-8318-8161-7
[19] Cardona A and Géradin M 1988 *International Journal For Numerical Methods in Engineering* **26** 2403–2438 ISSN 0029-5981
[20] 2015 Pointwise URL `www.pointwise.com`
[21] Jonkman J 2013 *51st AIAA Aerospace Sciences Meeting, Grapvine Texas* January ISBN 9781624101816

[22] Jonkman J, Butterfield S, Musial W and Scott G 2009 Definition of a 5-MW reference wind turbine for offshore system development Tech. rep.

Development of a Semi-submersible Barge for the installation of a TLP floating substructure. TLPWIND® case study.

Juan Amate [1], Gustavo D Sánchez [2] and Gonzalo González [1]

[1] Iberdrola Ingeniería y Construcción, Av. de Manoteras, 20, 28050 Madrid, Spain
[2] Iberdrola Engineering and Construction, Ochil House 10 Technology Avenue, High Blantyre, G72 0HT, United Kingdom

Abstract. One of the biggest challenges to introduce Tension Leg Platform (TLP) technology into the Offshore Wind market are the Transport & Installation (T&I) stages, since most of TLPs are not self-stable as semisubmersible or SPAR platforms, and consequently requires additional means to perform these operations. This paper addresses this problem that has been overcome through the development of a Semi-submersible "Transport & Installation" Barge (SSB) for Iberdrola's TLPWIND® floating support structure. The Semi-submersible Barge has been designed both through the use of numerical models and an extensive basin testing campaign carried out at the University of Strathclyde facilities. This paper also includes an estimation of the duration in time to carry out the installation process of a Floating Offshore Wind Farm, comprising 100x5MW TLPWIND® units in different scenarios.

1. Introduction

The development of offshore floating wind turbines is currently one of the toughest engineering challenges of wind energy. Numerous synergies have been established with Oil&Gas floating technologies (SPAR, semi-submersible and TLP) in order to determine the optimal concept designs that better matches with the conditions and constraints of the forecasted development areas worldwide.

In this sense, Oil&Gas floating technologies are designed like "one of a kind projects" that can be installed in reduced period of times and relatively short weather windows. On the contrary, the high number of units used for Offshore Wind developments require innovative systems that can perform an effective installation and operate at more severe sea states to increase the installation windows (or workable days per year). The possibility of assembling the wind turbine onshore has been determined as a major advantage for floating technologies, due to the development of a project lifecycle much more onshore based which reduces the cost and risks of a project.

SPAR designs cannot meet the requirement of assembling the wind turbine onshore due to its deep draught. TLPs have an outstanding dynamic behaviour in comparison to other floating technologies and also present a much lighter weight than Semisubmersible platforms, as well as much simpler construction processes and rated costs.

On the other hand, most of the TLP concepts present a lack of stability during the transport and installation stages, since the stability of this type of platforms is provided by the mooring system, once it has been connected. In order to solve this temporary drawback, aid systems which provide the required stability have to be used during T&I stages. The use of a Semi-submersible Barge is a solution which can efficiently address the abovementioned problem.

Previous developments for T&I based on the concept of a semi-submersible vessel have been proposed as presented in [1][2][3][4]. These concepts aim to reduce the cost of bottom fixed offshore wind technology, more specifically Gravity Based Foundations installation processes. Also, some companies have developed multipurpose semi-submersible vessels which can ease offshore operations as presented in [5]. For the floating wind sector, the development of a T&I barge for a TLP structure has been performed in the past as presented in [6], although this solution does not rely in the semi-submersible capacity but instead comprises a set of mechanical spuds to lower the platform (Figure 1).

Figure 1: Strabag Carrier (Top left) [1], Eide Barge 33 (Top centre) [2], Vinci GBF (Top right) [3], Optimus Pram (Bottom left) [4], Pelastar barge (Bottom Right) [5].

In this paper, the design philosophy and performance of a Semi-submersible barge, tailored to TLPWIND® technology, is presented continuing the work exposed in [7]. The results have proven the feasibility of the solution and the cost reduction potential that it offers not only for the installation of the wind farm, but also to perform maintenance actions and the decommissioning which represent a significant part of an Offshore Wind Farm life cycle.

2. TLPWIND® technology description

TLPWIND® is an innovative TLP concept designed specifically to support offshore wind turbines for very aggressive conditions in mid/large water depths. This concept is characterized by a simplified geometry (Figure 2) and light weight (900 – 1000 tons TLPWIND® 5MW) which drives to much simpler construction processes and lower costs, but it requires an ad-hoc system that provides stability during T&I phases to avoid the use of expensive Heavy Lift vessels.

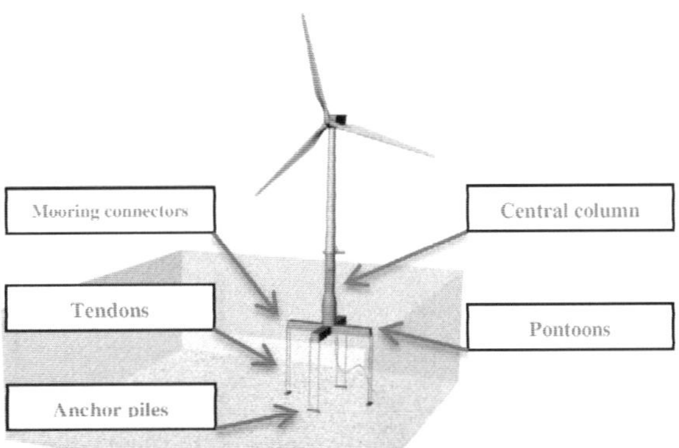

Figure 2: TLPWIND® technology main components

Since 2012, Iberdrola Ingeniería & Construcción has designed and tested at reduced scale three ad-hoc solutions for addressing this matter (Figure 3). An U-shaped barge, which can effectively perform the transport operations, a set of re-usable floaters which grant enough stability for performing both transport and installation operations, but also an U-shaped Semi-submersible barge (SSB); awarded PCT/2013/070697, that merged the advantages of the two systems formerly developed (High towing speed due to a significant low water resistance for transport operations & outstanding seakeeping capabilities at variable draughts).

Figure 3: TLPWIND® T&I solutions. Auxiliary floaters (left), semisubmersible barge (centre) and U-shaped barge (right)

3. TLPWIND® transport and installation operation

This section covers the development and detailed study of the Transport & Installation operations and the necessary means to perform these operations with the TLPWIND® floating platform in a safe manner.

This includes the characterization and analysis of the steps to follow in order to perform the installation of a platform, initially focusing in a single platform to later consider a full wind farm scale (100x5MW TLPWIND® units) and the impact of having repetitive operations. The results for this study are presented at the end of this paper.

In this regard, fifteen operations were identified for a single unit which can be summarised in six main stages (Figure 4) starting from the required port operations (1 & 2), followed by a transport stage (3), installation of the TLPWIND® platform (4) and final recovery of the SSB (5) and tow back to port (6). For this study a detailed estimation of the duration of each of these steps in ideal conditions (no weather delay) was additionally performed.

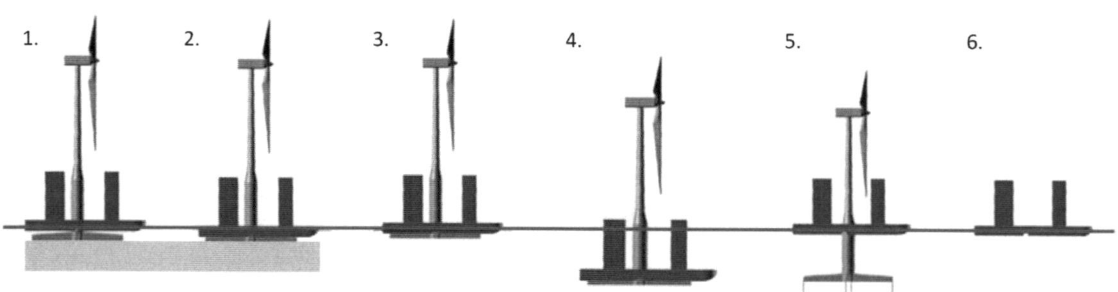

Figure 4: Transport and Installation stages

On a second step, the design of the ad-hoc SSB was carried out, taking into account the stability, towing resistance, dynamics and seakeeping characteristics of the vessel to comply with the requirements imposed by the wind turbine, the navigational limits and installation tolerances at each of the several draught conditions achieved during the whole installation operation.

4. TLPWIND® Semi-submersible Barge design & analysis methodology

Since the first stages of development of the TLPWIND® concept, and in parallel to the design of the TLP platform, the creation an auxiliary system that allows reducing the T&I operations complexity as well as the associated cost has been one of the key tasks of the TLPWIND® design team. Subsequently, the development of a design methodology that allows adjusting the characteristics of the SSB to each specific TLPWIND® geometry (defined depending on selected site characteristics) has been established.

Following the typical naval design spiral, the methodology that has been used for the first loop in the development of the SSB is shown in Figure 5.

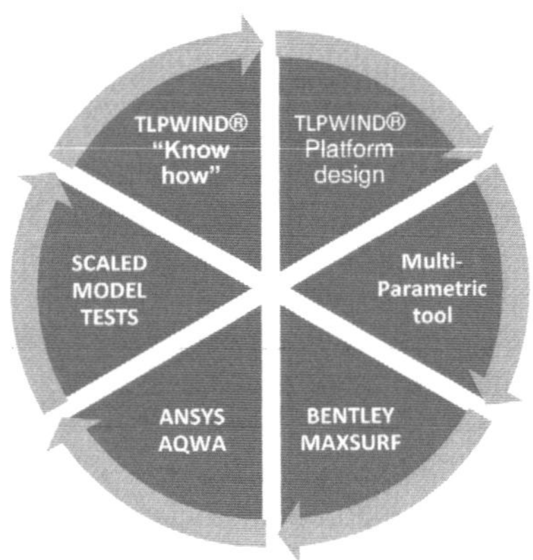

Figure 5: SSB Design Methodology scheme

On a first step an in-house Multi-parametric tool is used for achieving several preliminary configurations that meet the stability and dynamic performance requirements for a simple approach. Since the SSB is suited to a specific TLPWIND® geometry, the dimensions of the TLP platform have been taken into account to match both elements and the necessary footprint. Secondly, the use of commercial software like Maxsurf or Ansys AQWA allows verifying in the detail the stability and dynamic behaviour identifying the hydrostatic curves, static and dynamic stability, eigen periods and RAOs among many other parameters that define the performance of the system, thus this detailed analysis provides information that enables refining the design of the SSB.

For the selected design scaled model tests are carried out, which include a wide range of tests covering all the single stages that would happen during the T&I process apart from tests that help to characterize the SSB. A short list of the tests that have been performed is shown as follows:

- Free decay tests
- Semi captive towing tests in calm water and in waves for the SSB and SSB+TLP assembly
- Wired towing tests in calm water and in waves for the SSB and SSB+TLP assembly
- Seakeeping tests in regular and irregular waves

The results of the tests were correlated with the numerical simulations in order to validate the design and obtain additional information that allows once again performing a re-engineering process and improving even further, the design and performance of the SSB.

Consequently, it is worth highlighting that, the optimum solution will be reached by means of an iterative approach, where the design loop is repeated until the no significant improvement is achieved

More details on the results achieved from the development and scaled model testing of the SBB are presented in the following sections.

4.1. SSB behaviour in transport operation

The Semi-submersible Barge introduced in this paper presents a significantly low drag resistance, which leads into a reduction of the capacity needed to tow the SSB+TLP to the installation point (commonly known as Bollard Pull).

This low drag makes possible to perform the transport operation just using conventional tug boats and avoiding the need of the expensive vessels used currently used for the Oil&Gas sector.

To determinate the towing drag and dynamic behaviour of the SSB barge, towing tests were performed in Kelvin Hydrodynamic Laboratory (University of Strathclyde) by means of two different methods: an accurate semi-captive system, used for characterizing the towing resistance of the SSB+TLP assembly, and a wired towing system, which allowed to characterize in a more realistic way the seakeeping behaviour during the towing operation in still water, regular and irregular waves. These tests provided the technical background for the decision on operational limits for the towing operation. The towing resistance was evaluated both in the transport as well as in return condition, once the TLPWIND® platform has been installed and only the SSB is towed (Figure 6).

In both conditions, the SSB presents an extraordinary low towing resistance even at high towing speed, due to the optimized shapes developed, which minimizes viscous drag and wave formation components.

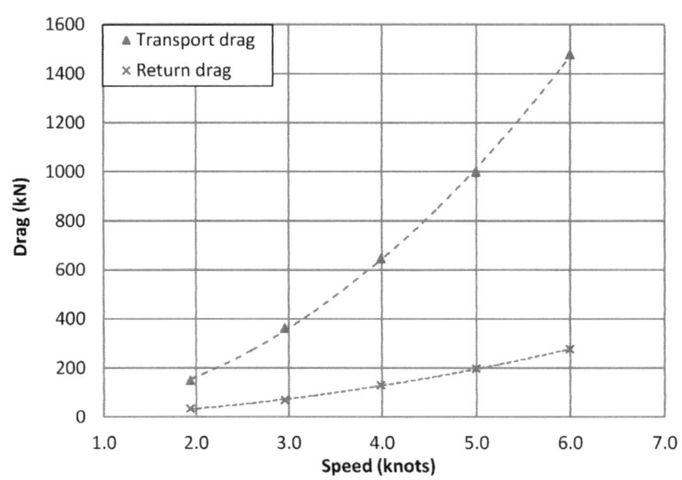

Figure 6: Calm water resistance with semi-captive system

As mentioned previously, this low towing drag leads into a reduction of the overall installation time, driving down the overall LCoE due to the shorter duration of the installation campaigns as well as the cost of the installation means needed.

Nevertheless, on a real scale operation, the transport operation would be performed with a towing wire (one line or various, depending on the specific transport scenario). In this sense, the position of the towage point may have a notable influence in towing resistance as well as in trim angles, which could lead into excessive tilting angles and loading over the nacelle, endangering the integrity of the wind turbine components.

Two different towing points were then considered for determining the best way to carry out this operation. Consequently, a precise tracking of the SSB's behaviour was required in these tests, which used an optical tracking system (*Qualisys*) to register the motions of the SSB. Also, the tension on the towing line was acquired by means of a load cell which connected the towing carriage and the towing wire that pulled the model.

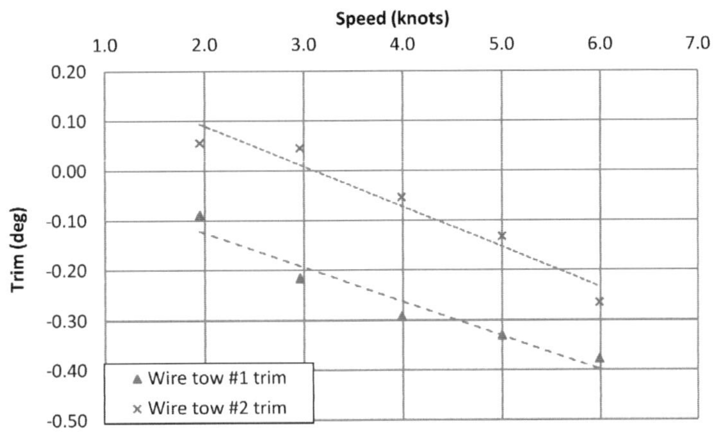

Figure 7: Calm water trim from wired tow tests

Difference between drag forces obtained using the different towing points are almost negligible; however, the SSB's trimming presents notable differences as shown in Figure 7. Towing point 1, located at the middle of the bow of the barge, develops almost twice trimming than the achieved when performed using tow point 2, which is located at the top of the bow of the barge.

The SSB has presented an extraordinary towing behaviour in calm waters; however, the transport operation tests have to be validated in realistic conditions, where waves are also taken into account.

The added resistance in waves was calculated for regular waves with a period ranging from 6 to 20 seconds (in real scale), with a nominal height of 1 meter.

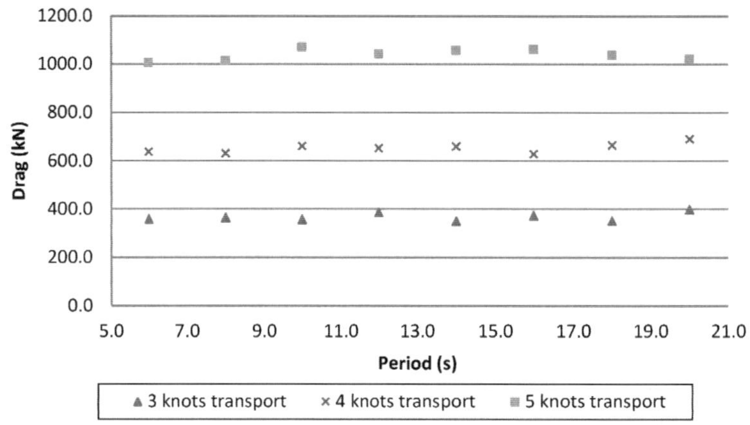

Figure 8: Transport operation resistance with waves using semi-captive system

The mean resistance is calculated in this section of the testing campaign integrating over a number of wave cycles, and the added resistance is obtained calculating the difference between this and the resistance calculated for calm water. Again, these tests were conducted for the transport and return condition. From these tests, it is obtained that the wave forming resistance increases the total resistance in between 2% and 6% depending on the specific wave period and towing speed as shown in Figure 8.

In the same way it was done for the calm water runs and aiming to represent a realistic behaviour of the barge on the towing operation, the tests for regular waves were repeated using a wired tow attachment.

With this wire towing configuration, it was noticed that the model develops a small directional instability and oscillates slightly from side to side (fish-tailing). This effect makes the resistance vary with the lateral oscillation. This phenomenon pointed that it seems to be more advisable to perform the towage of the TLPWIND® SSB platform using two tug boats and two towing lines rather than a single line.

On the other hand motions of the SSB in transport conditions seem to be appropriated for performing this operation in safe conditions. In the figure below (Figure 9), heave and pitch RAO resulting from testing the SSB in unitary height regular waves are plotted, presenting a good degree of reliability.

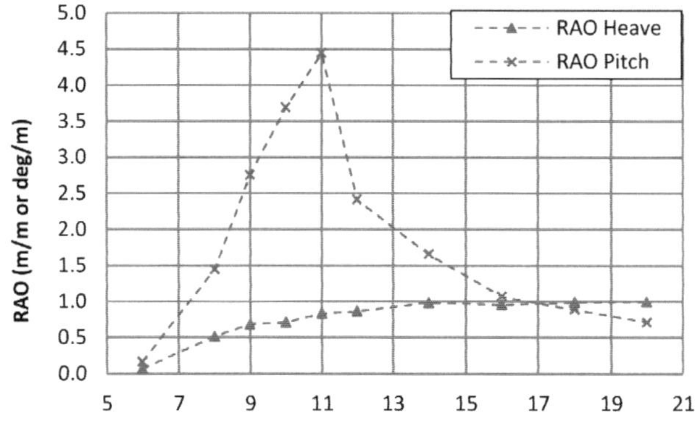

Figure 9: Towing test RAOs, regular wave with wire-tow system

As for most of the ships and conventional barges, heave RAO does not present any peak within tested range of wave periods, avoiding any undesired resonance and excessive motion in this degree of freedom. Therefore, the SSB+TLP assembly will just tag along the waves when towing in large period waves. On the other side, regarding pitch RAO, it is showed a maximum, corresponding to the pitch eigen period. In this sense, it is crucial to ensure this peak does not match with the most probable excitation periods that may occur during the transport operation, if not it is advisable to move this pitch eigen period above characteristic wave periods.

One of the aims of the design was to achieve pitch eigen periods above 10s, since sea state wave periods higher than that value are not expected during the transport operations for the selected location in this study, thus avoiding resonant phenomena.

4.2. SSB behaviour in installation operation

Installation operation in a TLP structure is probably one of the most critical and sensitive operations to be planned within its development. In this sense, designing an auxiliary installation mean that greatly eases this operation and increases the installation weather window is a crucial issue when developing a large scale Offshore Wind Farm.

A set of four elastic lines (soft moorings) were connected at each of the corners of the SSB to keep the model in place. The evaluation of the seakeeping behaviour of the SSB has been carried out through analysis of the motions RAOs over a range of different draughts that will occur during the installation operation and subsequent ballasting. This evaluation started from the transport draught (lowest draught) and ended up for the draught at which the connection of the mooring system is performed (deepest draught).

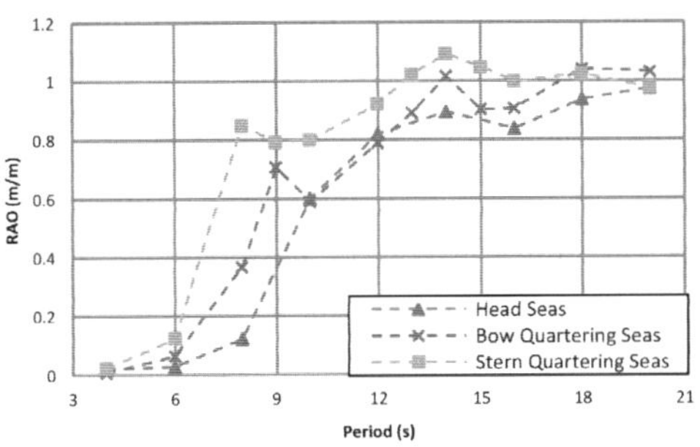

Figure 10: Zero forward speed Surge RAOs at transport draught (7m)

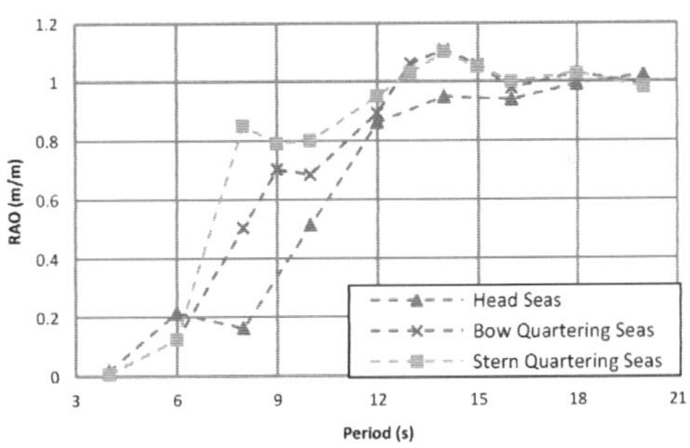

Figure 11: Zero forward speed Heave RAOs at transport draught (7m)

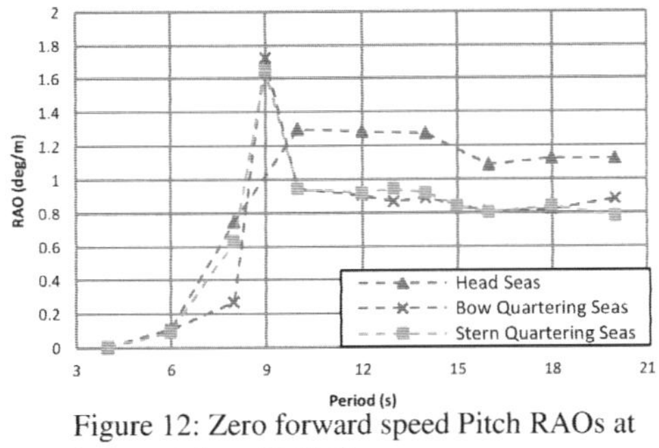

Figure 12: Zero forward speed Pitch RAOs at transport draught (7m)

These tests were carried out with three different wave headings (head, bow quartering and stern quartering waves) but same wave height of 1 meter and range of periods from 4s to 20s to make it comparable. Results for the initial installation stage (the one corresponding to the transport draught) are shown on Figures 10 to 12.

To ensure assure a good dynamic behaviour of the SSB and avoid undesired resonances or noise in the recorded signals, the mooring system is selected with a natural frequency far away from the generated wave frequencies and SSB natural frequencies.

The same evaluation of the stability and dynamic behaviour of the SSB was performed for the deepest draught or tendon connection draught (Figures 13 to 15). It is important to highlight that this condition requires a deeper analysis due to the sensitivity of this operation at which the platform is connected to the pre-installed mooring system.

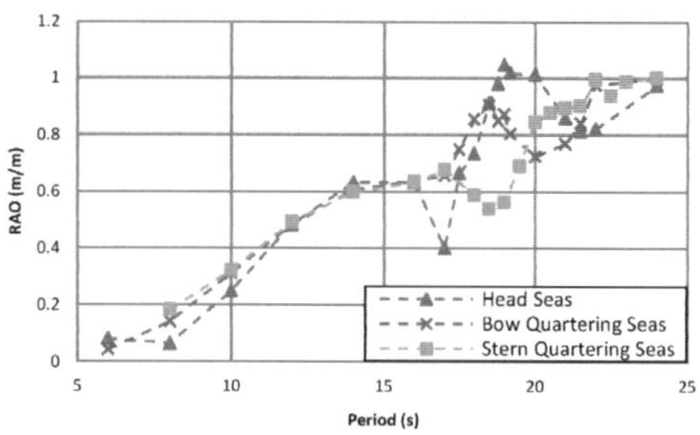

Figure 13: Zero forward speed Surge RAO at the connection draught (35.5m)

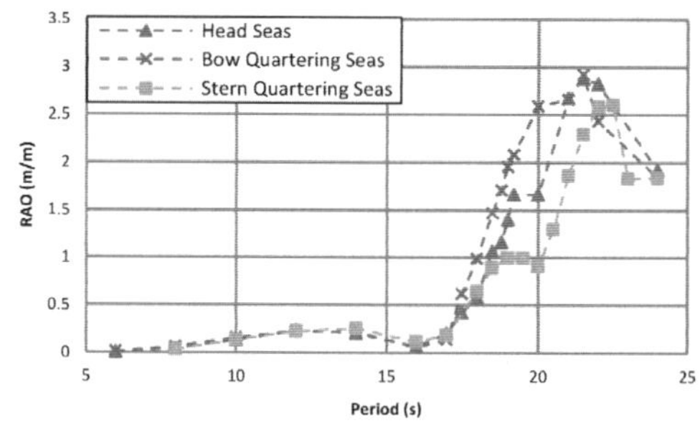

Figure 14: Zero forward speed Heave RAO at the connection draught (35.5m)

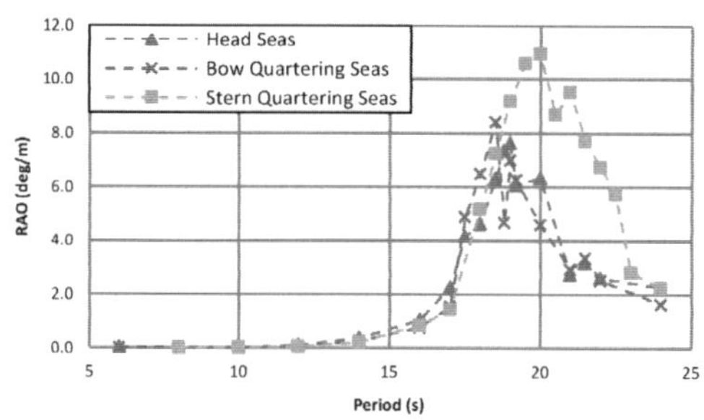

Figure 15: Zero forward speed Pitch RAO at the connection draught (35.5m)

For this tendon connection condition, other factors play an important role such as the need of accuracy in the positioning of the SSB+TLP assembly and the need of restricting as much as possible any displacement and rotation in the six degrees of freedom of the floating structure. Accordingly, the positioning methodology must be properly evaluated.

In order to evaluate this matter, a range of different stiffness for the elastic lines was tested, simulating the variation from a soft system (which was not expected to modify the behaviour of the SSB) to a stiff system (with approximately three times the stiffness of the abovementioned soft system).

Figure 16: Position keeping system arrangement during the installation operation.

It is important to highlight that although the model from Figure 16 does not have the columns fitted to make towing easier, the mass properties were reproduced accurately so that the model behaved in all aspects as if the columns were present.

From the results of the tested performed, it can be concluded that that the less stiff the position keeping system is, the larger the operational offset will be (meaning the SSB is displaced from the initial position). However, there is not a notable influence in the amplitude of motions of the system due to this stiffness change.

The only exception to this conclusion is the pitch rotation, where the position keeping system proved to have a notable influence, reducing the amplitude of the motion at the eigen period of the SSB considerably as shown in Figure 17.

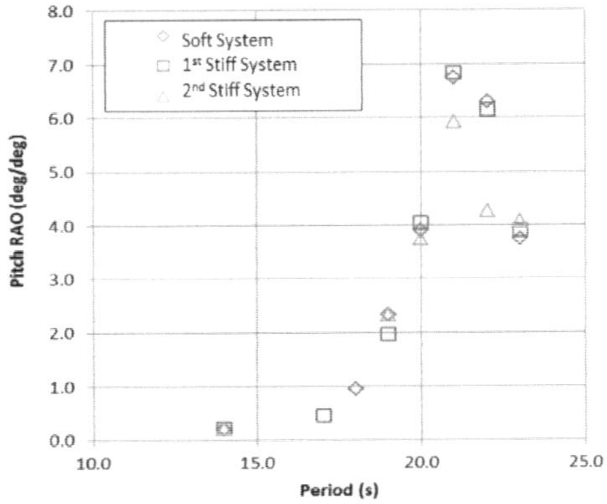

Figure 17: Pitch RAO comparison for three position keeping configurations tested

Nevertheless, since pitch eigen period of the SSB is intended to be higher than 20 s and well above the characteristic wave periods of the installation location (furthermore considering the operational limits for the installation operation), it can be concluded that the SSB at the connection draught is not very sensitive to variations introduced in the position keeping system.

5. Operational limits

The results obtained from the scaled model tests were used to evaluate and define realistic operational limits for the whole T&I operation.

This study showed that towing operations for the SSB+TLP can be performed up to a significant wave height (Hs) of 2.5 meters at a towing speed of 5 knots, while most of the remaining operations can be carried out in significant wave height sea state up to 2 meters except for the most critical tendon connections operations that may require a less severe sea state of around 1.5 meters of significant wave height. However, as this operation has a very short duration compared to the whole T&I process, the impact on the overall averaged operational limit is small. As an example in Scottish waters, the averaged operational limit for the whole T&I operation is close to 2.4 meters considering Aberdeen coast as the wind farm site and its associated wind & wave characteristics, besides of a towing distance of 217 km from a port in Cromarty Firth to the selected site.

6. Overall installation time and cost reduction

Having defined the operational limits for each of the T&I stages and calculated the ideal duration of each of them, weather climate effect was considered by completing a study of the weather conditions for three UK waters representative locations. The operational limits for each step along with the associated duration were contrasted against the sea state probability of exceedance for the selected sites, resulting in the estimation of the necessary weather windows and expected weather downtime. The three sites are represented through the corresponding three annual mean wave heights.

The insights obtained from the previous studies were implemented for a complete Offshore Wind Farm project scale considering 100 units. This thorough study allowed estimating the total duration for the installation process of a 500 MW floating Offshore Wind Farm based on 5MW TLPWIND® platforms in different scenarios varying the staging port to site distance, weather conditions and the use of one or two SSB for the T&I strategy. This study also takes into account that T&I procedures are to be repeated numerous times, involving many man-driven operations which will have a learning-by-doing improvement. This effect was captured with the application of learning rates for each different operation when applicable (ballasting operations would not have a variation in terms of the duration of the operation).

If a single SSB is considered and a continuous installation process starting in April is assumed for the installation of the whole Offshore Wind Farm, the total estimated duration obtained for the installation of 100 units is shown in Table 1.

100 units T&I duration (days)		**Annual mean wave height** 1.0 m	1.5 m	2.0 m
Port to site distance	100 km	225 days	241 days	264 days
	250 km	365 days	397 days	425 days
	400 km	492 days	524 days	577 days

Table 1: Overall installation time with one SSB, 100x5MW TLPWIND® units

Also, if two SSBs are considered for an equal continuous installation process starting in April for the installation of the whole Offshore Wind Farm, the total estimated duration obtained is reduced by more than half since some severe weather months could be avoided.

100 units T&I duration (days)		**Annual mean wave height**		
		1.0 m	**1.5 m**	**2.0 m**
Port to site distance	100 km	110 days	121 days	130 days
	250 km	175 days	185 days	197 days
	400 km	245 days	262 days	290 days

Table 2: Overall installation time with two SSB, 100x5MW TLPWIND® units

Due to the outstanding dynamic behaviour of the SSB even with seas up to wave height (Hs) of 2.5 meters, the overall installation time can be reduced which leads into a reduce of the overall LCoE of the wind farm project.

The application of TLPWIND® technology and the proposed T&I installation method has the potential to provide LCoE values of around £95/MWh by 2025 when considering a 500MW Offshore Wind Farm composed on 100 x 5MW TLPWIND® platform under Aberdeen conditions. An increase in the turbine size that would lower the number of installations to be performed could reduce the LCoE even further to values of £77/MWh by 2050 under the same conditions. Last, if additional reductions are considered like learning curves for the technology and market development LCoE values could reach £64/MWh by 2050.

7. Conclusions

The use of a semi-submersible barge for performing the transport and installation of TLP substructures supporting a wind turbine pre-assembled onshore is a technically and economically feasible solution. In this regard, a design of semi-submersible barge has been performed, which can efficiently carry out the abovementioned tasks reducing the overall installation process duration and costs. Also, the designed barge reduces the risks associated with the transport and installation operations besides of showing an outstanding performance in both transport and installation conditions.

The SSB has an exceptional navigation capability, even at seas with Hs=2.5m, with very low pitching angles, nacelle accelerations and towing drag (< 100 tons) at 5 knots or even higher. Towing back operations after the installation can be carried out at 8 knots or even higher if higher bollard pull vessels are used.

As for the installation operations, the TLP+SSB assembly shows very good seakeeping characteristics and dynamic behaviour, especially for the lowest draught which represents the most critical installation operation.

Installation with the SSB presents high number of workable days per year, since offshore operations are minimized. Installing a 100x5 MW windfarm in less than 8 months is possible in sites with mild weather conditions and relatively close to a staging port (100 km) with just one SSB.

Finally, all the above mentioned advantages of the SSB, will not just apply for the transport and installation phases, but also for the maintenance and decommissioning phases of the Offshore Wind Farm, turning the SSB into a new concept of an asset which can provide evident benefits throughout the whole project life cycle. Thus, the extended exploitation of the SSB can provide a further cost reduction and even lower LCoE figures.

Due to the SSB's low drag characteristics and the enhanced dynamic behaviour, marine operations can be done using standard vessel like an AHT (20-35 k€/day) assisted by standard tugs (8-15 k€/day) and a WROV (8-12 k€/day) to carry out tendon connections, instead of expensive specialised vessels (150-500 k€/day). This reduction on the installation costs compensates the fabrication of the Semi-submersible Barge, leading into an overall LCoE reduction.

The application of TLPWIND® technology and the proposed T&I installation method has the potential to break the £100/MWh barrier in the near future and could offer a significant cost reductions in the long term to achieve very competitive LCoE values of £64/MWh by 2050.

Aknowledgement

The studies shown in this paper have been carried out under the scope of works of the project: TLPWIND UK "Driving down the cost of Offshore Wind in UK waters". This project was possible thanks to a grant scheme awarded by Innovate UK, Infrastructure for Offshore Renewables R&D. (Grant 40145-288200, Project 101969). Moreover, the authors acknowledge the outstanding work performed by the University of Strathclyde and Kelvin Hydro Laboratory staff involved in this project as well as congratulate the ORE Catapult for the exceptional LCoE study performed.

References

[1] Innovative vessels: Concepts for heavy-lift carriers. *Ship & Offshore: The international publication for Offshore & Marine Technology*. July/August 2012. N°5. Page 10-15.

[2] Strabag Carrier. Strabag Offshore Wind. (http://www.strabag-offshore.com/fileadmin/Downloads/Flyer_STRABAG_OFFSHORE_eng.pdf) [Online]

[3] Vinci GBF Transport and Installation Barge TIB. (http://www.vinci-offshorewind.co.uk/VFOW.pdf) [Online]`

[4] Ahlstrom C. ; Andersson, A ; Blomgren, N. (2014) "Optimus Pram – Semi-submersible wind farm installation vessel for blekinge Offshore".

[5] EIDE BARGE 33 Technical specification. Eide Marine Servives AS. (http://ems.eidemarine.com/vessels) [Online]

[6] Pelastar Installation and Maintenance barge. The Glosten Associates (http://pelastar.com/?page_id=55) [Online]

[7] Amate J, Gómez P, de Diego V, Trigo P, Sánchez G, González G, Llana A, Giner L. 2015. How to install a TLP substructure for offshore wind? TLPWIND® case study. EWEA Offshore 2015. PO.ID028.

WindEurope Summit 2016 IOP Publishing
Journal of Physics: Conference Series **749** (2016) 012017 doi:10.1088/1742-6596/749/1/012017

Recommendations for load validation of an offshore wind turbine with the use of statistical data: experience from alpha ventus

Ricardo Faerron Guzmán, Po Wen Cheng

Stuttgarter Lehrstuhl für Windenergie, Allmandring 5b, 70192,Stuttgart, Deutschland
Email: faerron@ifb.uni-stuttgart.de , cheng@ifb.uni-stuttgart.de

Abstract. The present paper provides insight into the validation of computer models used for simulations of offshore wind turbines. The offshore turbines are affected by environmental conditions that must be logged during the measurement campaign of the prototype and used for the simulations during the validation process.
A simple generic methodology is presented to be used for the comparison of statistical data from the measurement campaign and the simulations. This allows a better analysis of the simulations and helps limit the apparition of outliers in the measurements.
An example of the use of the methodology is provided with the use of the data recorded for the AD5-116 5MW turbine at alpha ventus. For it, the turbine power, operational parameters and blade and tower loads are compared.

1. Introduction

The design of an offshore wind turbine (OWEC) requires the simulation of various design load cases according to the IEC-61400-3 [1]. Once a design has been determined appropriate, a prototype can be built. The IEC-61400-13 [2] guideline specifies how a measurement campaign of a full-scale wind turbine is to be carried out and used as a basis for the validation of computer models, which is exemplified in a general procedure shown in its appendix E. Despite this, it is still unclear what constitutes a validated simulation model, and furthermore the IEC-61400-13 is based on onshore turbines, providing only a few recommendations for the offshore case.

Early research focused on the verification of the aeroelastic and hydrodynamic codes in projects like OC3 [3] and OC4 [4]. For these projects, a range of different simulation codes are compared to one another with the use of load cases varying in complexity.

The research community has also addressed the issue of computer model validation with measurement data. Zierath et al. [5] have shown the validation of an OWEC with statistical and fatigue data, and the load spectrum. Seidel et al. [6] presented a time based comparison of simulation and measurements on the brace of a 5 MW offshore jacket mounted turbine within the project DOWNVInD. Kaufer [7] presented the validation of an 5 MW offshore jacket mounted turbine by comparing loads on the jacket during a nacelle rotation, as well as frequency domain comparisons of the blade, tower and jacket sensors, and the statistical comparison of 10 minute simulations.

Regarding the general subject of validation, Söker et al. [8] provided a description of the possible comparisons needed between measured data and simulation, outlining important steps to help guide the validation engineer.

Despite these efforts, the question still arises, how to compare measured data from a turbine that is affected by a combination of stochastic meteorological and oceanographic parameters such as: wave height, wave period, wave direction, wind-wave misalignment, wind shear and turbulence intensity. The aim of this work is thus to provide insight into the proper validation of an OWEC computer model. The work will as well outline a methodology for the validation of a computational model, with the use of statistical data from a measurement campaign by screening the data that is chosen for comparison. To exemplify the methodology, a validation of the statistics from simulations of a 5 MW AD5-116 tripod mounted turbine installed at the alpha ventus wind park is carried out.

2. On the use of the IEC 61400-13 with regards to offshore model validation

The validation of a simulation model used in the design of a turbine should be carried out with different methods. The aim is that the results of the simulation model need to be accurate enough to obtain a safe design. Söker et al. [8] mentions a number of validation steps which should be required to ensure a good agreement between measurements and model. This includes:

-Consistency of environmental conditions
-Consistency of turbine dynamic behaviour (frequencies)
-Consistency of turbine characteristic curves
-Consistency of behaviour of loads and operational parameters (time series and statistics)
-Consistency of fatigue characteristic behaviour

These validation steps require the analysis of the data in the frequency domain for certain time series, derivation of damage equivalent loads, load cycle spectrums, as well as comparison of statistical parameters such as the minimum, maximum, mean and standard deviation. The comparison of such data is also shortly mentioned in the Annex E of the IEC 61400- 13 [2]. These validation steps apply both to onshore and offshore structures. This paper will build on the idea of validation of the statistical data of the simulated load cases for an OWEC within normal power production.

The offshore environment and the complex array of support structures and foundations used by OWECs pose a challenge for a load measurement campaign. When considering the IEC 61400- 13 [2], there are several issues that are not discussed that would need to be taken into account for the validation of an OWEC. These include:

1. Calibration of the offset and sensitivity of strain gauges at the underwater structures is not possible for most cases due to the complexity of the support structure of some offshore turbines. The recommended procedure estimates the moment cause during stand still or nacelle rotation of the rotor-nacelle assembly on the tower, yet for a jacket or tripod structure the estimation of this moment for the statically indetermined structure is less reliable. As well, strain gauges located underwater are prone to failing with no chance for repair. This severely limits the possibility to compare the magnitude of the loads from simulations for the foundations.
2. Hydrodynamic loading needs to be modelled. These parameters involve: waves conditions, wind and wave directionality, sea currents, and water level. The procedure with which one could include such parameters into the capture matrix of a measurement campaign is unclear.
3. Marine growth and scouring could play important roles on the loading of the offshore structure. A quantification of these parameters from the measurement campaign will be needed for the simulation model.

The aforementioned issues have been taken into consideration as much as possible in the following sections, yet as mentioned, the calibration of the strain gauges at the underwater support structure of the modeled turbine was not successful and thus the results section of this paper doesn't show their comparison with the simulation model.

3. Validation of the statistical data

In an ideal measurement campaign, one would know all the environmental parameters affecting the turbine during its operation and be able to reproduce these exactly in the simulation. The response of the real turbine and the simulation model could then be easily validated. In reality, we have to simplify the environmental parameters, for example, generating an artificial wind field for simulations based on

Journal of Physics: Conference Series **749** (2016) 012017 doi:10.1088/1742-6596/749/1/012017

a mean hub height wind speed measured at a meteorological mast. Due to this fact, an exact time series of a loading signal from a simulation will not be the same as that from the one measured, yet it is still very important for the design and validation process that statistics, such as maximum loads and damage equivalent loads (DEL), are representative of what is really happening to the turbine.

The next section concentrates on how this statistic validation can take place.

3.1. Modelling of the wind turbine in Flex5/Poseidon

The presented results are based on the measurement campaign carried out at the Adwen AD5-116 OWEC located in alpha ventus. The tripod mounted OWEC has a rated power of 5MW at a wind speed of 12.5 m/s, a rotor diameter of 116m and a hub height of 90m. The investigated turbine is number 7, which is equipped with measurement devices (Figure 1). The measurement campaign at the offshore wind park provides statistical data for validation. The turbine is equipped with strain gauges at the tower base and blade root. SCADA data (50 Hz) and 10 minute statistical of the SCADA data are available, along with oceanographic and meteorological statistics from the Fino 1 metmast. The data used is taken for 13 months from 01-10-2010 to 31-10-2012.

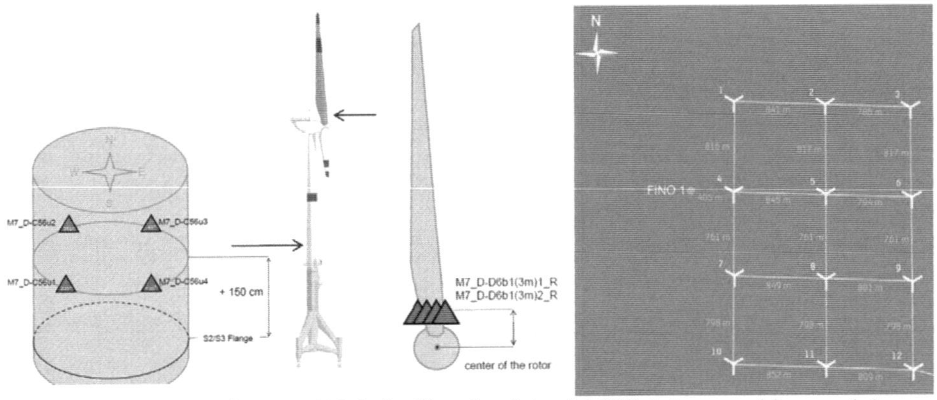

Figure 1: right- alpha ventus layout [9], left- Sketch of the 5 MW Adwen turbine at alpha ventus with location of blade root edgewise and flapwise, and tower bottom strain gauges (Copyright DEWI).

3.2. Modelling of the wind turbine in Flex5/Poseidon

The model data and controller of the 5MW turbine at alpha ventus are kindly made available by the turbine manufacturer, Adwen. For the simulation of the complete offshore wind turbine with substructure under the effects of hydrodynamic and aerodynamic loading, an integrated simulation environment was developed using Flex5 and Poseidon, which allows for a coupled simulation method, as exemplified by Kaufer [10]. The ability to use the coupled programs for OWECs simulations was shown in the OC3 project [3]. The rotor nacelle assembly and tower are simulated in Flex5 whereas the substructure and foundation are modelled in Poseidon.

The pile foundation is modeled as a spring damper at the three connection points of the tripod with the mudline. All tubular elements in Poseidon are modelled as Bernoulli beam elements with 6 degrees of freedom at each node.

For a 10 minute simulation, 660 seconds are simulated and the first 60 seconds are cut out to try to eliminate the effect of transients during start up. Turbulent wind fields are set up with VindSim. The 3-D wind vectors at predefined nodes in a polar grid are saved in defined time intervals. This wind field describes the stochastic wind through the use of the Kaimal spectrum as proposed in [11]. Waves are modelled in Poseidon. For it the Jonswap spectrum for waves is utilized, also as proposed in [1].

3.3. Data Screening Methodology

The goal of screening of data is to pick certain events and their measured statistics (from the ten minute time series), and reproduce them with the simulation model. Still, the use of statistics of the measurements from an offshore campaign poses a challenge due to the stochastic behaviour of the wind and wave parameters. The stochastics leads to large scattering of the data points along with the appearance of outliers as exemplified in Figure 2. Thus a comparison of simulation data with the measurements would be more meaningful if the wind and wave parameters have been chosen to be representative of the conditions simulated. This is done through a data screening process.

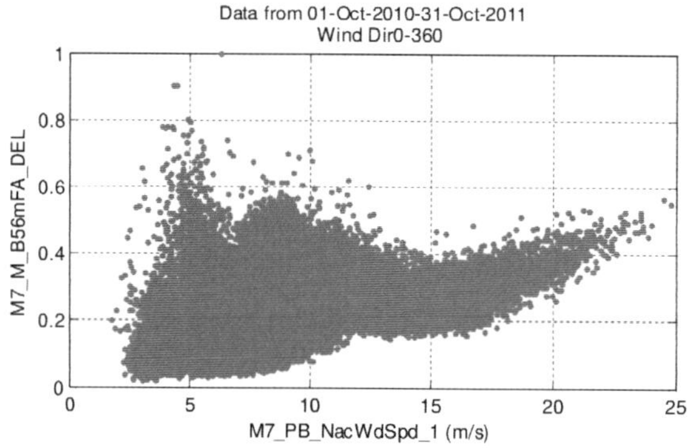

Figure 2: plot showing scattering and outlier in the tower for-aft DEL.

When screening the data, some parameters will be more important than others since they affect the loads to a greater extent, for example wind speed and turbulence intensity. Furthermore wave height and direction, along with wind/wave correlation and the wave spectrum should be reported by the measurement campaign and included in the screening.

For the validation engineer to be able to carry out the validation procedure of statistical data, the following methodology is recommended for the screening of the data:

1. List all meteorological and oceanographic parameters that are needed for the validation, along with other parameters such as marine growth thickness and density
2. Define which parameters need to be binned. For example wind speed
3. Define which parameters will be constrained. For example only data from 5-7% turbulence intensity
4. Define which parameters are, for the purpose of the validation, constants. For example marine growth thickness or mean sea level

Additionally, simulation parameters need to be set based on the screened data. This can be done the following way:

1. Determine a mean value representative of the bin or constraint
2. In the case that a bin or constraint is large, there might be the need to divide the constrained parameter into several different mean values. For example, if the wind wave misalignment is filtered from +60 to -60 degrees, the simulated misalignment can be +45, 0 and -45 degrees.
3. Determine the amount of seeds that are appropriate for each simulation

Using this procedure, the alpha ventus data for the 5MW turbine is screened. An example of the selection and filtering of parameters can be found in

Table 1. As is usual, the wind speed is binned. Furthermore, as in Figure 3, the turbulence intensity is constrained to an area where most of the data points can be found. Similarly, the significant wave height is also plotted in Figure 3 as a function of the wind speed and therefore defined by a fitted curve. The constraint for the filtering is based on this fitted function. The different meteorological and oceanographic parameters would have to be similarly analysed.

After applying the screening of the data, a smaller data pool is obtained which better mirrors the simulated parameters to the measurements and helps limit the appearance of outliers in the data. The advantage of this procedure is that since the simulation parameters are now clearly defined, later it will be easier to compare these simulations with the measurements.

Table 1: Meteorological and Oceanographic parameters and their respective values for the alpha ventus measurement campaign

	Parameters	**Constraint or Binning**	**Values for data screening**	**Value for simulations**	**Seeds**
Wind condition	Wind direction	Free stream	207-275 degrees	270 degrees	-
	Mean hub wind speed	1m/s bins	3.5-23.5m/s	4-23m/s	-
	Turbulence intensity	Constrained	5.5%-6.5%	6%	9 turbulent seeds per wind bin
	Wind shear	Not binned or constrained.	None	0.14 power law exponent (design basis)	-
Wave conditions	Significant wave height	Constrained as function of wind speed	Bin is defined as a function of the fitted curve. This gives the relationship between the significant wave height and the binned wind speed. The bin will be +/- 0.5 m of the fitted significant wave height value for a given wind speed	For each wind speed bin, a significant wave height value is given by the best fit curve	-
	Peak spectral period	Constrained	6-8 seconds	7 seconds	-
Wind and wave	Misalignment	Constrained	-30 to +30 degrees	-30,0, +30 degrees	3
Sea currents	Current velocity	Not binned or constrained.	None	0 m/s	-
	Direction	Not binned or constrained.	None	-	-
Water level	Mean sea level	Not binned or constrained	None	27m	-
Air Properties	Density	Not binned or constrained	None	1.225 kg/m^3	- -
Marine growth	Thickness	Not binned or constrained	None	0.05 m	-
	Density	Not binned or constrained	None	1325 kg/m^3	-
Wind/Yaw	Misalignment	Not binned or constrained	None	-5,0,+5 degrees	3
Soil parameters	Scour	Not binned or constrained	Not available	None	-
	Stiffness and Damping	Not binned or constrained	Not available	Provided by manufacturer	-

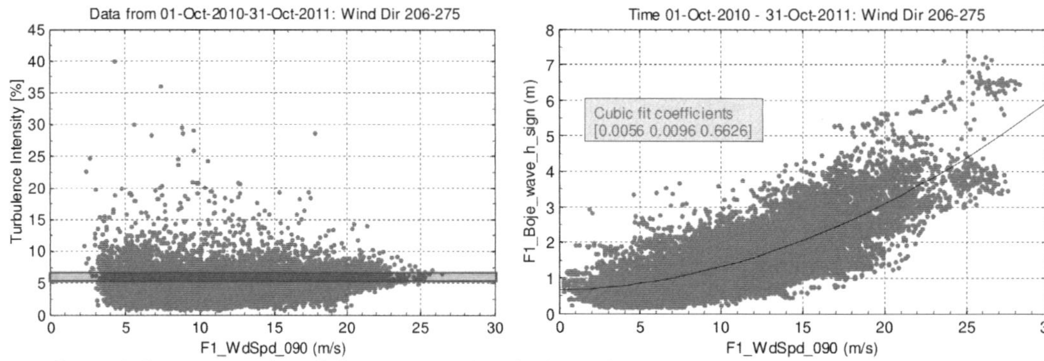

Figure 3: (a) left- scatter plot of the measured turbulence intensity (%) in relation to the 90 meters, 10 minute averaged wind speed at Fino 1. The constraint for data screening is the shaded area, (b) right - scatter plot of the measured significant wave height (m) in relation to the 90 meters, 10 minute averaged wind speed at Fino 1. The curve has been fitted to the data to establish a simulation value.

The method of data screening is not fixed in this case but left open for the validation engineer due to other factors such as data availability and measurement errors. It also leaves the possibility for the measurement and validation engineer to investigate different effects on the simulation model. For instance, high turbulence intensities or the effects of yaw-wave misalignment could be picked.

4. Comparison of results from measurements and simulations

To validate a computer model, different sensors or measured quantities have to be compared. Examples are shown in the following section. Here, the methodology for selection of data leads to a better representation of the measurements with simulation and thus higher certainty in the validation of the model.

As stated in section 2. , the initial comparison of the dominant frequencies and masses of the turbine and the computer model should be one of the first steps in the validation.

Table 2: Comparison of frequencies

	Simulation model eigenfrequency (Hz)	Measured excitation frequency (Hz)	Measurement conditions
1st edgewise	1.57	1.661	Normal operation
1st flapwise	0.835	0.91	Idle @ 3m/s wind
1st tower for-aft	0.333	0.335	Stop procedure

Since 3 seeds for wind and wave misalignment as well as 3 seed for wind- yaw misalignment have been chosen, their combination leads to 9 different seeds per wind bin, and a total of 180 simulations. Simple statistical values such as the mean, minimum, maximum and standard deviation can be compared. As well, for the case of the loads, the DELs can be computed in order to get a better understanding of the fatigue loading of the simulations and measurements. The following equation has been used to calculate the DEL,

$$S_{r,eq} = \left(\sum_{i=1}^{n} \frac{S_{r,i}^{m}}{N_{eq}} \right)^{1/m}$$

where S_r is the range of a load cycle and -1/m is the slope of the S-N line on log-log scale for the material of the measured component, N_{eq} is the number of equivalent cycles. This simplified equation does not carry a correction for the mean value of the load data.

4.1. Turbine characteristic curve

Before considering actual loads on the turbine, operational characteristics of the turbine, such as the power produced can be compared.

The left graph in Figure 4 shows good agreement in the simulations for the underrated region when compared to the measurements. Nevertheless, the simulations slightly under-predict the produced power for the region above rated. This can be explained due to the fact that the simulation model takes into account further losses, which are not measured by the exact positioning of the sensor in the alpha ventus turbine since the measurement point is directly at the generator. On the other hand, the standard deviation of the 10 minute measured generator power seems to fit very well above the rated wind speed, while at around rated wind speed there seems to be larger scattering of the measured data meaning more fluctuations of the produced power.

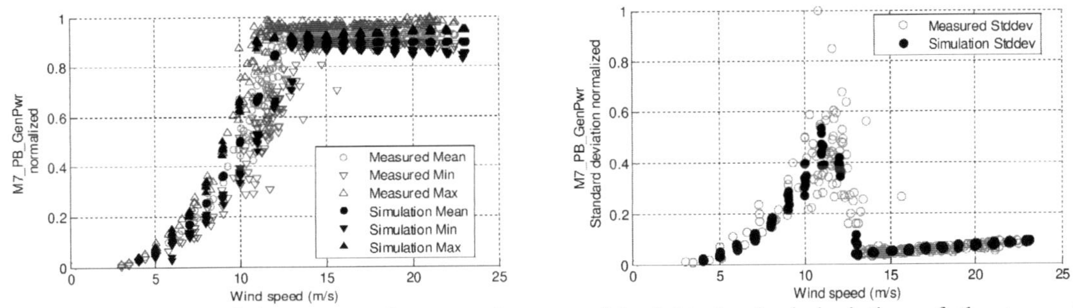

Figure 4: (a) left-minimum, maximum and mean, (b) right-standard deviation of the ten minute statistics for the power production.

4.2. Operational data

A good way of quickly checking the behaviour of the model is with the use of the operational parameters.

4.2.1. Generator Speed. The generator speed comparison shows the behaviour of the controller with regards to the wind speed.

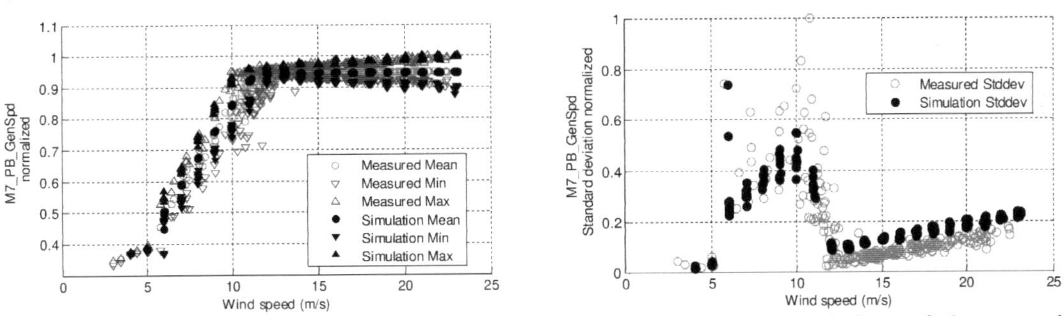

Figure 5: (a) left-minimum, maximum and mean, (b) right-standard deviation, of the ten minute statistics for the generator speed.

The minimum, maximum and mean from the statistics in Figure 5-left show very good agreement. The standard deviation above rated shows however that the fluctuations in the rotor speed are higher for the simulations. Also, the transition region programmed into the controller which tries to avoid the first excitation frequency of the tower is visible in the 6m/s wind speed simulation bin. In this case the screening of the data is too thorough and therefore at low wind there are few data points available for comparison.

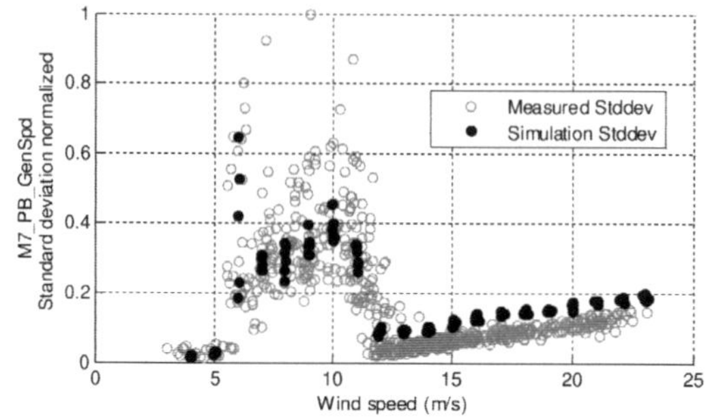

Figure 6: standard deviation of the ten minute statistics for the generator speed when wave spectral period screening criteria has been removed from the screening procedure.

In this case the validation engineer might decide to change some of the screening criteria in order to obtain more data for analysis. As an example the screening of the peak spectral period has been (which in hindsight should not have a large effect on the rotational speed of the rotor) is removed for the data plotted in Figure 6. Here it is seen that the simulations do behave as the real turbine does at wind speeds around 6m/s.

4.2.2. Pitch Angle. The pitch angle comparison also helps show the behaviour of the controller, especially at the region above rated wind speed. Figure 7-left shows again good agreement between measurements and simulations for the mean, minimum and maximums.

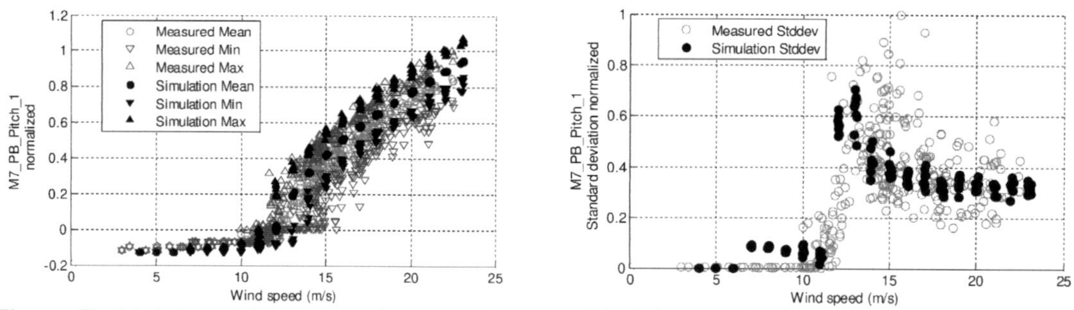

Figure 7: (a) left- minimum, maximum and mean, (b) right- standard deviation, of the ten minute statistics for the pitch of blade 1.

When taking a look at the results from the standard deviation, it is apparent that the pitch activity of the real turbine has a slightly different behaviour than that given by the simulations. At around 7 m/s wind speed there is a small amount of pitch activity in the simulations and above the rated wind speed the pitch activity of the real turbine can be more or less than that of the simulations. In this particular case, since the controller has been obtained by the manufacturer and is encrypted, a closer look at these results is not possible.

4.3. Loads

The most important part of the whole validation process is to show that the loads of the simulations are representative of the real loads seen by the turbine.

4.3.1. Fore-aft bending moment. The thrust of the rotor plays an important role on the fore-aft loading. Also, the wave loading on the underwater structure can influences the dynamics of the tower and substructure. Figure 8 shows that the simulations can reproduce the minimum, maximum and mean of the measurements very well, even though there is still some deviation at the region between 14-16 meters per second wind speed, which is clearly seen from the plot of the standard deviation.

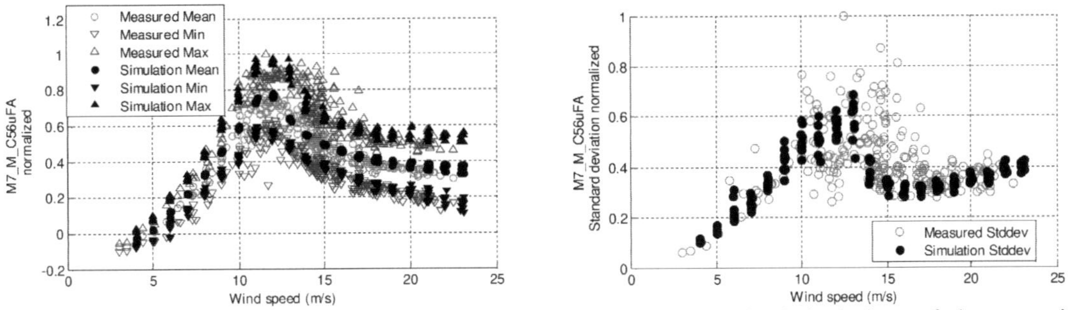

Figure 8: (a) left- minimum, maximum and mean, (b) right- standard deviation, of the ten minute statistics for the tower fore-aft bending moment.

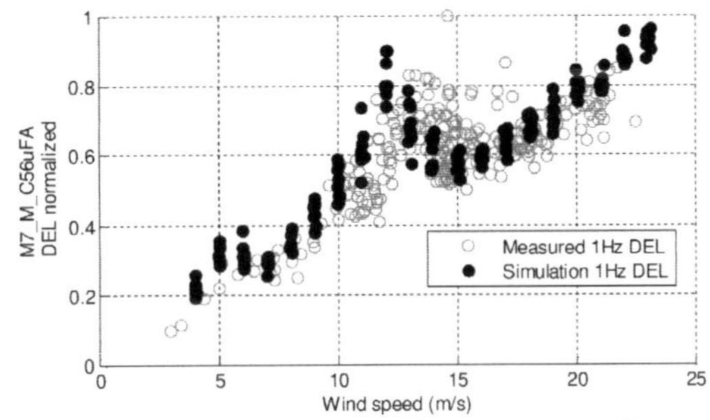

Figure 9: DEL of the ten minute statistics for the tower fore-aft bending moment.

The damage equivalent load for the fore-aft bending moment of the tower seem to agree well, with a slight over prediction of the DEL at rated wind speed and an under-prediction for the 14-16 m/s wind speeds.

4.3.2. Flapwise bending moment. The lift and drag caused by the aerodynamic profile of the blade influence the flapwise loading at the blade root. As the blade is pitched these force are changed and also the loading. Furthermore, turbulence also strongly influences the flapwise bending moment fluctuations.

The analysis of the flapwise bending moment shows the example of a situation where the validation engineer would have to carefully look at the simulation model to find the source of the discrepancies. In Figure 10- left there is a slight under-prediction of the statistics from the simulation at the region above rated wind speed, which could be linked to the slightly higher pitch angles of the simulations. The standard deviation also is over-predicted and does not show the scattering that is obtained in the measurements. This is also mirrored in the DELs, where the simulations cannot reproduce the scatter of the measurements in Figure 11.

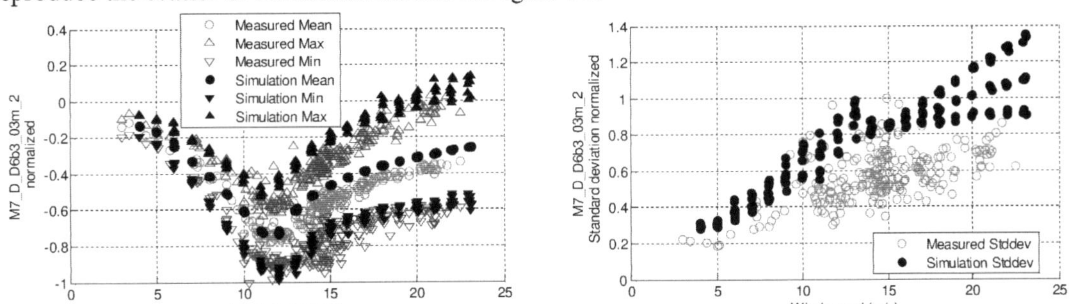

Figure 10: (a) left - minimum, maximum and mean, (b) right- standard deviation, of the ten minute statistics for the flapwise root bending moment.

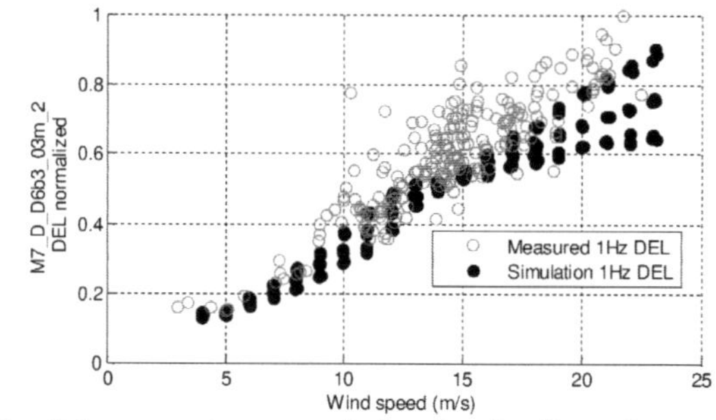

Figure 11: DEL of the ten minute statistics for the flapwise root bending moment.

5. Further Considerations

Particularly for the estimation of fatigue, the wake conditions in a wind park are important. Schmidt et al. [12] presented the validation of a wake meandering model with the data from alpha ventus. Here, an increase in the measured DELs is clearly seen for the case when turbine 7 is in the wake of other turbines. The main question when dealing with wake is, how to determine the inflow wind speed and turbulence intensity, since the nacelle anemometer of the turbine provide a distorted value of the measurements. It seems almost necessary in this case to carry out a measurement campaign with a nacelle Lidar in order to properly characterise the wind field of the wake as it approaches the wind turbine to be analysed, before one can simulate wake inflow on the computer models.

Weitjens et al. [13] also showed the increased in the measure loads on a monopile as a wave hits it. A validation for such cases might be better suited in the time domain with use of high resolution data, instead of statistics as in this paper. The methodology in this paper however, could be used to investigate the influence of the wave parameters, similar to how section 4.3. shows the trends with respect to the wind speed. Initial results from alpha ventus show, however, that the relationship between the statistics of the wave parameters and the loads is not apparent for the sensors at the splash zone. This is shown in Figure 12, where data is binning for wind speed and the statistics are plotted as a function of the significant wave height. The other filters mentioned in Table 1 are left unchanged.

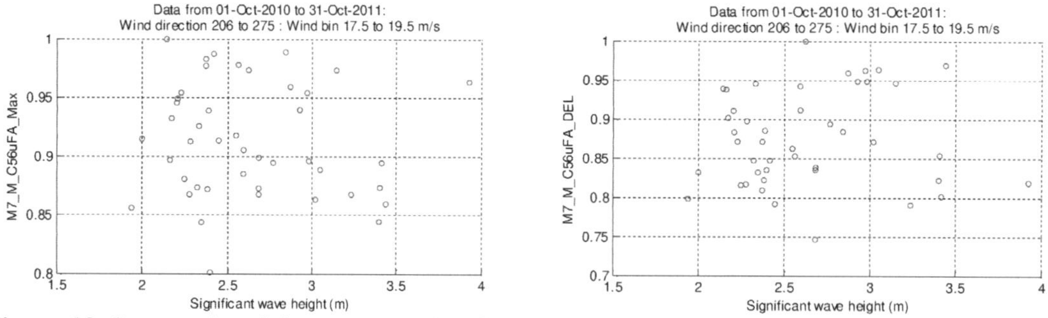

Figure 12: Scatter plot of the measured significant wave height against the for-aft maximum loads and the DELs for tower sensors at the splash zone.

6. Conclusions

Model validation of offshore wind turbines presents specific challenges which have to be addressed. Oceanographic parameters add complexity to the loading of the turbines and need to be properly modelled by the simulations tools. The work thus described the necessary meteorological and oceanographic parameters that are needed from an offshore measurement campaign. These are then used to implement a methodology of data screening for the statistical comparison of measurements and simulations. The methodology is shown to be helpful to define the simulation parameters that will be used and limits the influence of measured outliers as well as limits the spreading of the measured data created by the stochastic environment. The methodology itself is a flexible procedure since not all offshore turbines prototypes will be the same, and their measurement campaigns will not capture the same environmental conditions since these vary from location to location.

The Flex5-Poseidon coupled model of the alpha ventus 5 MW tripod mounted turbine agrees well with the measurement data, and serves as an example of the use of the methodology. The process thus shows how better management of the available data and data screening helps improve the validation process. The plotting of the screened statistics from the turbine data help the validation engineer determines the strengths and weaknesses of the simulation model.

Acknowledgments
The presented work was funded by the German Federal Ministry for Economic Affairs and Energy (BMWi) within the scope of the research project OWEA Loads as part of a Research at alpha ventus (RAVE) initiative. The support is gratefully acknowledged.

7. References

[1] International-Electrotechnical-Commission, *IEC 61400-3 Wind Turbines – Part 3: Design requirements for offshore wind turbines*, First Edition, ISBN 978-2-88910-514-4, 2009.

[2] International-Electrotechnical-Commission, *IEC 61400-13 Wind Turbines – Part 13: Measurement of Mechanical Loads, First Edition,* ISBN 978-2-8322-3087-9, 2015.

[3] J. Jonkman and W. Musial, "OC3 Task 23 Subtask 2 Final Report," National Renewable Energy Laboratory, IEA Wind Task 23, 2010.

[4] W. Popko, et al., "Offshore Code Comparison Collaboration Continuation (OC4), Phase I - Results of Coupled Simulations of an Offshore Wind Turbine with Jacket Support Structures," in *22nd ISOPE Conference*, Greece, 2012.

[5] J. Zierath, R. Rachholz, C. Woernle and A. Müller, "Load calculation on wind turbines: validation of Flex5, Alaska/Wind, MSC.Adams and Simpack by means of field tests," in *ASME International Design Technical & Computers and Information in Engineering Conference*, New York, 2014.

[6] M. Seidel, A. Curvers, D. Kaufer, M. Kühn and C. Böker, "Validation of Offshore load simulations using measurement data from the DOWNVinD project," in *European Offshore Wind*, Stockholm, 2006.

[7] D. Kaufer and P. W. Cheng, *Validation of an Integrated Simulation Method with High Resolution Load Measurements of the Offshore Wind Turbine REpower 5M at alpha ventus,* Journal of Ocean and Wind Energy Volume 1, No. 1, 2014.

[8] H. Söker, M. Damascke, C. Illig and N. Cosack, "A guideline to load validation," in *Deutsche Windenergie Konferenz*, Wilhelmshaven, 2006.

[9] Unknown, "alpha ventus," 2015. [Online]. Available: https://www.alpha-ventus.de/technik/. [Accessed 09 06 2016].

[10] D. Kaufer, N. Cosack, C. Böker, M. Seidel and M. Kühn, "Integrated Analysis of the Dynamics of Offshore Wind Turbines with Arbitrary Support Structures," in *EWEC*, Marseille, 2009.

[11] International-Electrotechnical-Commission, *IEC 61400-1 Wind Turbines - Part 1: Design requirements Third Edition,* Geneva, Zwitzerland, 2005.

[12] B. Schmidt, U. Smolka, S. Hartmann and P. W. Cheng, "Validation of the Dynamic Wake Meander Model with AREVA M5000 Load Measurements at alpha ventus," in *EWEA*, Vienna, 2013.

[13] W. Weijtjens, A. Iliopoulos, J. Helsen and C. Devriendt, "Monitoring the consumed fatigue life of wind turbines on monopile foundations," in *EWEA Offshore*, Copenhagen, 2015.

Monitoring Wind Turbine Loading Using Power Converter Signals

C A Rieg[1], C J Smith[1], C J Crabtree[1]

[1]School of Engineering and Computing Sciences, Durham University, Lower Mountjoy, South Road, Durham, DH1 3LE, UK

c.j.crabtree@dur.ac.uk

Abstract. The ability to detect faults and predict loads on a wind turbine drivetrain's mechanical components cost-effectively is critical to making the cost of wind energy competitive. In order to investigate whether this is possible using the readily available power converter current signals, an existing permanent magnet synchronous generator based wind energy conversion system computer model was modified to include a grid-side converter (GSC) for an improved converter model and a gearbox. The GSC maintains a constant DC link voltage via vector control. The gearbox was modelled as a 3-mass model to allow faults to be included. Gusts and gearbox faults were introduced to investigate the ability of the machine side converter (MSC) current (I_q) to detect and quantify loads on the mechanical components. In this model, gearbox faults were not detectable in the I_q signal due to shaft stiffness and damping interaction. However, a model that predicts the load change on mechanical wind turbine components using I_q was developed and verified using synthetic and real wind data.

1. Introduction

Extreme wind conditions such as gusts can lead to very large loads on the turbine that cause fatigue, shut-downs and damage to components such as the gearbox [1]. In response the condition of wind turbine components is monitored so that a developing fault can be detected and appropriate action taken. This allows maintenance to be scheduled before the impact on the system has become too large, resulting in lower downtimes and lower cost of energy (CoE) [2].

Condition monitoring (CM) techniques such as vibration and strain measurement require expensive sensors that are often impractical in the high-torque applications of wind turbines [3]. Using readily available signals from other areas of the turbine could prove an inexpensive alternative CM approach.

The power converter could provide this information for CM applications; the converter should respond to any disturbances and therefore its signals should show the drive train response. For example, the quadrature-axis component of the machine side converter (MSC) current signal (I_q) controls the real power flow and contains torsional information from the drive train. Monitoring I_q could provide useful information about torsional loads on components that could be used for early fault detection without extra sensors.

This investigation focuses on whether power converter signals can be used for CM with a focus on two potential applications:

1. Gear tooth failure detection.
2. Mechanical load estimation from damaging gusts.

2. Approach

To carry this work out a drive train model was required. The model developed at Durham in Simulink [4] was used. It is a drivetrain model of a fully rated, direct-drive 2MW permanent magnet synchronous generator (PMSG) wind turbine with two voltage sources connected to ground simulating the DC link. To make this model suitable for this study the following modifications were made:

1. A full grid-side converter (GSC) was added for a more realistic converter model.
2. A gearbox was added.
3. A gearbox fault model was used to provide fault conditions.
4. A gust model was added to provide data for load prediction.

A schematic of the final model is shown in figure 1. This section outlines how these aspects were modelled. Modifications to the PMSG and MSC as a result of including a gearbox are also detailed.

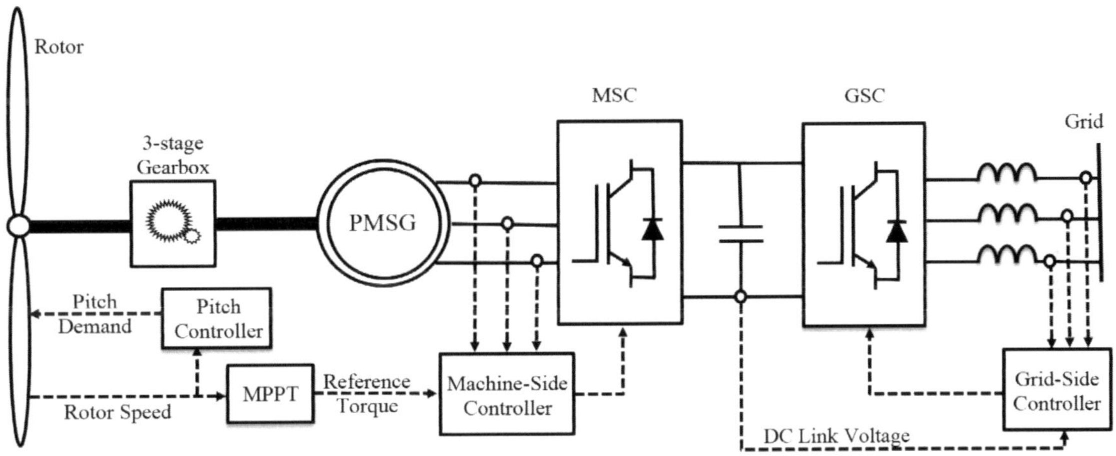

Figure 1. Schematic of the 2MW geared PMSG wind energy conversion system. MPPT stands for maximum power point tracking.

2.1. Grid-Side Converter

The main objective of the GSC is to control power flow between converter and grid to maintain a constant DC link voltage regardless of the power input from the MSC (figure 1). In this configuration the GSC acts as an inverter and the MSC acts as a rectifier. The GSC was modelled as a 2-level insulated-gate bipolar transistor (IGBT)/diode pair active inverter. In the model the 'Universal Bridge' block from the Simulink library was used with the power electronic device set to 'IGBT/Diodes'. It is controlled using the 'PWM Generator (2 level)' block that takes the voltage from the grid side controller as the modulating input signal. The DC-link voltage is $1150V_{DC}$. The grid is represented as ground connected to a three phase programmable voltage source connected to the GSC via inductors.

To control the GSC, vector control was chosen as it is able to respond to transient events more robustly than load angle control [5]. Figure 2 outlines the control schematic for the GSC. I_d is the direct-axis current, V_d is the direct-axis voltage, ω is the grid frequency (rad/s), L is the grid inductance, V_{DC_link} is the DC-link voltage, V_q is the quadrature-axis voltage, $V_{d,r}$ is the converter reference V_d, $V_{q,r}$ is the converter reference V_q, and V_0 is the 0-component voltage. To convert between 3-phase sinusoidal and direct-quadrature-zero (dq0) reference frames the Park and inverse Park transforms were used.

2.2. Gearbox Model

The gearbox is connected to the hub via the low-speed shaft and to the generator via the high-speed shaft. It increases the speed of the incoming turbine speed to the desired generator speed while reducing the torque by a gear ratio N_{GB}. The dynamic interactions of the rotor, gearbox and generator were modelled as a 3-mass model. Higher order models were considered, however no data was found and the 3-mass model represents the dynamic interactions of the rotor, gearbox and generator adequately for this project. The 3-mass model is shown in figure 3. J_R is the rotor moment of inertia, J_{GB} is the gearbox moment of inertia, $J_{m1,2}$ are the equivalent moments of inertia for the low and high speed gear sections respectively, $T_{m1,2}$ are the equivalent mechanical torques for the low and high shafts respectively and J_g is the generator moment of inertia.

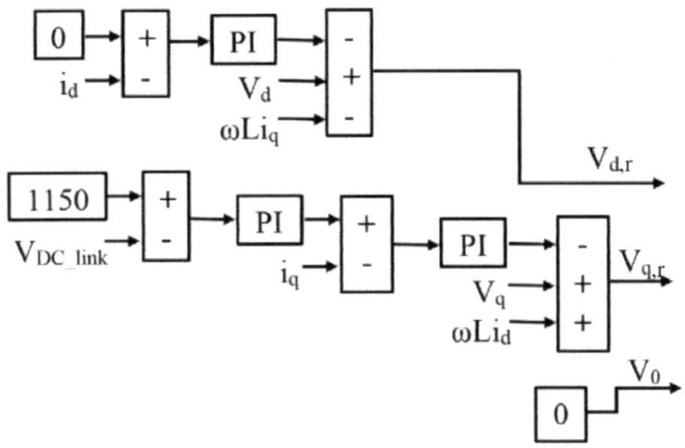

Figure 2. Vector control scheme for the GSC.

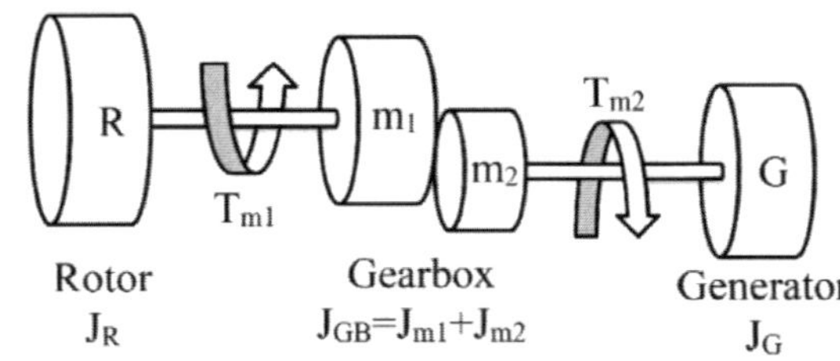

Figure 3. Schematic of the 3-mass model dynamics.

The first mass in the rotor, the second mass is the gearbox and the third mass is the generator. The model uses the principles of a mass-spring-damper system where each mass has inertia J and each shaft a stiffness K and viscous damping B. The second mass (gearbox) is divided into two parts that are related through N_{GB} to represent the difference in speeds of each gear. As such the resulting $T_{m1,2}$ and rotational speeds of the various components (ω) can be represented using equations (1-6).

$$T_{m1} = B_1(\omega_r - \omega_{m1}) + K_1\int(\omega_r - \omega_{m1})dt \tag{1}$$

$$T_{m2} = B_2\left(\omega_{m2} - \omega_g\right) + K_2\int\left(\omega_{m2} - \omega_g\right)dt \tag{2}$$

$$\omega_r = \int\frac{T_r - T_{m1}}{J_r}dt \tag{3}$$

$$\omega_{m1} = \int\frac{T_{m1} - T_g N_{GB}}{J_{GB}}dt \tag{4}$$

$$\omega_{m2} = \int\frac{T_{m2} - T_r / N_{GB}}{J_{GB}}dt \tag{5}$$

$$\omega_g = \int\frac{T_e - T_{m2}}{J_g}dt \tag{6}$$

Where $B_{1,2}$ are the viscous damping of the low and high-speed shaft respectively, $K_{1,2}$ are the shaft stiffnesses of the low and high-speed shafts respectively, ω_r is the rotational speed of the rotor, $\omega_{m1,2}$ are the rotational speeds of the low and high-speed gear components respectively, ω_g is the generator rotational speed. T_r is the rotor torque, T_g is the generator torque, and T_e is the electromechanical torque.

The torque and speed across the rotor and generator are related through the gearbox ratio, N_{GB} using equation (7).

$$N_{\mathrm{GB}} = \frac{\omega_{\mathrm{g}}}{\omega_{\mathrm{t}}} = \frac{T_{\mathrm{r}}}{T_{\mathrm{e}}} \tag{7}$$

Due to the new torque and speed in the generator from [4], changes of the PMSG were made to accommodate the current and voltage requirements. To keep the current and voltage outputs the same equations (8) and (9) were used. The number of poles was reduced to 4 because the generator has a rotational speed of 1500rpm. The flux density was changed to 1.611Vs and the armature inductance was changed to 0.4mH.

$$I_{\mathrm{q}} = \frac{4T_{\mathrm{m2}}}{3p\varphi} \tag{8}$$

$$V_{\mathrm{d},m} = L_{\mathrm{d}}\omega I_{\mathrm{q}} - I_{\mathrm{q}}R_{\mathrm{s}} \tag{9}$$

p is the number of generator poles, φ is the generator flux linkage, $V_{d,m}$ is the MSC direct-axis voltage, L_d is the direct-axis generator inductance, and R_s is the stator resistance. The data for the 3-mass model has been taken from research papers and is given in Appendix B.

2.3. Gearbox Fault Model

The most severe gearbox failure modes that arise from extreme wind conditions have been identified as fretting corrosion and high cycle bending fatigue [6]. Fretting corrosion is the deterioration of contacting gear tooth surfaces as a result of vibratory motion between teeth and is this study's focus.

The gear friction coefficient varies according to three different types of surface structure: adhesion, unevenness and wear [7]. Friction losses in the gears are part of the normal force exerted by each gear at the point of contact F_{N} as a friction factor μ. Due to difficulties involved in the estimation of the μ with lubrication it is often assumed constant [8] and was not used in this project due to a lack of relevant experimental data in the literature. As gears are well lubricated this assumption was deemed satisfactory.

Instead, the gear wear impact on stiffness was considered. The effect of tooth wear on the mechanics of the system has previously been examined and it was found that gear tooth wear causes a reduction in the stiffness of the gear. It was found that it can be modelled as a rectangular pulse wave or a half sine function. The half sine wave function is used in detailed gearbox models that include the gear meshing process in their calculations [9]. For this model the rectangular pulse function was chosen as it represents the fault accurately for the purpose of this investigation.

The reduction of the gear tooth stiffness can be calculated according to equation (10).

$$K_{\mathrm{wear}} = K_{\mathrm{g}} l_{\mathrm{w}} A \tag{10}$$

Where K_{wear} is the wear stiffness, K_{g} is the hertz contact stiffness, l_{w} is the wear length and A is the amplitude of wear.

Typical values of l_{w} are between 1 and 2mm and A typically has a value between 0 and 1 [9]. The contact stiffness with a wear fault present, $K_{\mathrm{g,wear}}$, is given as the difference between the non-faulty gear stiffness and the wear stiffness as in equation (11) [9].

$$K_{\mathrm{g,wear}} = K_{\mathrm{g}} - K_{\mathrm{wear}} \tag{11}$$

The relationship between the contact stiffness of the gears and the stiffness of the shaft can be modelled as springs connected in series. The total stiffness K_{Total} is calculated from K_{g} and the shaft stiffness K_{S} as in equation (12).

$$K_{\mathrm{Total}} = \frac{1}{\dfrac{1}{K_{\mathrm{g}}} + \dfrac{1}{K_{\mathrm{s}}}} \tag{12}$$

The effect of tooth wear in the gearbox was modelled as a reduction in the total stiffness every time there is contact with a worn gear tooth as shown in figure 4. The total stiffness is applied across the shaft in the model.

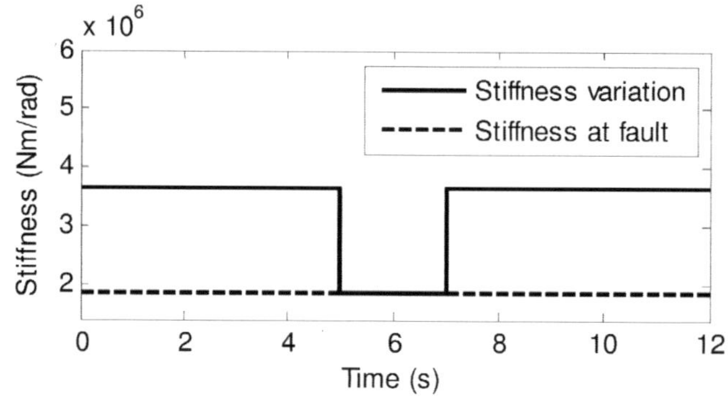

Figure 4. The stiffness relationship in a worn gear.

Other faults such as gear cracks have been modelled using a periodic cosine based variation in shaft stiffness given in (13) where K_{crack} is the reduction in shaft stiffness due to the crack that can be calculated using finite element analysis [9, 10]. The underlying calculations for a crack fault and a wear fault are very similar as they both rely on a periodic reduction in the shaft stiffness due to a fault.

$$K_s = \frac{1-\cos(\omega t)}{2} K_{crack} \tag{13}$$

Gearbox faults are often modelled as a periodic variation in tooth stiffness to indicate the presence of a fault. As a widely used, well-established method of modelling faults and experimental data available, it was chosen in this project.

A typical gearbox in a wind turbine has 3-stages with a planetary gearbox at the first stage, coupled to two parallel gearboxes at the second and third stage [11]. Due to the speed dependency of the gear fault model, faults can be introduced into any of the gear stages. Appendix A gives a summary of the gear ratios and output speeds corresponding to the individual stages.

2.4. *Gust Model*

Existing gust models rely on real wind data to model the amplitude, duration and gust shape introduced along with a running average wind speed [12, 13]. These wind gust profile characteristics can be extracted and applied using square or cosine shaped wind profiles that have a gust amplitude, duration and frequency. The maximum gust speed ($U_{G,max}$) in a given time period is calculated from the gust factor $G(t)$ in equation (14). An expression for the gust factor is given in equation (15) [14].

$$U_{G,max} = G(t) U_W \tag{14}$$

$$G(t) = 1 + 0.42 I_u \log_e \left(\frac{3600}{t_G} \right) \tag{15}$$

Where U_w is the mean wind speed, I_u is the longitudinal turbulence intensity, and t_G is the gust duration.

The International Electrotechnical Commission (IEC) has divided the value for turbulence intensity into three categories - higher, medium and lower turbulence characteristics with values of 0.16, 0.14 and 0.12 respectively [15]. The underlying square wave gust characteristic was used as the basis for all gust analysis.

For the load prediction model gust, 10 gust categories were defined, each representing a reduction in the gust wind speed (table 1).

Journal of Physics: Conference Series **749** (2016) 012018 doi:10.1088/1742-6596/749/1/012018

Table 1. Gust category assignment.

Gust Category	U_G	Gust Category	U_G
1	$(U_G+U_W) + U_W$	6	$[(U_G + U_W)/2.25] + U_W$
2	$[(U_G + U_W)/1.25] + U_W$	7	$[(U_G + U_W)/2.5] + U_W$
3	$[(U_G + U_W)/1.5] + U_W$	8	$[(U_G + U_W)/3] + U_W$
4	$[(U_G + U_W)/1.75] + U_W$	9	$[(U_G + U_W)/4] + U_W$
5	$[(U_G + U_W)/2] + U_W$	10	$[(U_G + U_W)/6] + U_W$

3. Results

The section presents and discusses the results of gearbox fault detection using converter signals (section 3.1), and estimating turbine drive train loads from gusts using converter signals (section 3.2).

3.1. Gearbox Fault Detection

Gearbox wear faults were introduced using the method outlined in section 2.3. The first fault was introduced as a wear fault with wear amplitude 0.5 and wear length 1mm present on every other tooth, giving a fault frequency of 1.72Hz. The incoming wind speed was constant at 7m/s. The parameters used to introduce the first fault in the second gearbox stage of the gearbox are detailed in Appendix B.

By taking the Fast Fourier Transform (FFT) of the MSC I_q signal the frequency spectrum was computed to identify differences between the 'healthy' (no fault) and faulty spectrum. Figure 5 shows the frequency response of the MSC I_q signal in its 'healthy' and faulty state as well as the amplitude difference between the healthy and faulty state. It can be seen that there is no clear difference in the spectrum at the fault frequency. There is a small difference at 2Hz, where both the healthy and the faulty spectrum show a spike due to control errors.

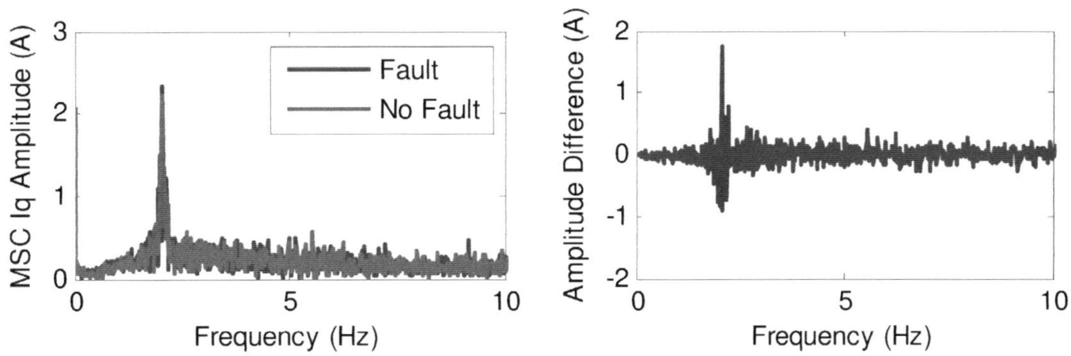

Figure 5. MSC I_q frequency response

It was investigated why the fault does not appear in the MSC I_q frequency spectrum by looking at the frequency spectrum of the relevant torque components. The torque across the high speed shaft is an input to the PMSG and is used to determine the MSC I_q and is result of the addition of the torque due to stiffness (T_K) and the torque due to damping (T_B). Figure 6 shows the frequency spectrum of each of these individual torque signals in their 'healthy' state and their faulty state. It can be seen that the fault is visible in the frequency spectrum of T_K and T_B (figure 6), yet is no longer visible in the resulting total torque spectrum (figure 5).

To understand the impact of the damping and stiffness components on the fault frequency response, the time sequence of T_K and T_B was monitored with the fault present (figure 7). The time sequences showed that the oscillatory motion of T_K due to the fault is counteracted by an opposite oscillatory motion from T_B removing the oscillation due to the fault from the frequency spectrum.

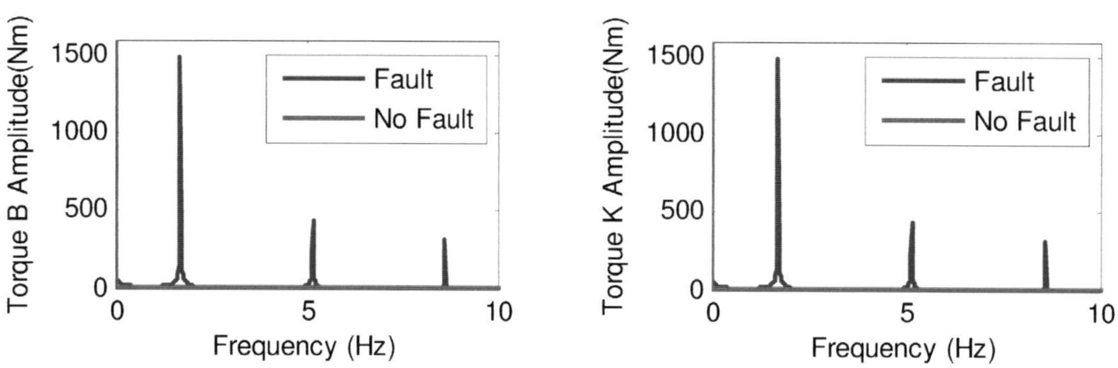

Figure 6. Frequency spectrum of mechanical torque components.

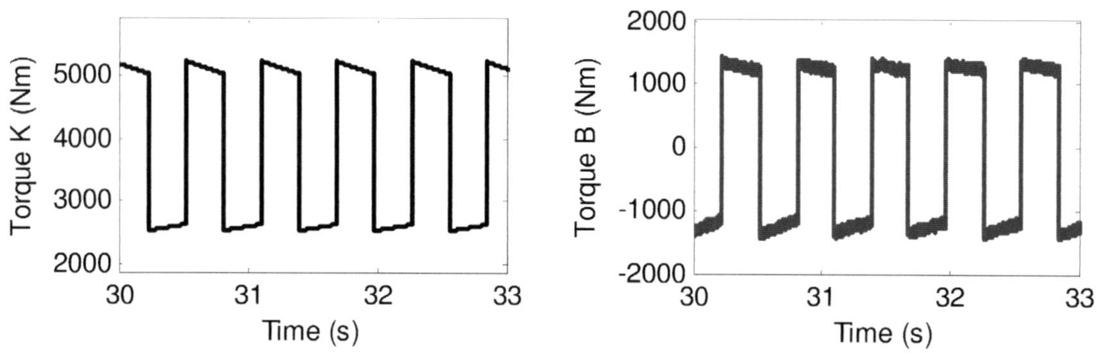

Figure 7. Temporal spectrum of mechanical torque components.

The amplitude of the torque due to damping counteracts the amplitude of the torque due to stiffness exactly, resulting in a critically damped system. The gearbox was modelled analogous to a mass-spring-damper system. In this system the role of the damper is to reduce or prevent oscillations. The fault amplitude was varied in the full range of 0 to 1 and the wear length was varied in the full range of 1mm to 2mm and the input speed was varied. However in each case the system remained critically damped, resulting in the fault not appearing in the MSC signals.

In a real gearbox the torque due to damping and torque due to stiffness cannot be measured separately as they have been in this model. In a real gearbox the system parameters might not be as perfectly balanced as in this modelled system and the damping might not have the same effect as in this model. Thus there is a possibility that faults can be detected in the MSC I_q signal of a real gearbox system where the components and parameters are not as balanced as in this drive train model.

3.2. Load prediction on mechanical components using MSC signals

The MSC converter signal spectrum changes with the incoming wind speed and wind pattern. Wind gusts at varying frequencies appear clearly on the spectrum and can be monitored using the MSC signals. Figure 8 shows the variation of the frequency spectrum as the gust frequency of the incoming wind is varied at a mean wind speed of 7.5 m/s using the maximum gust speed.

Simulations were done at different speeds and constant gust frequency of 3Hz. A relationship between the MSC I_q amplitude and the difference in rotor torque magnitude was derived for each gust category using simulation results as data points. The result for the first, second and third gust categories are shown in Figure 9 with equations (16-18) representing their relationship respectively. ΔT_r is the change in mechanical load on the rotor in kNm.

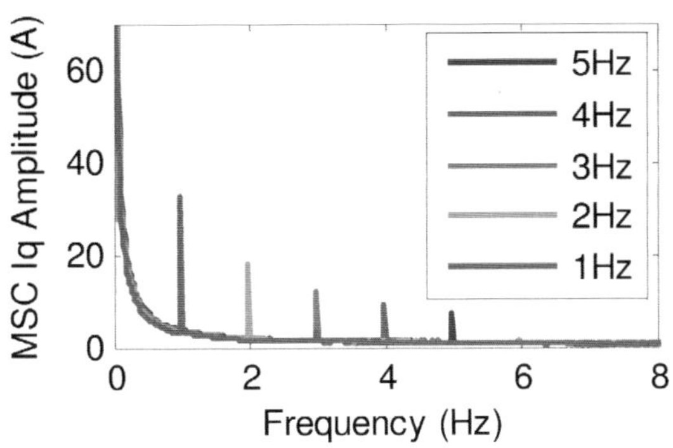

Figure 8. MSC I_q frequency spectrum for different gust frequencies.

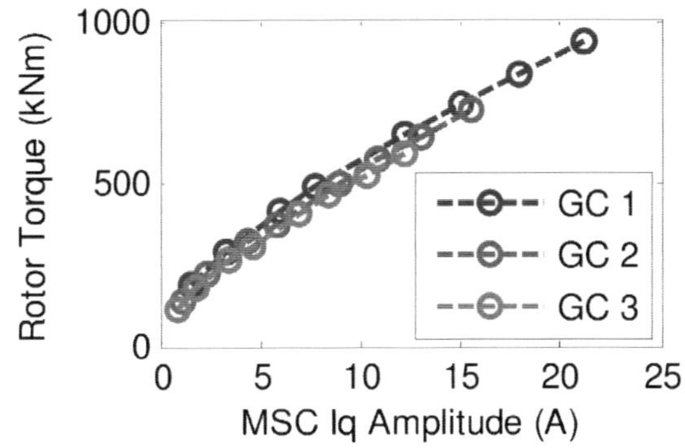

Figure 9. ΔT_r vs MSC I_q amplitude for different gust categories.

$$\text{GC 1: } \Delta T_r = 137.7 I_q^{0.6238} \tag{16}$$

$$\text{GC 2: } \Delta T_r = 130.21 I_q^{0.6206} \tag{17}$$

$$\text{GC 3: } \Delta T_r = 123.92 I_q^{0.6176} \tag{18}$$

With this information, a load prediction model can be constructed. The proposed model works on the basis that the wind speed and MSC signal amplitudes can be measured. A flowchart of its operating principle is illustrated in Figure 10. The wind is monitored and depending on the mean wind speed and gust magnitude it can be assigned a gust category. Each gust category has an equation relating the change in torque and the MSC current amplitude for an assigned frequency range.

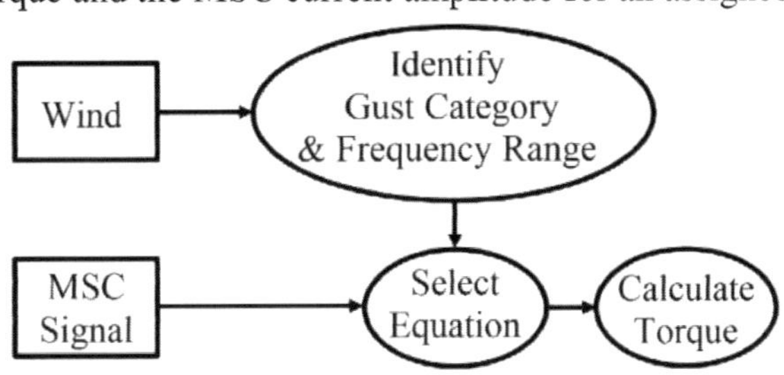

Figure 10. Flowchart of the load prediction model.

Depending on the gust category and gust frequency, an equation is selected from which ΔT_r can be calculated. The frequency ranges become smaller as the gust frequency decreases because the change in I_q amplitude increases. Severe load changes can then be counted to estimate the mechanical fatigue.

In order to verify the functionality of this method a variety of ideal category 1 gusts with different mean wind speeds were inputted into the model. The MSC I_q FFT amplitude was measured and the expected load on the rotor was calculated according to equation ($\Delta T_{r,est}$) (16). $\Delta T_{r,est}$ was compared to the measured torque from the simulation ($\Delta T_{r,sim}$) using the percentage error.

The results are summarised in table 2. The percentage error between the measured and the calculated error is very small, below 1%. This shows that the model is able to predict the load on the mechanical components in the wind turbine drive train through MSC signal measurements adequately.

The model was tested using real wind data from the anemometer on a 1.5MW variable speed wind turbine in order to investigate the accuracy of the model using a real, non-ideal wind characteristic. The data was identified as GC 10 and frequency 0.29Hz. The equation relating $\Delta T_{r,est}$ and MSC I_q in this case is given by (19). Table 2 gives a summary of the results.

Table 2. Model Verification and response to real wind input.

U_W (m/s)	MSC I_q (A)	$\Delta T_{r,est}$ (Nm)	$\Delta T_{r,sim}$ (Nm)	% Error
		Verification (GC 1)		
5.5	4.536	353636.5	350370	0.92
6.2	6.648	448868.5	445450	0.76
7	9.84	573264.2	568160	0.89
8.2	16.344	786715	780330	0.81
		Real wind input (GC 10)		
8.4	11.82	98978.5	104860	5.9

$$\Delta T_{r,est} = 15.739 I_q^{0.7445} \quad (19)$$

The percentage error for the real wind is higher than for the ideal wind. This is expected as the real wind gusts have a larger variation in duration and magnitude. The frequency of the gusts in the real wind characteristic is not as clear as in the ideal characteristic. The frequency categories allow for some variation that increases the percentage error. For a mean wind speed of 8.5m/s with gusts of frequency 0.29Hz the difference between the maximum (GC 1) and minimum (GC 10) change in rotor torque is 555910Nm. The difference between the calculated and measured rotor torque from Table 3 is 5881.9Nm, which is 100 times smaller than the difference between GC1 and GC10. This indicates that the model has the ability to estimate the change in load using real wind characteristics well.

4. Conclusion

CM of wind turbine components allows appropriate action to be taken to minimise the impact of developing faults but currently requires expensive sensors and data acquisition devices. This paper investigates whether converter signals, which are already monitored by turbine controllers, can be used for CM.

A drive train model was modified to include a gearbox, GSC and gearbox fault model to determine whether gearbox faults could be detected in the converter signals. Gusts were also modelled to determine if drive train mechanical loading could be predicted using converter signals. The conclusions from this study are:

- Gear wear cannot be detected in the MSC signals due to the model damping effects. However, physical testing should be carried out to explore the impact of non-ideal dynamics.
- A model using MSC signals successfully predicted the load changes in the turbine with a percentage error < 1% under ideal wind conditions, and <6% for a real wind speed case.

Further investigations into the magnitude of load changes that cause mechanical component damage could lead to the application of this accurate MSC-based load prediction model to prevent gearbox faults through turbine shutdown during damaging wind conditions.

Appendices

Appendix A. 3-stage gearbox gear ratios.

	Stage 1	Stage 2	Stage 3
Gear type	Planetary	Parallel	Parallel
Gear ratio	1:16.667	1:2	1:2
Output speed	375rpm	750rpm	1500rpm

Appendix B. Data for 3 mass model and gear faults.

Parameter	Value	Ref	Parameter	Value	Ref
J_r	2.92×10^6 kgm^2	[16]	K_2	2.29×10^8 Nm/rad	[18]
J_g	200 kgm^2	[17]	K_g	3.715×10^6 Nm/rad	[17]
J_{GB}	190 kgm^2	[17]	K_{wear}	1857.5 Nm/rad	(10)
B_1	6.72 Nms/rad	[4]	$K_{g,wear}$	3713143Nm/rad	(11)
B_2	6.72 Nms/rad	[4]	K_{Total}	3655585 Nm/rad	(12)
K_1	4.00×10^7 Nm/rad	[16]	$K_{Total,wear}$	1842070 Nm/rad	(12)

References

[1] Kanev S and van Engelen T 2010 Wind turbine extreme gust control *Wind Energy* **13** pp 18-35
[2] Han Y and Song Y 2003 Condition monitoring techniques for electrical equipment-a literature survey *IEEE Trans. Power Del.* **18**(1) pp 4-13
[3] Yang W, Tavner P, Crabtree C and Wilkinson M 2010 Cost-effective condition monitoring for wind turbines *IEEE Trans. Ind. Electron.* **57**(1), pp 263-71
[4] Smith C, Wadge G, Crabtree C and Matthews P 2015 Characterisation of electrical loading experienced by a nacelle power converter *EWEA Annual Event 17-20 Nov. (Paris)*
[5] Anaya-Lara O, Jenkins N, Ekanayake J, Cartwright P and Hughes M 2009 *Wind Energy Generation: Modelling and Control* (West Sussex: Wiley)
[6] Sheng S and McDade M 2011 Wind turbine failure modes *IJTC 23-26 Oct. (Los Angeles)*
[7] Mahdavian S and Mai Y 1984 Further study in friction, metallic transfer and wear debris of sliding surfaces *Wear* **95** pp 35-44
[8] Brethee K, Gao J, Gu F and Ball A 2015 Analysis of frictional effects in the dynamic response of gear systems and the implications for diagnostics *ICAC 11-12 Sep. (Glasgow)*
[9] Cui L and Cai C 2015 Nonlinear Dynamics of a Gear-Shaft-Bearing System Breathing Crack and Tooth Wear Faults *Open Mech J* **9** pp 483-491
[10] Kidar T, Thomas M, El Badaoui M and Guibault R 2013 Control of phases by ESPRIT and WLSE methods for the early detection of gear cracks *3rd International Congress on Engineering Industrial Risk 3-5 Jul. (Riems)*
[11] Feng Y, Li J, Qiu Y, Yang W and Infield D 2014 Study on order analysis for condition monitoring wind turbine gearbox *RPG 24-25 Sep. (Naples)*
[12] Sheridan P 2011 Review of techniques and research for gust forecasting and parameterization *Forecasting Research Technical Report 570* (Exeter: Met Office)
[13] Verheij F, Cleijne J and Leene J 1992 Gust modelling for wind loading *J WIND ENG IND AEROD* **42**(1-3) pp 947-958
[14] Burton T, Jenkins N, Sharpe D and Bossanyi E 2011 *Wind energy handbook* (West Sussex: Wiley).
[15] IEC 2005 Wind turbines-Part 1: design requirements *IEC 61400-1:2005(E)*
[16] Akhmatov V 2003 Analysis of dynamic behaviour of electric power systems with large amount of wind power *Electric Power Engineering Ørsted-DTU Thesis*
[17] Jiang B 2010 Modelling of dynamics of driveline of wind stations *Chalmers University of Technology Master's Thesis*
[18] Ackermann T 2005 *Wind power in power systems* (West Sussex: Wiley)

Modelling the failure behaviour of wind turbines

S Faulstich, V Berkhout, J Mayer and D Siebenlist

Fraunhofer Institute for Wind Energy and Energy System Technology, Königstor 59, Kassel, Germany

stefan.faulstich@iwes.fraunhofer.de

Abstract. Modelling the failure behaviour of wind turbines is an essential part of offshore wind farm simulation software as it leads to optimized decision making when specifying the necessary resources for the operation and maintenance of wind farms. In order to optimize O&M strategies, a thorough understanding of a wind turbine's failure behaviour is vital and is therefore being developed at Fraunhofer IWES. Within this article, first the failure models of existing offshore O&M tools are presented to show the state of the art and strengths and weaknesses of the respective models are briefly discussed. Then a conceptual framework for modelling different failure mechanisms of wind turbines is being presented. This framework takes into account the different wind turbine subsystems and structures as well as the failure modes of a component by applying several influencing factors representing wear and break failure mechanisms. A failure function is being set up for the rotor blade as exemplary component and simulation results have been compared to a constant failure rate and to empirical wind turbine fleet data as a reference. The comparison and the breakdown of specific failure categories demonstrate the overall plausibility of the model.

1. Introduction

Operation and Maintenance (O&M) for wind turbines has evolved as a sector with important levers to increase availability, energy yield and thereby increasing financial profits of wind farm projects. At the same time O&M costs are to be lowered constantly, mainly by shifting towards a more preventive approach on turbine O&M.

Especially in the offshore sector a wide range of O&M simulation and modelling tools are available on the market to support decisions on O&M strategy and resources like ships and personnel to be allocated in order to achieve the most efficient operational setup. The development of these tools is being supported by public research projects and industry partnerships.

While these tools strive hard to model the full range of the complexity of offshore O&M processes including weather and wave conditions and restrictions for vessel operations on the wind turbine service side there remains a large uncertainty on the demand side which is the characteristics of failure behaviour of wind turbines.

2. Existing failure models in offshore O&M tools

Existing offshore O&M tools have been compared in their capabilities and results by [1]. In the reference case for the comparative study the failure behaviour was modelled by applying a constant failure rate for five different failure categories. This simplified approach is the basic starting point for modelling the failure behaviour, which enables comparing different tools.

Journal of Physics: Conference Series **749** (2016) 012019 doi:10.1088/1742-6596/749/1/012019

ECN O&M Cost Estimator

The ECN O&M Cost Estimator uses the mean time to failure (MTTF) on a component level to represent the failure characteristics of wind turbines. Failure modes are grouped by components and main systems based on the RDS-PP structure [2]. Maintenance actions are categorized into different categories and fault type classes with individual resource use and costs associated. The overall number of failures is taken is required as input from the user and may be taken from internal experience or published literature. These input parameters may be determined or updated by using the O&M data analyser, a complementary tool, which analyses the operational data structured in an event list and provides information on the frequency of failure events per components [3]. Additionally it shows confidence intervals for the failure rates that have been derived from empirical data [4].

Far Offshore Wind Interactive Tool - University of Strathclyde

The Far Offshore Wind Interactive Tool (FOWIT) developed by the University of Strathclyde references a study by Raademakers and Braam [5] for failure rates of offshore turbines. The failure model allows choosing between a subsystem or a five repair categories approach with different durations for repair actions and logistic efforts. Statistical distributions such as power law and Poisson processes in order to model the bathtub curve with early failures, intrinsic failures and deterioration may be applied [6].

NOWIcob / SINTEF

The NOWIcob model by SINTEF assigns maintenance strategies to maintenance tasks and includes corrective and time-based maintenance. It considers the main systems of wind turbines with a set of maintenance tasks each. Poisson processes are used to distribute the failure rate over time [7].

MAINTSYS™ / Shoreline

The Offshore Wind Simulation Model by the University of Stavanger, which has been taken forward as the commercial software MAINTSYS™ also use a homogenous Poisson function to model the dynamic behaviour of the failure rate. The wind turbine is subdivided into 19 components according to the RDS-PP structure and occurring failures are categorized by using the failure type classification [8].

SIMLOX / SYSTECON

The commercial software Simlox from the Swedish company Systecon also applies a Poisson process to model the failure characteristics. The wind turbine is subdivided into subsystems with failure modes and related failure mode frequencies for condition based and preventive maintenance. Calendar time and operational time are the underlying factors. Maintenance actions such as inspection, repair or replacement for each failure modes are defined and may consist of several sub processes. For preventive maintenance measures anticipation and deferment times are set [9].

Discussion

Although these approaches succeed in varying the failure rate over time they neglect a number of known characteristics of failure behaviour. They rely on time-based distribution only. Thus they do not take into account the physical mechanisms in the modelled failure modes and do not reflect the characteristics visible from empirical operation and event data.

Results of FMEA studies have been published by [10], [11], [12] and show the most critical failure modes and components. Among these are converter, pitch and yaw systems and the rotor blades with their respective failure modes.

Empirical data from the University of Strathclyde [13], Reliawind [14] and the WInD-pool [15] also provide data on the affected components. Moreover, the German scientific measurement and evaluation programme (WMEP) also shows statistics of failure causes giving shares of component failures and external causes like storm, lightning or icing. There is a seasonal pattern for the occurrence of failure due to these causes and it is also a known correlation that the failure probability increases at higher wind speeds [16], [17], [18].

These effects are relevant for offshore O&M modelling as they affect the accessibility of turbines at the time the failure occurs. Therefore are more detailed approach is proposed in this abstract. Recent research papers have already indicated that further parameters on the underlying physical failure mechanisms are to be included in the failure modelling:

Holmstrøm has shown a validation of the rate of occurrence of failures used in the non-homogenous Poisson process models with real farm data on a yearly basis over a 20-year lifespan and found a time dependent rate to match better [19]. A detailed approach including covariates, unobserved heterogeneity and seasonality has been outlined by [20] based on data analysis of the WMEP.

Building the link to the physical failure mechanisms Asgarpour and Sørensen suggest applying deterioration curves for structural components, discrete degradation functions for mechanical components and bathtub curves for electrical components [21]. Jürgensen has developed a model adding condition monitoring data to calculate individual failure rates for components [22].

3. Modelling different failure mechanisms of wind turbines

The proposed approach builds on a reliability block diagram of the wind turbines sub-assemblies and components. Additionally an extra layer is inserted into the block diagram to represent the failure categories for different failure mechanisms. The overall framework is illustrated in the following figure.

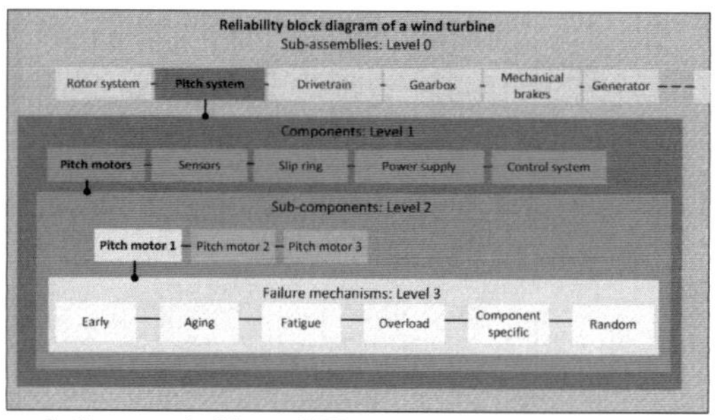

Figure 1: Framework for modelling different failure mechanisms of wind turbines

3.1. Turbine structure

A wind turbine is a system of over one thousand components which can be divided into component groups. Therefore, the system of the wind turbine is described as a combination of individual components and for the simulation of a wind turbine, a commonly used method for network analysis is applied. Reliability block diagrams (RBD) are used in several studies for modelling the failure behaviour and the reliability of wind turbines (for example [23]). The structure of a RBD defines the logical interaction of failures within a system. Each block is to represent both components and subsystems. In various literature the series connection of components or groups of components is chosen for modelling a wind turbine. Since part failures usually lead to a stop of the entire WT, this principle is also used in the described approach. However, redundancy of components means that a failure of one of these components does not lead to failure of the entire system as a functionally identical component ensures continued operation and a parallel connection is used in the case of redundant components. In a further breakdown, each component group is regarded as a series connection of individual components at the next level.

For the breakdown the wind turbine is to be divided into different areas of functional relationships and an unambiguous assignment of the components is needed. With the Reference Designation System for Power Plants, short RDS-PP® [24], a standard was introduced to designate parts of the system.

The structuring of a wind turbine according to RDS-PP® is classified according to functions and functional units.

3.2. Failure Categories

To simulate the failure behaviour of a component a categorization of failures based on the phases of a bathtub curve and the various types of damage has been performed.

The development of failure intensity with time for non-repairable systems is well known and is often described by the bathtub curve, which divides the lifetime of a technical system into three phases. The first phase is marked by falling failure intensity due to 'early failures or teething problems'. This is followed by a longer second phase, when failure intensity is constant due to 'intrinsic or random failures', which can be called failure rate. This is followed by a period of rising failure intensity as damage accumulates with operational age due to 'wear out'.

In addition, there are several factors that influence the WT failure rate, such as wind speed, turbine concept and climatic conditions, which should be part of any appropriate reliability analysis.

The failure categories are derived from the bathtub curve and a general classification of failures of technical systems in wear and break mechanisms. Early failures are used to represent the infant mortality of turbines as a decreasing failure rate over the first years. Wear effects include aging and fatigue failures whereas overload refers to break mechanisms which may be due to heavy gusts of wind. Component specific mechanisms allow to model individual risks to a component that may be related to the specific time or place of the site e.g. lightning damages to the blades. Finally, there remains a random block with a constant failure rate that cannot be tied to other mechanisms.

- **Early failures**: failures that occur at the beginning of the start-up phase, for example, due to installation or manufacturing errors of individual components.
- **Random failures**: failures, which is a constant risk based
- **Aging failures**: failures of a component that occur due to aging effects, e.g. corrosion.
- **Fatigue failures**: failures that occur because of fatigue or wear.
- **Overload failures**: failures of a component due to an overload of a component.
- **Component specific failures**: this category allows modelling additional effects. This could also comprise natural events such as lightning strikes, earthquakes and damage due to bird strike

3.3. Parameter estimation

In the following section the procedure of modelling the failure behaviour is described stepwise.

3.3.1. Determine the total number of failures for the simulation period. In the first step of modelling the value of the average failure rate of the modelled subsystem or component needs to be selected from a suitable study. This value determines how often the component fails in a selected period.

3.3.2. Weighting of failure modes. In the next step the total number of component failures with respect to the weighting of the different failure categories must be allocated. Depending on the component selected the number of failures per category can vary greatly.

After having all failure modes divided according to the presented categories, the next step is to describe a corresponding function for the failure behaviour of the different categories.

3.3.3. Functions of time-dependent failures. Time-dependent failures are modelled by using Weibull distributions. The function of the failure rate is

$$\lambda(t) = \frac{\beta}{\eta - t_0}\left(\frac{t-t_0}{\eta-t_0}\right)^{\beta-1} \quad for \quad t \geq 0 \qquad (1)$$

where ß is the corresponding form factor, η the characteristic lifetime and t_0 the failure-free period. These parameters need to be adjusted to fit with the weighted empirical failure rates for the different components.

Random failures are characterized by constant failure rates over time. In this case the form factor has the value 1 and the Weibull distribution turns into an exponential distribution.

For modelling the early failures a form factor of 0.4 << 0.9 is used, so that the lifetime distribution is described with a decreasing failure rate. For design-related or installation failures it is assumed that the probability of a component failure due to these causes is high at the beginning and after a component specific period decreases to nearly zero and the probability of failure is only very low. Therefore an additional limitation of the lifetime distribution period, e.g. failures occurring only in the first 2 years, is necessary.

In the case of aging failures the failure rate increases over the age of the component. The shape parameter is assumed to be <1. Additionally, the beginning of the aging phase and therefore the failure-free period, which is strongly dependent on the component, needs to be estimated. Since wind turbines are usually designed by the manufacturer with a minimum service life of 20 years, for most components the beginning of the aging phase will occur after at least 10 years.

3.3.4. Function of fatigue failures. The modelling of the fatigue failure is not described as a function of the "calendar" age because wind turbines are exposed to the effects of fatigue only in operation and a different lifetime characterizing variable needs to be introduced. For many components, the operating time or the hours of full load are more meaningful values for indicating the development and the state of fatigue. Therefore, fatigue can be modelled by tracking the turbine's working hours and capacity factor. However, the overall failure function as well as the choice of influencing parameters is strongly component specific. A different approach to implement the effects of fatigue would be to model the function depending on the strain gradient or the number of load cycles of a component.

Similar to the aging failures, the failure behaviour described by the corresponding lifetime characteristic is expressed by a Weibull distribution. For the failure-free period it can be assumed that fatigue effects act on the component from the start-up. With the additional assumption of an increasing failure rate, a value of the shape factor must be estimated. Here fore it is necessary to translate the results coming from failure statistics to the changed lifetime variable (for example, energy generated since commissioning of the component).

3.3.5. Function of overload failures. For the modelling of failures due to overload another lifetime characterizing variable needs to be introduced. In general, strong mechanical forces, temperatures or high electrical voltages to components are possible causes of overload failures.

In the case of a wind turbine, it is obvious that the most dominant factor for overload failures for the majority of the components can be seen in the wind, which introduces forces to the wind turbine. The dependence of reliability on wind speed is analysed in general in [17]. The relation between failure rate and wind energy index is shown for a population of Danish WTs in [16], with failure rate increasing at higher wind speeds and electrical subassembly failure rates showing the strongest dependency.

A reasonable assumption is that an increase in the risk is proportional to the wind speed. At too low wind speeds no overload failures will occur. At high wind speeds the wind turbine is shut down and it is believed that the risk of failure for most of the components due to overload is then negligibly small. Therefore, the corresponding function for overload failure can be based on the current wind speed with a progressive function beginning at cut-in wind speed ending just after the cut-out speed.

The overload of components due to much stronger storms (for example hurricanes) will not be displayed in this function and the resulting risk of failure can be found in the category of random failures or in the group of natural disasters such as earthquakes and lightning strikes.

3.3.6. Component specific failures. Even though the majority of reported failures are attributable to defective or loose components in many cases the faults were caused by external influences such as storms, lightning, ice accretion or network outages. These failure causes are represented in the model as "component specific", as shown in Figure 1). While as a cause of breakdowns network outages are not dependent on season or location, the other external conditions show both a clear seasonal and geographic dependency. For example, most damage and interruptions to operations caused by direct or indirect lightning damage – i.e. voltage surge damage following lightning strikes into the electricity network – are reported during the summer (in Germany). As most operational problems caused by ice accretion arise between December and March (a period with normally good wind conditions), a significant loss of yield can sometimes be expected in the case of such breakdowns. Therefore it is essential to include those effects in a suitable reliability model.

3.3.7. Combined failure function. The failure categories are weighted system or component specific and then summed up to build an overall failure function. This requires modelling the different failure categories independently from one another. The combined function looks like

$$\lambda(t, X_1, X_2, X_3) = \lambda_{konstant} + \lambda(t) + \lambda_{fatigue}(X_1) + \lambda_{overload}(X_2) + \lambda_{specific}(X_3) \quad (2)$$

with

$$\lambda(t) = \lambda_{Aging}(t) + \lambda_{Early}(t) \quad (3)$$

4. Simulation of the reliability behaviour

With the described framework for modelling the reliability behaviour it is possible to set up a reliability module which can be implemented in an overall software tool. Fraunhofer IWES has developed an in-house O&M modelling tool with a multi-agent approach throughout the project MAS-ZIH [26]. The tool serves as the development environment for the analysis and simulation results presented in this paper. It is based on a time-step Monte Carlo simulation and includes individual failure transitions for specific components and failure impact factors.

In the following the reliability module is described. It contains the information about the reliability behaviour of the different wt components. As described before it can be seen as a serial connection of the different components. This allows the flexibility needed since the natural characteristics of the components can be taken into account to the extent needed for the simulation processes. The flexibility also comprises the level of detail chosen for the different components. Since the input for estimating the parameters describing the failure function comes from failure statistics, it is necessary to have some default values for those components for which the failure characteristic is not well known. The parameter estimation follows the steps described before and a model with estimated parameters for the rotor blades is shown as an example within this subchapter.

The failure behaviour of the rotor blades of wind turbines with a capacity in MW range is subject in several statistics, such as the ReliaWind study [14] and the WMEP [25]. These statistics provide some indispensable information about the most common causes of failure of rotor blades. The possible failure causes shown in Figure 2 were assigned to the various event categories. Since no direct information about certain categories is available, the assignment needs to be conducted using some additional weighting.

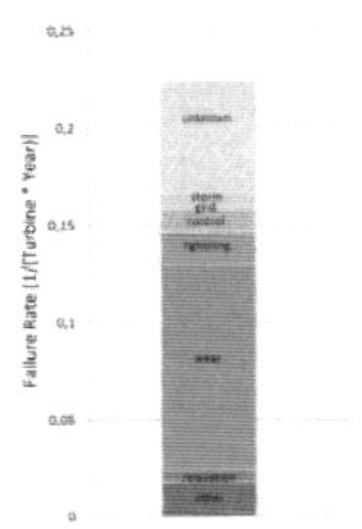

Average Rotor Blade Failure Rate:
λ = 0.225 /(Turbine * Year)

Failure Mechanism	Share of total Failure Amount [%]
Early	5
Random	15
Aging	15
Fatigue	55
Overload	10

Figure 2: Failure statistic for failure categories of rotor blades

The rate of random failures makes a total of about 15% of the total failure rate of the rotor blade, about 5% of the total losses correspond to the early failures and with the assumption that about 20% of failures "wear" and about 10% of failures due to "relaxation" are counted in the category aging, a share of 15% of total losses results for the aging failures.

With these shares the function parameters need to be estimated using some further assumptions. It is believed that early failures on rotor blades occur only in the first two and a half years and that the expected value of the probability density function lays between a half and one year. With the assumption of a declining failure rate, a form-factor is β = 0.5 estimated. For describing the function for the aging failures the assumptions are made that an increasing failure rate exists, that aging failures occur only after 10 years after installation but at the latest after 40 years and that the expected value of the distribution density function is at approximately 20 years. This leads to

$$\boldsymbol{f(t) = 0,25 * \left(\frac{t}{2}\right)^{-0,5} e^{-\left(\frac{t}{2}\right)^{0,5}}} \tag{4}$$

for the early failures and for the aging failures to

$$\boldsymbol{f(t) = 0,09 * \left(\frac{t-10}{17}\right)^{0,3} e^{-\left(\frac{t-10}{17}\right)^{1,3}}} \tag{5}$$

The dominant failure group for the rotor blades can be seen in the fatigue failures. For the calculation of f(x) for the distribution density function a run-time variable needs to be chosen and being translated out of the existing statistics. The energy generated or the cumulated number of full load hours is taken as life-describing parameter for components that are constantly running during time of energy production.

Because components are subject to fatigue failure over the entire operating period, the average failure rate of fatigue failures can be expressed with 0.55 times the average failure rate of the rotor blade. This in turn would mean a mean time to failure due to fatigue of about 8 years. With an average wind speed of 8.5 m/s, a 4MW plant produces about 1,84MW, so that the energy produced in 8 years corresponds on average to approximately 129GWh. It is believed that the expected value of the failure due to fatigue is at that time at which a wind turbine has produced this energy. The value of the energy delivered, approximately 129GWh, corresponds to the expected value of the associated probability density function. With an increasing failure rate for fatigue failures, the form factor is selected with 2.2 so that the failure rate slowly rises at the beginning and in the course faster. The value of the characteristic lifetime of a rotor blade can be calculated from this data and is approximately 160GWh. With the assumption that the fatigue effects are applied on the rotor blades from the beginning and occur at any later time with probability greater zero, the distribution function is valid for all times t>0. With these assumptions, the distribution density function of fatigue failures can be described completely.

$$f(t) = \begin{cases} 0 & \text{for } t < 0 \\ \frac{\beta}{\eta}\left(\frac{t}{\eta}\right)^{\beta-1} e^{-\left(\frac{t}{\eta}\right)^{\beta}} & \text{for } t \geq 0 \end{cases} \tag{6}$$

leading to

$$\boldsymbol{f(t) = 0,01375 * \left(\frac{t}{160}\right)^{1,1} e^{-\left(\frac{t}{160}\right)^{2,2}}} \tag{7}$$

For the overload failures an additional parameter describing the failure probability is needed. In most cases the average failure rate for overload failures can be assigned to the value of the mean wind speed at hub height of the wind turbine.

With the assumptions described afore the profile of the failure rate of the category of overload failures is described as a linear function with the slope

$$m = \frac{\lambda(v_{average})}{v_{average} - v_{min}} \tag{8}$$

Resulting in the function

$$\lambda_{Overload}(v_{Wind}) = \begin{cases} 0 & for\ v < v_{min} \\ m * (v - vmin) & for\ v_{min} < v < v_{max} \\ 0 & for\ v > v_{max} \end{cases} \tag{9}$$

The overall function describing the failure behaviour can be expressed combining the aforementioned single functions for the different categories.

5. Results

For the assessment of the model quality a comparison has been conducted. Three distributions have been compared. The first distribution can be seen as the starting point for the simulation of the reliability behaviour. A constant failure rate with no dependencies is simulated for a large offshore wind farm over the whole lifetime. The second distribution comes from the simulation using the developed reliability module. A clear shift towards a broader distribution with smaller peaks throughout the lifetime can be seen. For the reference case a set of event data, which has not been used for the establishment of the model, has been analysed. The distributions are illustrated in Figure 3.

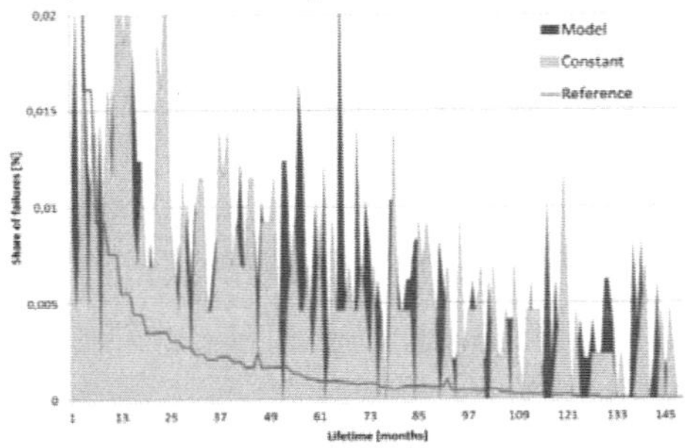

Figure 3: Empirical distributions for a) constant failure rate b) reliability model and c) reference case

Even though the curves appear to be quite similar, some differences can be seen. In order to demonstrate the better fit of the reliability model the difference to the reference scenario has been calculated. The RMSE for the constant failure rate is at 2,01 while the RMSE only slightly decreases to 2 for the developed model. The main difference can be seen for both simulations in the incredible higher amount of failures in the very beginning of the components life. This might be a statistical artefact due to the fuzziness of the failure definition. In the reference scenario the larger amount of early failures might be explained by the fact that the simulation covers only failures leading to a turbine downtime whereas the event database also comprises smaller component defects. Nevertheless, the comparison demonstrates the overall plausibility and proves that the outcome is in the same order as the initial simulation. Therefore, for showing the real value of the model a deeper investigation of the results is needed.

In a first step the breakdown according to the different failure categories is presented and the desired output is discussed. The breakdown of the failures for the different failure categories is shown in Figure 4.

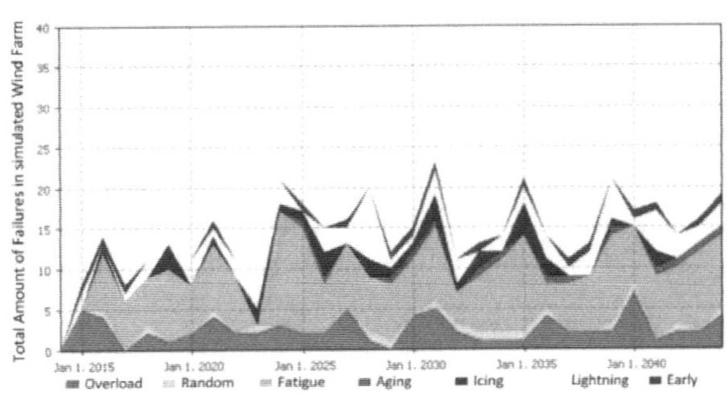

Figure 4: Breakdown of simulated failures according to the failure categories

It can be seen that the share of the different categories show the proportion of the initial assignment. The main share is at the fatigue failures and the simulation results confirm the proper behaviour of the model. Since the simulation is done by simulating a wind farm over the whole lifetime and by having component exchanges included, we can see that the early failures, which are due to the components life and not the turbine life, occur over the whole period. The aging failures however only appear to happen after a certain amount of time.

For a clearer picture the dependencies regarding the external effects are investigated in order to validate the simulation results. Icing is being modelled using statistical information on icing events. Icing season is from November to March with a clear maximum in January. The distribution over the other months may vary.

The characteristics of lightning failures are site specific. For this analysis the events have been modelled with reference to a study on lightning climatology in the German federal state of Saxony. [27]. For lightning events a strong seasonality on yearly and daily time scale are well known from experience and statistical data.

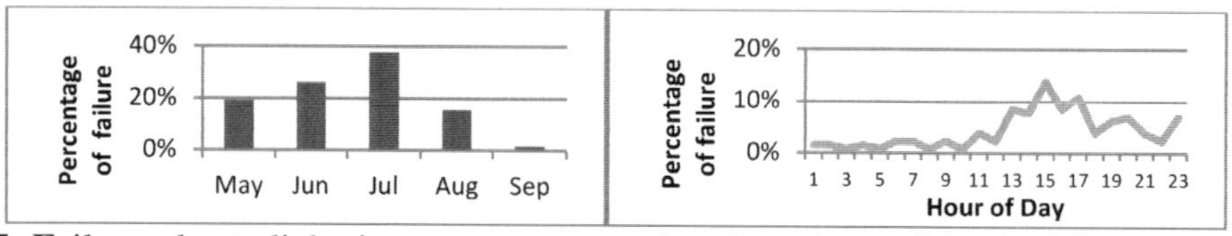

Figure 5: Failures due to lightning events per month and per hour of day in failure model output.

The wind dependent module in the failure function in the model produces a dynamic failure rate that follows the fluctuations of wind speed. As wind speed has seasonal and daily patterns as well these are also considered in the model. Figure 7 shows the difference in the general rate level between summer and winter season.

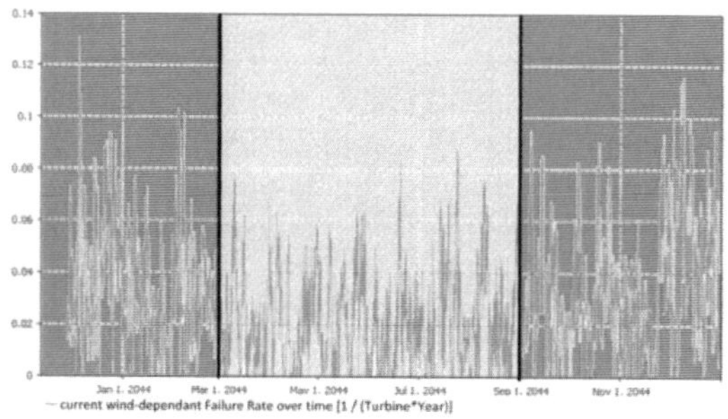

Figure 6: Wind speed dependent failure rate in failure model over one year

6. Conclusions

Modelling the failure behaviour of wind turbines is an essential part of offshore simulation software as it forms the demand side for all O&M processes and related costs incurred. State-of-the-art O&M tools use advanced models to represent the failure behaviour of wind turbines but come short when considering characteristic impact factors on specific components or failure causes. Important effects not yet modelled include increased failure rates at higher wind speed and seasonal effects on failures due to lightning or icing.

To address these issues a failure model based on a reliability-block-diagram incorporating different failure mechanisms has been proposed. It described the failure behaviour based on failure categories derived from the bathtub curve and fundamental wear and break mechanisms and has been fitted with available data to match empirical failure rates. The failure model has been implemented in Fraunhofer IWES multi-agent-system for modelling O&M processes.

The development of a detailed failure function for the rotor blade has been described and simulated. The fit to empirical data proves the general plausibility of the concept. The definition of failures in model and operational data needs further analysis to reduce existing uncertainties especially in the early stages of component life.

The proposed framework for a more detailed model of failure behaviour is essential for better including preventive maintenance strategies to wind farms and a step towards a more functional modelling of component failures. A key factor to benefit from the more detailed model is the availability of high quality O&M data from operational experience which is needed to identify the input data.

References

[1] I Dinwoodie et al 2015 Reference Cases for Verification of Operation and Maintenance Simulation Models for Offshore Wind Farms *Wind Engineering* pp. 1–14

[2] H Braam, T H Obdam, R van de Pieterman and L Rademakers 2011 Properties of the O&M Cost Estimator (OMCE)

[3] M Asgarpour and R van de Pieterman 2014 O&M Cost Reduction of Offshore Wind Farms - A Novel Case Study

[4] L Rademakers 2013 How to assist operators? The development of low cost load monitoring and decision support rules

[5] L Rademakers and H Braam O&M Aspects of the 500MW Offshore Wind Farm NL7

[6] University of Strathclyde: 'The Tool. URL: http://www.esru.strath.ac.uk/EandE/Web_sites/14-15/Far_Offshore_Wind/our_work/our_work_files/pages/the_tool.html#TUT'

[7] M Hofmann and I B Sperstad 2013 A Tool for Reducing the Maintenance Costs of Offshore Wind Farms *Energy Procedia* pp. 177–186

[8] J O Endrerud, J P Liyanage and N Keseric 2014 Marine logistics decision support for operation and maintenance of offshore wind parks with a multi method simulation model *Proceedings of the 2014 Winter Simulation Conference* pp. 1712–1722

[9] J Johansson 2013 Operational Validation of SIMLOX as a Simulation Tool for Wind Energy Operations and Maintenance (O&M), Master thesis

[10] M Spring et al 2015 Top 30 Charts of wind turbine failure mechanisms *Proceedings of the EWEA Annual Event*

[11] M G Bharatbhai 2015 Failure Mode and Effect Analysis of Repower 5M Wind Turbine *International Journal of Advance Research in Engineering, Science & Technology (IJAREST)*, 2 (2015) 5

[12] H Jung et al 2014 Abschlussbericht für das Verbundprojekt „Erhöhung der Verfügbarkeit von Windkraftanlagen"

[13] J Caroll et al 2015 Offshore Wind Turbine Sub-Assembly Failure Rates Through Time *Proceedings of the EWEA Annual Event*

[14] J B Gayo 2011 ReliaWind final report. More efficient and reliable wind turbines by 2011'

[15] S Faulstich 2015 Performance and reliability benchmarking using the cross-company initiative WInD-Pool *Proceedings of the rave international conference*

[16] P J Tavner et al 2012 Study of weather and location effects on wind turbine failure rates *Wind energy*

[17] B Hahn 1997 Zeitlicher Zusammenhang von Schadenshäufigkeit und Windgeschwindigkeit'

[18] P J Tavner et al 2006 Influence of Wind Speed on Wind Turbine Reliability *Wind energy*

[19] K M Holmstrøm 2014 How can more advanced failure modelling contribute to improving life-cycle cost analyses of offshore wind farms?

[20] V Slimacek, B H Lindqvist 2016 Reliability of wind turbines modeled by a Poisson process with covariates, unobserved heterogeneity and seasonality *Wind energy*

[21] M Asgarpour, J D Sorensen 2016 O&M Modeling of Offshore Wind Farms – State of the Art and Future Developments

[22] J H Jürgensen 2016 Condition-based Failure Rate Modelling for IndividualComponents in the Power System: Licentiate Thesis

[23] I Athamna 2015 Zuverlässigkeitsberechnung von Offshore-Windparks

[24] RDS-PP – Application Guideline: 'Part 32: Wind Power Plants', 2014

[25] Fraunhofer IWES 2014 *Windenergie Report Deutschland*

[26] V Berkhout et al 2015 MAS-ZIH - Einsatz von Multi-Agenten-Systemen als Unterstützung für eine zuverlässigkeitsorientierte Instandhaltung. Abschlussbericht : Laufzeit des Vorhabens: 01.10.2011 - 31.03.2015.'

[27] A Schucknecht, J Matschullat 2014 Raum-zeitliches Blitzaufkommen im Freistaat Sachsen – Ursachen, Phänomene, Risiken (BlitzSn)

3D augmented reality for improving social acceptance and public participation in wind farms planning

S Grassi* & T M Klein°

*Institute of Cartography and Geoinformation – ETH Zurich

Stefano-Franscini-Platz 5, 8093 Zurich, Switzerland

°Planning of Landscape and Urban Systems – ETH Zurich

Stefano-Franscini-Platz 5, 8093 Zurich, Switzerland

Corresponding author: sgrassi@ethz.ch

Abstract. Wind energy is one of the most important source of renewable energy characterized by a significant growth in the last decades and giving a more and more relevant contribution to the energy supply. One of the main disadvantages of a faster integration of wind energy into the energy mix is related to the visual impact of wind turbines on the landscape. In addition, the siting of new massive infrastructures has the potential to threaten a community's well-being if new projects are perceived being unfair. The public perception of the impact of wind turbines on the landscape is also crucial for their acceptance. The implementation of wind energy projects is hampered often because of a lack of planning or communication tools enabling a more transparent and efficient interaction between all stakeholders involved in the projects (i.e. developers, local communities and administrations, NGOs, etc.). Concerning the visual assessment of wind farms, a critical gap lies in effective visualization tools to improve the public perception of alternative wind turbines layouts. In this paper, we describe the advantages of a 3D dynamical and interactive visualization platform for an augmented reality to support wind energy planners in order to enhance the social acceptance of new wind energy projects.

1. Introduction

Wind energy has been a source of debates and arguments in the last decades due to general accepted willingness of renewable energy sources to reduce the CO2 emissions. On the other hand, one of the main concerns about the siting of new wind farms is the visual impact on the surroundings and their noise and impact on birds and wildlife [1]. The acronym NIMBY ('Not-In-My-Back-Yard') describes resistance to siting specific projects close to one's area of residence while exhibiting acceptance of similar projects elsewhere [2]. The local participation with respect to the installation of wind turbines has been investigated in the last decades [3, 4] showing that the acceptance is influenced by the perception of the landscape and the value given to the aesthetic impact on the environment. Previous studies showed the importance of the visual impact in different countries such as Sweden [5], Netherlands [6] and USA [7]. In addition, according to previous work, the social acceptance is a key aspect for a successful market development [8, 9] and stakeholders admit the lack of useful instruments to support social acceptance [10, 11].

The importance of involving local communities when planning wind energy projects has been investigated in many regions of the world such as Australia [12], France and Germany [13]. Overall, factors such as quality of communication, public participation in the planning process are demonstrated to play a critical role in the acceptance of local communities with respect to new wind projects [14, 15].

This issue has been addressed in last decades in order to support planners and investors in increasing the social acceptance of wind projects and thus to reduce the construction time and investment risk.
With the technology progress, IT solutions and processes were developed aimed at supporting planners and decision makers in the planning process. Typical approaches of superimposition of images from given points of view have been adopted for years [16] to show the visual impact from different distances for offshore and onshore wind projects [17] showing limitations due to the lack of visualization from other subjective perspectives.
The more realistic the image or the visualization, the higher is the achieved credibility [18]. In addition, previous work demonstrated that the interactivity of the 3D visualization of wind farms in the landscape is a benefit for users [18-20]. Visualization software program pushed by computer games gave a significant contribution for the interactive visualization enabling for real-time navigation through virtual environment [21].
Approaches using Geographic Information Systems (GIS) have been developed to enhance the public participation and acceptance of wind energy projects [22, 23] for visualization [24] and integrating also the acoustic 3D simulation [25]. The technological advances in digital landscape visualization techniques allow using digital 3D visualizations for landscape design, planning and management [26].
In this paper, we describe the advantages of a 3D dynamical and interactive visualization platform for an augmented reality to support wind energy planners to enhance the social acceptance of new energy projects. The developed process consists of a combination of GIS processes aimed at identifying the regions suitable for wind energy projects whose outcome is then used in the 3D visualization for augmented reality. The advantage of the proposed generalized method is the ability of combining the planning phase with the 3D interactive visualization into a unique tool enabling users to have an intuitive and real-time understanding of possible scenarios of wind turbines immersed in a virtual landscape. The additional benefit consists of the flexibility of tool in being adapted for public participation and thus giving investors and administrations a tool for a better communication with local communities. The tool can be also applied for planning for example transmission lines which are infrastructures necessary both for wind farm interconnection and the enhancement of the existing power grid.

2. Materials and Methods

2.1. Dataset

The data required for a 3D visualization with augmented reality consist of the geospatial data of the landscape (i.e. topography, satellite images). The higher the resolution the better the landscape is represented. The objects in the landscape are added in the exact location and with their corresponding characteristics (e.g. the type and the height of trees).

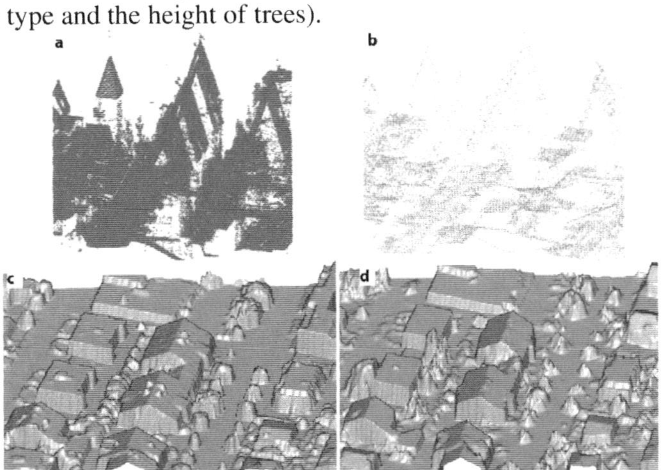

Figure 1. Building detection using different LiDAR data point clouds (Buffat R,2016)

LiDAR (Light Detection and Ranging or Laser Imaging Detection and Ranging or simply 3D scanning) data can be also integrated and used to better describe the objects in the landscape. LiDAR consists of a surveying technology collecting 3D data points cloud (x, y, z) of the landscape with high-resolution (up to several ten hundreds of points per square meter) collected using a device installed on an airplane flying over a region. LiDAR is applied in many fields ranging from agriculture to transport, archeology and meteorology. It allows reconstructing 3D objects on the Earth such as buildings [27] and vegetation [28]. Figure 1 shows an example of how LiDAR data can be used in order to describe in details real objects such as buildings by combining satellite images [29].

2.2. Method

The layout of wind turbines is optimized in order to maximize the economic profit using different IT solutions based on CFD but neglecting the analysis of their visual impact. Often visual assessments are carried out by superimposing wind turbines on fixed images taken from few viewpoints limiting thus a full perspective of the whole landscape. The developed platform uses, first, GIS data and geoprocessing (figure 2) to identify suitable locations of wind turbines far from natural and anthropological constraints (figure 3). The suitable regions are identified by excluding areas not suitable for wind turbines installation and using a multi-criteria decision analysis (MCDA)[30] which classifies areas using a matrix of weights corresponding to the land suitability as done in previous work [31, 32] and applying buffer distances [33]. These weights are set by planners based on their experience and by experts' opinions and by regulation.

Then, each suitable land is discretized in cells of arbitrary spatial resolution corresponding to potential suitable wind turbines locations (figure 4).

The obtained suitable regions are integrated into a 3D visualization (web)-platform that allows to place 3D wind turbines models in the suitable areas. Each cell contains information and data of the wind resource and the corresponding geographic position.

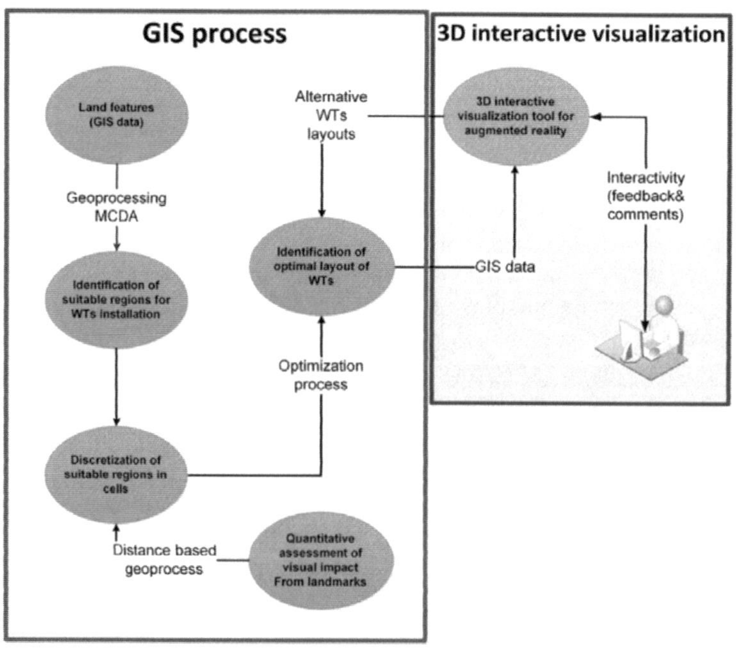

Figure 2. Geoprocessing of the developed platform for an augmented 3D interactive visualization of wind turbines

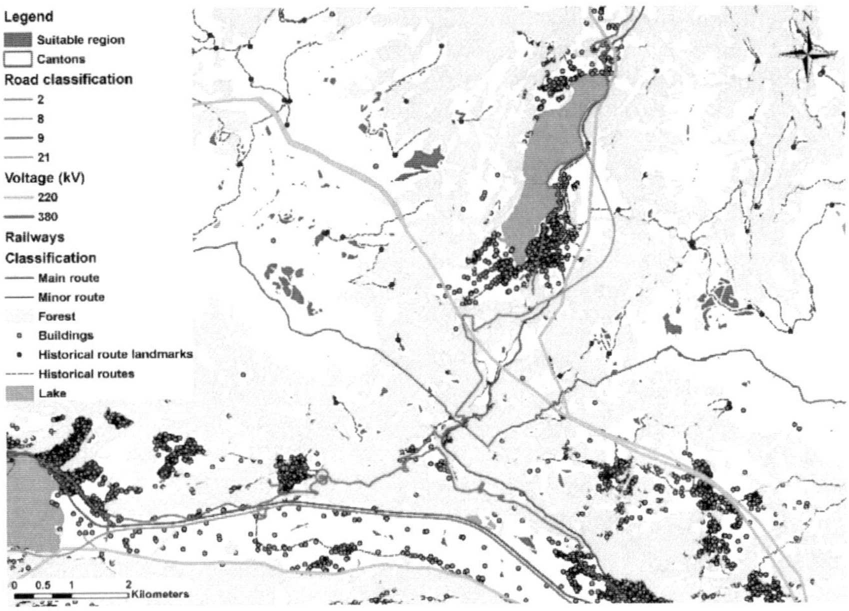

Figure 3. Identification of suitable areas far from constraints

As additional information, the visibility of the wind turbines from selected landmarks of the landscape can be also estimated using a GIS-based process. In Switzerland, when planning large infrastructures, such as wind farms or transmission lines, the visual impact plays a critical role since the landscape is considered as part of the well-fare and a quantitative assessment of the visual impact from given observer points, such as buildings or landmarks, gives planners additional information about the potential locations where issues can raise.

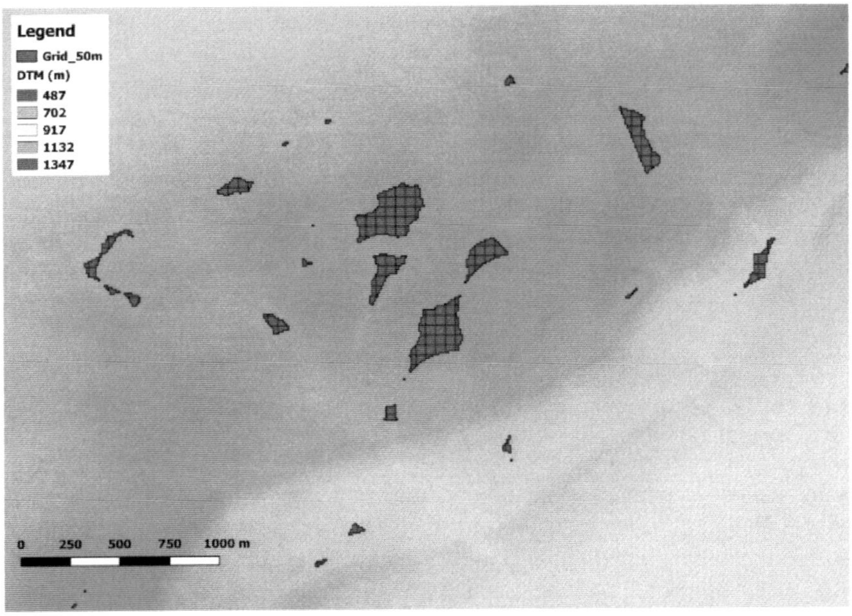

Figure 4. Discretization of suitable region in cells of 50m resolution.

Journal of Physics: Conference Series **749** (2016) 012020 doi:10.1088/1742-6596/749/1/012020

As investigated in previous studies [34], for instance, the visual impact of such massive infrastructures can be quantified as function of the distance from the buildings in the landscape within a certain radius from the wind farm area.

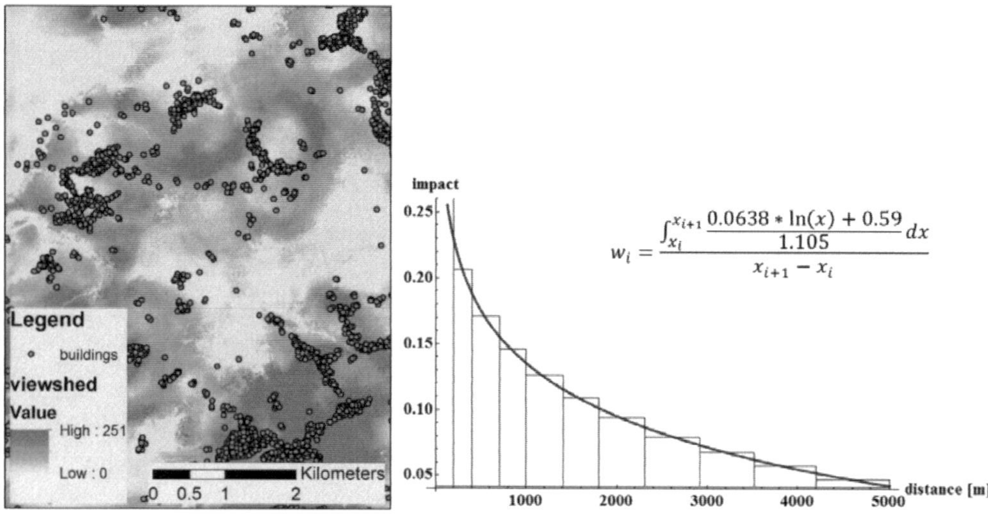

Figure 5. Spatial distribution of the visibility of transmission line towers from the buildings (left), approach used to estimate the visibility depending on the distance (right).

The buildings are used as observer points since they are the locations where people spend part of their private life and thus considered – with the surroundings – fundamental element of well-being. Nevertheless, other elements such as specific landmarks in the landscape or mountains (e.g. hiking trails, panoramic spots, etc.) can be selected or included in the assessment of the visual impact.
In figure 5 (left) the estimate of the spatial distribution of the visibility is shown: with this approach the area around each cell is divided into multiple concentric rings and to each of them a weight is assigned depending on the distance from the center of each cell.
The width of the histograms corresponds to the width of the rings and is determined by discretizing the function expressed in the figure 5 (right). The weight of a circular ring equals the mean value of the impact between the inner and the outer radius and is calculated with the relation in figure 5 (right). The difference between the inner and the outer radii increases with the distance whilst the corresponding weight decreases. The formula describing the visual impact depending on the distance has been defined in previous work when assessing the impact of transmission line towers [35]. In previous work [36] the perceived impact wi of TLs depending on the distance x from the TLs (in this case limited to 5000m) has been defined with the relation in figure 5(right). The same approach is used for wind turbines where the height of the observed point is adapted depending on the wind turbine model.
The weight of each ring is roughly equal to 0.85 times of the weight of the previous ring. This map can be generated for any region and any type of wind turbine and to assign a value of visual impact to each cell (figure 6). This map can be used as input data to optimize the wind turbines locations in order to find possible layouts which are a trade-off between two possible wind turbine layouts: the one that maximizes the return on investment and the one that minimizes the visual impact on the surroundings.
Once the information and the data are collected, the layout of wind turbines is identified in the space, the 3D visualization can be created (figure 7). The possible layouts can be interactively visualized by switching on-off the corresponding layers, which enable the user to assess each layout from different perspectives.

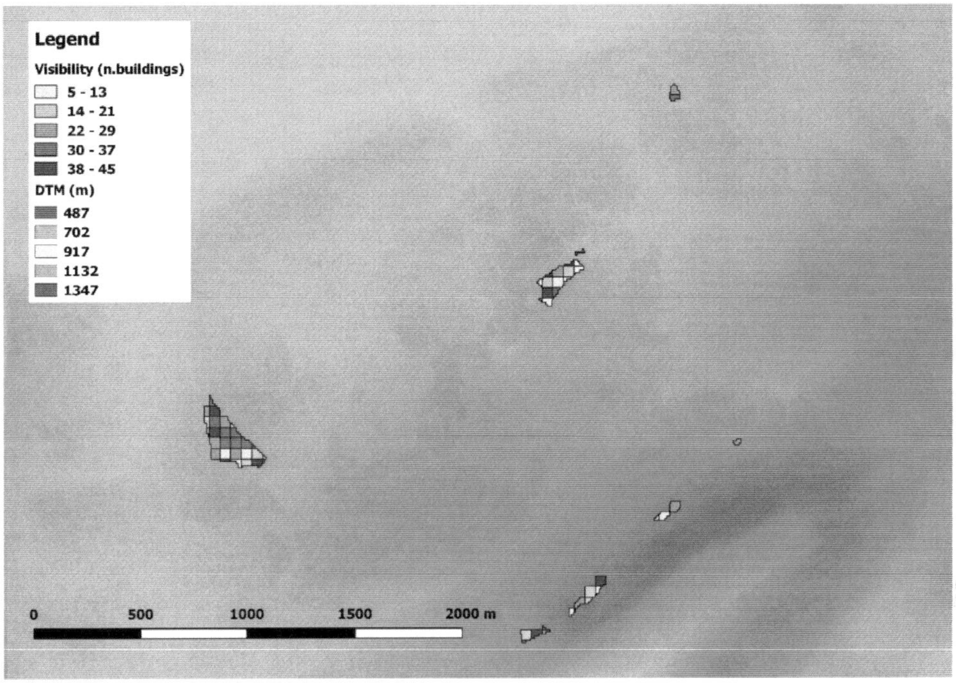

Figure 6. distribution of the number of visible buildings from each potential cell suitable for wind turbines.

A user can navigate through a given project using customized functions in order to get an augmented and more realistic perception of the impact of wind turbines on the landscape (figure 8). The user can both simulate a walk at the ground level and fly over the project region to gain a personal perspective. This allows having a realistic subjective impression of the impact of wind turbines from arbitrary locations or landmarks in the landscape. In countries such as Switzerland, the environment and the landscape are highly considered as element of the welfare therefore such massive infrastructures are seen as an issue for the beauty of valleys and mountains. Overall, the landscape threatened by massive infrastructures, such as wind turbines, is one of the common and frequent reason raising local protests and opposition in many other areas worldwide.

This reasonable concern of local population stems from both a lack of knowledge of wind turbines technology and the lack of efficient form of communication and visualization allowing one to get a visual interactive experience of how a project will look like. Aware of this issue, the platform is designed in order to assess the visual impact from different points such as trails in the mountains, ridges or any other given location and can be applied to any region worldwide.

Figure 7. Snapshot of the 3D visualization of wind turbines from the ground.

Since multiple locations are suitable for wind turbines, the interactivity feature of the platform allows modifying the positions of wind turbines within the allowed areas identified with cells. This enables identifying different scenarios both in terms of number of wind turbines and layout in order to assess the different visual impacts. At the same time, the tool performs the estimate of the long-term electricity generation and thus allows for comparing different scenarios.

The advantage of the developed platform is its flexibility of being adapted for public communication in a form of a web-platform where stakeholders can exchange feedback and comments about a given wind farm project (figure 2).

The driving concept for the development of this platform is the need of giving a more realistic and intuitive experience of a future project. The superimposition of fixed images from few viewpoints is a great limitation when showing new wind farm projects since it does not show their impact from different perspectives. Since the local population is familiar with their landscape and has a high consideration of the landscape and natural values as part of their welfare, they look at the wind turbines as a threat for their wellbeing. Any sort of environment and landscape can be reproduced with high resolution and fidelity (figure 8).

Additional features of the 3D visualization are the shadowing and flickering of the tower and the blades of wind turbines depending on the position of the sun and the simulation of the different weather conditions.

The flexibility of the 3D interactive platform allows integrating functionalities useful for wind farm planners in order to obtain a quick overview and visualization of a wind energy layout. The platform is useful to enhance the communication with the local population and to identify critical aspect of a wind farm layout (e.g. wind turbines position not accepted by the population).

Previous work demonstrated how a 3D dynamic visualization can support users in better understanding the impact of wind turbines on the landscape [25]. The platform can be also used as a useful tool for public participation by integrating and exchanging comments and feedback between all stakeholders involved in the projects. Suitable locations for wind turbines can be graded depending on the individual perspective and finally ranked based on their suitability.

Figure 8. Assessment of a wind farm layout on site by augmented reality

3. Results and Discussions

The social acceptance of wind energy projects can be improved by using 3D interactive visualization platform which allows a user to navigate through a more realistic representation of the landscape. The 3D platform realistically reproduces 3-dimensional objects of the environment using GIS data and high-resolution satellite images. It helps users to experience the visual impact of wind energy projects from different viewpoints and to have a better understanding and perspective of the impact of wind farm projects on the landscape. The developed platform can be applied to any region and for any wind farm size, layout and wind turbine model overcoming the limitations of fixed images usually used to reproduce the visual impact from a limited number of viewpoints. The flexibility of the platform allows integrating also additional tasks to support planners during the planning phase of wind farms, in particular:

- the identification of the optimal layout of wind turbines in suitable regions for wind farms,
- a preliminary quantitative assessment of the visual impact of wind turbines from landmarks or buildings in the surroundings
- the assessments of visual impact of other elements of wind energy projects such as transmission line interconnection
- the creation of a platform for enhancing the public participation and interaction between all stakeholders of the project and the local communities

References

[1] Tabassum-Abbasi, Premalatha M, Abbasi T and Abbasi S A 2014 Wind energy: Increasing deployment, rising environmental concerns *Renew Sust Energ Rev* **31** 270-88

[2] Petrova M A 2013 NIMBYism revisited: public acceptance of wind energy in the United States *Wires Clim Change* **4** 575-601

[3] Devine-Wright P 2005 Beyond NIMBYism: towards an integrated framework for understanding public perceptions of wind energy *Wind Energy* **8** 125-39

[4] Wolsink M 2007 Wind power implementation: The nature of public attitudes: Equity and fairness instead of 'backyard motives' *Renew Sust Energ Rev* **11** 1188-207

[5] Carlman I 1986 Public opinion on the use of wind power in Sweden. In: *European Wind Energy Association Conference and Exhibition, Rome,* pp 7-9
[6] Wolsink M 1989 Attitudes and expectancies about wind turbines and wind farms *Wind engineering* **13** 196-206
[7] Thayer R L and Freeman C M 1987 Altamont - Public Perceptions of a Wind Energy Landscape *Landscape and Urban Planning* **14** 379-98
[8] Hall N, Ashworth P and Devine-Wright P 2013 Societal acceptance of wind farms: Analysis of four common themes across Australian case studies *Energy Policy* **58** 200-8
[9] Pepermans Y and Loots I 2013 Wind farm struggles in Flanders fields: A sociological perspective *Energy Policy* **59** 321-8
[10] Huber S H, Robert 2010 Technical Report Results of IEA Wind Task 28 on Social Acceptance of Wind Energy. IEA pp 1-91
[11] Strub P and Ziegler C 2009 Code of Conduct for wind-power projects-Feasibility study. No. SFOE-290007
[12] Gross C 2007 Community perspectives of wind energy in Australia: The application of a justice and community fairness framework to increase social acceptance *Energy Policy* **35** 2727-36
[13] Jobert A, Laborgne P and Mimler S 2007 Local acceptance of wind energy: Factors of success identified in French and German case studies *Energ Policy* **35** 2751-60
[14] Wolsink M 1996 Dutch wind power policy - Stagnating implementation of renewables *Energy Policy* **24** 1079-88
[15] Wolsink M 2007 Planning of renewables schemes: Deliberative and fair decision-making on landscape issues instead of reproachful accusations of non-cooperation *Energy policy* **35** 2692-704
[16] Bishop I D and Miller D R 2007 Visual assessment of off-shore wind turbines: The influence of distance, contrast, movement and social variables *Renew Energ* **32** 814-31
[17] Molnarova K, Sklenicka P, Stiborek J, Svobodova K, Salek M and Brabec E 2012 Visual preferences for wind turbines: Location, numbers and respondent characteristics *Applied Energy* **92** 269-78
[18] van Lammeren R, Momot A, Loohuis R O and Hoogerwerf T 2005 3D Visualization of 2D Scenarios
[19] Warren-Kretzschmar B and Tiedtke S 2005 What role does visualization play in communication with citizens?–A field study from the interactive landscape plan *Buhmann, E.: Trends in Real-Time Landscape Visualization and Participation, Wichmann Verlag* 156-67
[20] Lange E and Hehl-Lange S 2005 Combining a participatory planning approach with a virtual landscape model for the siting of wind turbines *Journal of Environmental Planning and Management* **48** 833-52
[21] Herrlich M 2007 A tool for landscape architecture based on computer game technology. In: *Artificial Reality and Telexistence, 17th International Conference on*: IEEE) pp 264-8
[22] Berry R, Higgs G, Fry R and al. 2011 Web-based GIS Approaches to Enhance Public Participation in Wind Farm Planning *Transactions in GIS* 15(2): 147–72
[23] Aitken S 2002 *Community participation and geographic information systems,* (London: Taylor & Francis: ed. W. Craig, T. Harris, and D.Weiner) pp 357–66.
[24] Chias P and Abad T 2013 Wind farms: GIS-based visual impact assessment and visualization tools *Cartogr Geogr Inf Sc* **40** 229-37
[25] Manyoky M, Hayek U W, Heutschi K, Pieren R and Gret-Regamey A 2014 Developing a GIS-Based Visual-Acoustic 3D Simulation for Wind Farm Assessment *Isprs Int Geo-Inf* **3** 29-48
[26] Lange E 2011 99 volumes later: We can visualise. Now what? *Landscape and Urban Planning* **100** 403-6
[27] Buffat R 2016 Feature-Aware Surface Interpolation of Rooftops Using Low-Density Lidar Data for Photovoltaic Applications *Geospatial Data in a Changing World,* ed T Sarjakoski, et al. Springer pp 337-50

Journal of Physics: Conference Series **749** (2016) 012020 doi:10.1088/1742-6596/749/1/012020

[28] Wasser L, Day R, Chasmer L and Taylor A 2013 Influence of Vegetation Structure on Lidar-derived Canopy Height and Fractional Cover in Forested Riparian Buffers During Leaf-Off and Leaf-On Conditions *Plos One* **8**

[29] Zhou Z, Gong J and Guo M 2016 Image-Based 3D Reconstruction for Posthurricane Residential Building Damage Assessment *Journal of Computing in Civil Engineering* **30** 04015015

[30] Malczewski J 1999 *GIS and multicriteria decision analysis.* (John Wiley & Sons, New York)

[31] Hansen H S 2005 GIS-based Multi-Criteria Analysis of Wind Farm Development. In: *ScanGIS'2005 - The 10th Scandinavian Research Conference on Geographical Information Science,* ed H Hauska and H Tveite (Stockholm, Sweden pp 75-87

[32] Van Haaren R and Fthenakis V 2011 GIS-based wind farm site selection using spatial multi-criteria analysis (SMCA): Evaluating the case for New York State *Renewable and Sustainable Energy Reviews* **15** 3332-40

[33] Grassi S, Chokani N and Abhari R 2012 Large scale technical and economic assessment of wind energy potential with a GIS tool: case study Iowa *Energy Policy* **45** 58-73

[34] Grassi S, Friedli R, Grangier M and Raubal M 2014 A GIS-Based Process for Calculating Visibility Impact from Buildings During Transmission Line Routing *Connecting a Digital Europe Through Location and Place* Springer International Publishing pp 383-402

[35] Paul H-U, Uther D, Neuhoff M, Winkler-Hartenstein K, Schmidtkunz H and Grossnick J 2004 GIS-gestütztes Verfahren zur Bewertung visueller Eingriffe durch Hochspannungsfreileitungen. In: *Naturschutz und Landschaftsplanung: Zeitschrift für angewandte Ökologie,* pp 139-44

[36] Nohl W 1993 Beeinträchtigungen des Landschaftsbildes durch mastenartige Eingriffe. (Deutschland, München: Ministeriums für Umwelt Raumordnung und Landwirtschaft des Landes Nordrhein-Westfalen)

Wind offering in energy and reserve markets

T Soares[1a], P Pinson[2a], H Morais[3b]

[a]Technical University of Denmark, Elektrovej B. 329A, 2800 Kgs. Lyngby, Denmark
[b]Polytechnic Institute of Porto, Porto, Portugal

[1]tiasoar@elektro.dtu.dk, [2]ppin@elektro.dtu.dk, [3]hugvm@isep.ipp.pt

Abstract. The increasing penetration of wind generation in power systems to fulfil the ambitious European targets will make wind power producers to play an even more important role in the future power system. Wind power producers are being incentivized to participate in reserve markets to increase their revenue, since currently wind turbine/farm technologies allow them to provide ancillary services. Thus, wind power producers are to develop offering strategies for participation in both energy and reserve markets, accounting for market rules, while ensuring optimal revenue. We consider a proportional offering strategy to optimally decide upon participation in both markets by maximizing expected revenue from day-ahead decisions while accounting for estimated regulation costs for failing to provide the services. An evaluation of considering the same proportional splitting of energy and reserve in both day-ahead and balancing market is performed. A set of numerical examples illustrate the behavior of such strategy. An important conclusion is that the optimal split of the available wind power between energy and reserve strongly depends upon prices and penalties on both market trading floors.

1. Introduction

In last two decades electricity markets have been evolving in different ways with the aim to improve the competition among the different players without compromising the required reliability and stability in the electric system. In this scope, electricity markets are composed by different market stages for different commodities. Besides the energy commodity traded in energy auctions, there are ancillary services commodities (usually traded in reserve markets) that are used by power system operators to ensure proper levels of reliability, stability and security in the power system.

With the continuous introduction of wind generation in the electricity market, the behavior of electricity market participants has been changing. Currently, multiple methodologies for optimizing the strategic behavior of wind power producers (WPP) in the energy market have been proposed, accounting for expected costs from the balancing market [1–8]. Part of this work has been conducted based on the assumption that the strategic behavior of wind power plants does not have a significant impact on the market equilibrium, thereby, assuming price-taker behavior, i.e. the WPP does not exert market power [1–5]. In an opposite direction, several works exist, claiming that WPP may have a significant impact in the market equilibrium, and somehow may exert market power – yielding a price-maker assumption [6–8].

Nevertheless, wind power generators are now able to provide ancillary services, such as frequency and voltage control [9]. Namely, wind power plants can control active power injection in a few seconds; injecting/consuming reactive power while maintaining proper voltage levels, as well as providing virtual inertia to the system [10–12]. Thus, new business models may emerge, stimulating the willingness of wind power producers to participate and take advantage of reserve markets to increase their profit, as detailed in [13–15]. A analytical approach based on probabilistic forecast for wind power participating in the energy and reserve market is proposed in [13,14]. In [13], a simplistic strategy for splitting the available wind power in energy and reserve is applied, while [14] uses two different control strategies (proportional and constant wind power control) for WPP to participate in both energy and reserve markets.

We place ourselves under the proportional control strategy used in [14], contributing with a new stochastic methodology that maximizes the expected revenue of the WPPs in the day-ahead energy market and in the reserve market, while accounting for expected costs from failing to provide the energy and reserve products at the balancing stage. Besides that, this work contributes with a new perspective of facing the lead time of the WPP between the day-ahead and balancing stage by considering that energy and reserve bids submitted in the day-ahead market by the WPPs can be changed in the balancing market, i.e. the use of more accurate forecast of the wind power production in the balancing stage reduces the deviation between the power production committed in day-ahead stage and the effective production during the energy delivery. This may allow WPPs to bid in both market stages with more precise information about their wind power production, thereby, reducing expected energy and reserve costs in the balancing market. The results show that allowing a change in the proportionality of energy and reserve between day-ahead and balancing market, improves the expected revenues of the WPP, as well as, reduces the expected power deviation between the day-ahead and the energy delivery.

The paper is structured as follows. Section 2 describes the market structure for wind power producers participating in energy and reserve markets. Section 3 presents the detailed mathematical formulation of the optimal offering strategy for wind power producers. Section 4 numerically evaluates the offering strategy in expectation under different prices and penalties schemes that may occur in the market. Section 5 assembles the most important conclusions.

2. Wind power in electricity markets

2.1. Day-ahead and balancing participation

Currently, wind power producers can participate in the wholesale market by submitting their power bids (usually, their expected production) in the day-ahead market. The uncertainty of the wind power production is usually mitigated through the balancing market (the last mechanism for correcting the system and market participant imbalances), where the deviations of the wind power producers (the difference between their day-ahead market bids and the expected power production close to real-time) may induce some penalties for the wind power producers by failing to provide their day-ahead bids (either in deficit or surplus of power production) [16,17].

In that context, the expected costs to the wind power producers depend on the energy imbalance of the power system and of the difference between the sell and the delivered energy by the WPP. Furthermore, two different penalty mechanisms (one-price settlement and two-price settlement) can be applied depending on the characteristics and of the market rules [18,19]. For instance, the two-price system is assumed in the balancing mechanism in Denmark [20].

In what concerns the price bids of wind power plants in the day-ahead market, usually, WPPs places their power bids in the market at zero price or even negative price. This behavior depends on the internal rules of each market, as well as, on the different incentive schemes that wind power producers are submitted in each country. For instance, in Denmark, wind power producers are remunerated based on a scheme that lies on a combination of market participation (negative prices are allowed in NordPool) plus a premium [14]. In Portugal, similar schemes have been followed, yet most of wind power producers are still under fixed feed-in tariffs [21]. Besides, the Iberian market does not allow for negative bids [22,23].

2.2. Energy and reserve markets model

WPPs are willing to provide some ancillary services, since in their perspective, providing reserve (even with uncertainty and under high penalties when failing to provide the service) can somehow increase their revenue. Thus, the development of a methodology for wind power participation in energy and reserve markets at the day-ahead market, while accounting with expected costs in the balancing market is proposed and illustrated in Figure 1. The energy and reserve markets assume different characteristics, so different considerations are taken. On the one hand, wind energy bids submitted in the day-ahead market should account for potential imbalance situations and their asymmetric penalties. On the other hand, bids submitted in the reserve market should take into account the possibility to fail in providing the service.

Nevertheless, this model allows WPPs to submit bids into the energy and reserve market at day-ahead stage, following a proportional strategy for the split of the available power into energy and reserve (a share parameter is obtained by the split between energy and reserve). The bidding strategy for the day-ahead market assumes an expected energy market price, while the reserve market participation strategy takes into account the capacity reserve price.

At the balancing stage, expected costs for energy and reserve deviations are considered. On the one hand, expected costs for energy surplus or deficit of the WPP are considered. In contrast, reserve costs are only accounted for deficit of reserve, since the reserve surplus is not detrimental to the system. Additionally, this models assumes that the share parameter (split between energy and reserve) at the balancing stage can assume a different value from the one used for the day-ahead market decision. Thus, WPPs have the opportunity to reduce some energy or reserve deviations, thereby, increasing its expected revenue.

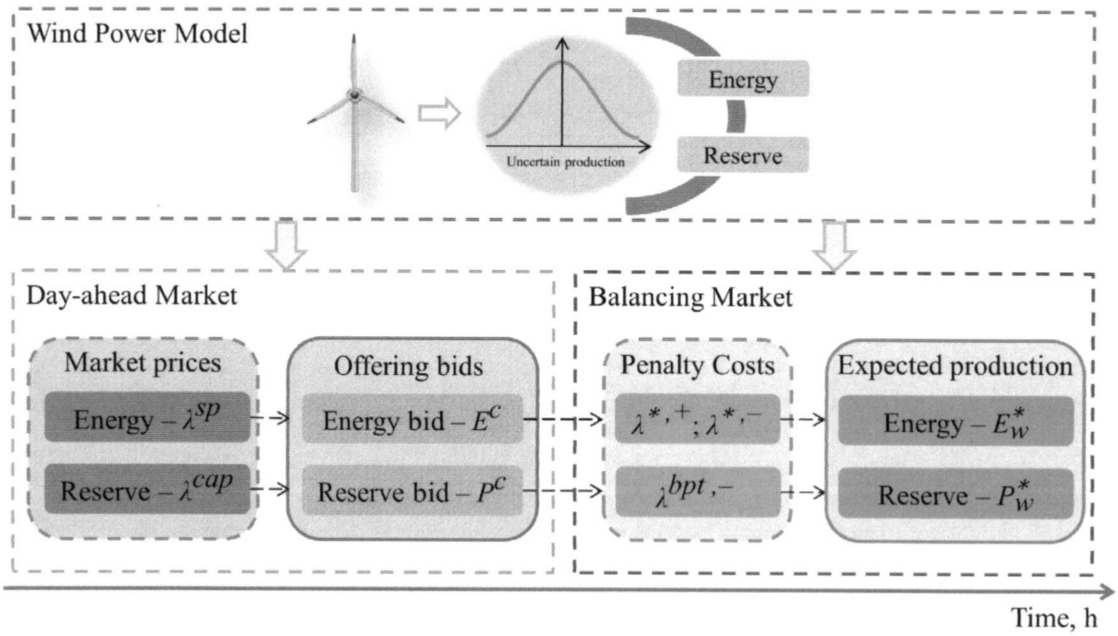

Figure 1. Wind power participation model in the energy and reserve market.

2.3. Wind power control to provide energy and reserve

Currently, wind power plants have developed several ways of active power control to provide energy and reserve, thereby, ensuring the stability of the power system. The use of these controls has been required by the system operators in different countries with high penetration of wind power, thereby, updating the grid-codes with new active power controls methodologies. System operators may require the use of such controls in cases of excess of wind power, to decrease congestion or even just for reserve provision. In this context, four methods for active power control of wind power for providing energy and reserve are described in detail.

2.3.1. Proportional wind control. This control mechanism consists in the proportional split of the available active power in energy and reserve, as illustrated in Figure 2. In terms of market strategy, the proportional wind offering strategy is used to define the share of energy E^c and reserve P^c to be submitted in the market [14,24], where Q is the total power bid and α^c the strategy parameter controlling the share of energy and reserve bids at day-ahead stage, which varies between 0 and 1.

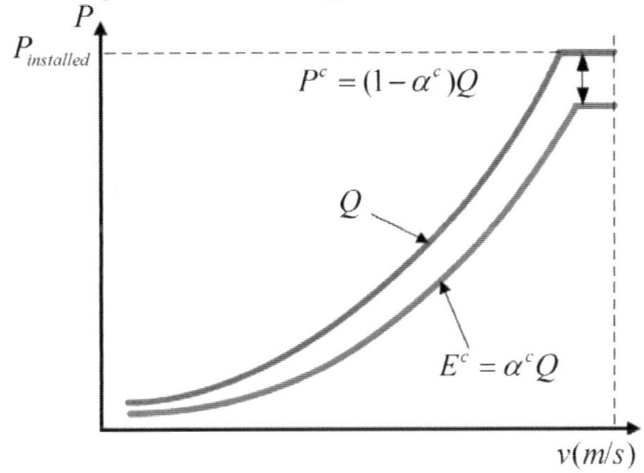

Figure 2. Proportional wind offering strategy (reproduced with authorization from [24]). The blue line Q stands for the available wind power production, while the red curve E^c comprises the energy part of Q to offer in the energy market. The reserve bid P^c to offer in the reserve market is equal to area between the blue and red curve. α^c is the proportional share parameter that splits the available wind power into energy and reserve assuming a value between 0 and 1.

2.3.2. Constant wind control. The constant wind power consists in a constant curtailment of energy in case that the expected forecast is bigger than a specified level of wind power (Figure 3) [14,24]. The strategy, reserves a fixed amount of power reserve to face system imbalances. The remaining active power is dispatched for the energy service. In this control, wind power plant has a fixed amount of power reserve for ancillary services, when the wind power available is above of a certain percentage X% of the installed wind power. Otherwise, the available wind power is offered to the energy market.

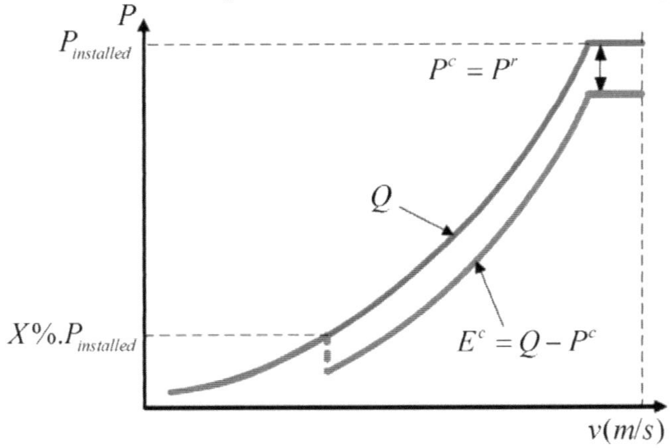

Figure 3. Constant wind offering strategy (reproduced with authorization from [24]). The blue line Q stands for the available wind power production, while the red curve E^c is the remaining part of Q to offer in the energy market assuming a certain fixed amount of reserve P^c.

2.3.3. ΔP control. This control is similar to the constant wind power control. The control curtails a certain and fixed amount of the maximum available power in function of the system operator requirements [15,25,26]. The biggest difference between the ΔP control and the constant control is the use of a minimum threshold (X% of the installed wind power) in the constant control to allocate part of the available wind power as power reserve.

2.3.4. Output cap. The output cap is an active power control for wind turbines establishing a maximum level of active power that can be provided to the energy service [15]. In cases of the available wind power higher than the output cap, the power that exceeds the output cap is curtailed by the wind turbine. Thus, system operator can require low levels of output cap to decrease the variability of the wind power production. However in the WPP standpoint, this control most likely reserves a significant part of the available energy to power reserve, which may result in a small but somehow constant provision of the energy service.

3. Wind offering methodology

3.1. General optimization framework

A methodology for the optimal offering of a WPP in energy and reserve markets at the day-ahead stage, while accounting for the expected penalties in the balancing market is proposed. A two-stage stochastic approach is used to optimize the revenue R for a given WPP, and is expressed as

$$\underset{Q,\alpha^c,\alpha^*_w}{Maximize}\ R = \lambda^{cap} P^c + \sum_{w\in\Omega} \pi_w \left[\lambda^{sp} E^*_w - T^*_w - W^*_w \right] \tag{1}$$

where λ^{cap} is the capacity price for primary reserve allocation, P^c is the reserve contracted (offered) in the day-ahead market, λ^{sp} is the spot price, E^*_w is the delivered energy in scenario w, T^*_w is the regulation costs from the regulation market, and W^*_w is the penalty cost for wind power plant failing to provide the scheduled reserve.

Additionally, it is assumed that the WPP acts as a price-taker. Thus, the market prices and penalties are independent of the WPP production. Then, the regulation costs from the regulation market can be defined as

$$T^*_w = \begin{cases} \lambda^{*,+}\left(E^*_w - E^c\right), & E^*_w - E^c \geq 0 \\ -\lambda^{*,-}\left(E^*_w - E^c\right), & E^*_w - E^c < 0 \end{cases} \tag{2}$$

where (E^*_w- E^c) is the difference between the delivered energy E^*_w and the amount of energy offered at day-ahead market E^c. The variables $\lambda^{*,+}$ and $\lambda^{*,-}$ are the regulation unit costs for positive and negative deviations, respectively,

$$\begin{aligned} \lambda^{*,+} &= \lambda^{sp} - \lambda^{c,+} \\ \lambda^{*,-} &= \lambda^{c,-} - \lambda^{sp} \end{aligned} \tag{3}$$

where $\lambda^{c,+}$ is the unit down-regulation price for being long, while $\lambda^{c,-}$ is the up-regulation price for being short. In addition, a two-price settlement rule (as in NordPool) is assumed [1]. Thus, when the power system imbalance is negative, there is a need for downward regulation (energy surplus), which is given by

$$\begin{aligned} \lambda^{c,+} &\leq \lambda^{sp} \\ \lambda^{c,-} &= \lambda^{sp} \end{aligned} \tag{4}$$

On the other hand, when the power system imbalance is positive, there is a need for upward regulation so the prices and penalties hold that

$$\begin{aligned} \lambda^{c,+} &= \lambda^{sp} \\ \lambda^{c,-} &\geq \lambda^{sp} \end{aligned} \tag{5}$$

The penalty costs for reserve imbalance is given by

$$W_w^* = \begin{cases} \lambda^{bpt,+}\left(P_w^* - P^c\right), & P_w^* - P^c \geq 0 \\ -\lambda^{bpt,-}\left(P_w^* - P^c\right), & P_w^* - P^c < 0 \end{cases} \tag{6}$$

where (P_w^*- P^c) is the reserve power imbalance between the deployed level of reserve P_w^* in real-time and the reserve offered, $\lambda^{bpt,+}$ is a unit penalty when wind producer generates more power than the contracted (surplus), and $\lambda^{bpt,-}$ is the unit penalty cost when the WPP generate less than contracted. These are given by

$$\begin{aligned} \lambda^{bpt,+} &= \lambda^{cap} - \lambda^{pt,+} \\ \lambda^{bpt,-} &= \lambda^{pt,-} - \lambda^{cap} \end{aligned} \tag{7}$$

hence $\lambda^{pt,+}=0$ since (extra) positive reserve is not detrimental to the system's reliability. $\lambda^{pt,-}$ is the penalty for negative reserve imbalance, weighted by the probability that reserve is needed.

3.2. *Proportional wind offering strategy*

In this work and by simplicity, the proportional wind offering strategy is applied for splitting the available wind power for energy and reserve, as illustrated in Figure 2. The objective function is subject to the following constraints regarding the proportional strategy split of energy and reserve. The proportional wind offering strategy is used to define the share of energy E^c and reserve P^c to be submitted in the market [14,24].

$$E^c = \alpha^c Q \tag{8}$$

$$P^c = (1-\alpha^c)Q \tag{9}$$

$$1 \leq Q \leq E^{\max} \tag{10}$$

Under some support schemes, the WPPs are required to participate in the day-ahead market, thereby, the bounds of the total power bid Q reflects the minimum power bid to participate in the market (1 MW in most of electricity markets) and the installed capacity of the WPP.

Equations (11) and (12) concerns the wind offering strategy under the balancing power market

$$E_w^* = \alpha_w^* E_w^{obs} \quad ,\forall w \in \Omega \tag{11}$$

$$P_w^* = (1-\alpha_w^*) E_w^{obs} \quad ,\forall w \in \Omega \tag{12}$$

where E_w^{obs} donates the eventually observed wind power production, composed by energy E_w^* and reserve P_w^* share actually available. α_w^* is the strategy parameter for the splitting in real-time operation.

3.3. *Fixed and relaxed approach of wind strategic split in day-ahead and balancing market*

Under the fixed approach (problem with "non-anticipativity" constraints), it is assumed that the share of energy and reserve established in the balancing stage cannot be different from the day-ahead stage. This ensures that perfect information on real-time cannot be used to change the share of energy and reserve decided on the first-stage problem, thereby avoiding the decision process to play with full degree of freedom. Equation (13) represents the "non-anticipativity" constraint of the wind offering problem.

$$\alpha_w^* = \alpha^c \quad ,\forall w \in \Omega \tag{13}$$

On the other hand, a simplification of the proportional strategy in the stochastic problem can be performed, assuming that the wind power producer can change the share of energy and reserve in both day-ahead and balancing stages (relaxed approach). This means that the wind power producer can adjust the share of energy and reserve in real-time, accordingly with the expected power production in each scenario w. Thus, the WPP can improve their revenue by changing their bid according with better information of their production when closer to real-time operation. The mathematical formulation for that case relies on equations (1) to (12).

The wind power producer problem presented here has been modelled as two-stage stochastic approach in GAMS [27] modelling language and carried out with CONOPT [28] as a NLP solver on an Intel Core i5 2.70 GHz processor with 8 GB RAM.

4. Evaluation of offering strategies

A wind power plant with 15 MW of installed power is considered. The wind total bid offer is subjected to a minimum amount of power to participate in the markets. Currently, electricity markets settle 1 MW as the minimum power for the bidding process. A set of 100 wind power scenarios for a single period presented in [29], has been considered for evaluating the proposed methodology. It is assumed that all the scenarios have equal probability.

The evaluation of the proposed strategy is performed according with a set of prices and penalty costs combination allowing us to test the behavior of the strategy for different assumptions, such as $\alpha_w^*=\alpha^c$ and allowing that α_w^* can be free (i.e., α_w^* can be equal or different of α^c) – stochastic approach with and without "non-anticipativity" constraint.

4.1. Normal operation

Under normal operation in the electricity market, adequate price signals for wind participate in both energy and reserve markets should be ensured. In this scope, the capacity price in the day-ahead market should be higher than the spot price ($\lambda^{cap} \geq \lambda^{sp}$). Besides that, the reserve penalties in the balancing stage for failing to provide the bid offered in the day-ahead stage should be higher than the penalty for failing to provide the energy ($\lambda^{bpt,-} \geq \lambda^{*,-}$), since for the power system is much worse a unit failing to provide reserve rather than energy.

Thus, the prices for energy and reserve, and the unit penalty costs for up and down deviations under normal operation for the power system (our base case) are presented in Table 1.

Table 1. Prices and penalty costs in energy and reserve market for base case.

Energy	Price (€/MWh)	Reserve	Price (€/MW)
λ^{sp}	40	λ^{cap}	41
$\lambda^{c,+}$	30	$\lambda^{bpt,+}$	0
$\lambda^{c,-}$	50	$\lambda^{pt,-}$	96

Under the normal operation case, it is expected that both strategies may behave differently, since the allocation of the available energy to one of the markets is not straightforward. One can expect the strategy with fixed share parameter base their decision with the information available in the day-ahead stage, while the approach with the flexible share parameter may use the better information of the balancing stage to reduce expected costs. Figure 4 illustrates the energy market participation for both stochastic approaches with standard and flexible share parameter relationships between day-ahead and balancing market. The standard approach chooses to participate only in the energy market, since the gain from participating in the reserve market is not much higher than participating in energy-only, and account with a high penalty when failing to provide the offered power reserve (risk adverse behavior). In contrast, the flexible approach (without "non-anticipativity" constraint) presents a different behavior (closer to the risk neutral), since participating in both energy and reserve markets. The participation in both energy and reserve can in one way give flexibility to the wind power producer to increase the expected revenue while taking the risk of getting penalties from failing to provide energy and reserve. Thus, for lower levels of available wind power (until wind available power equal to P^c, as can be seen in Figure 5) it is allocated all the available power to the reserve market (E^*=0, in Figure 4), where the penalty for failing is higher.

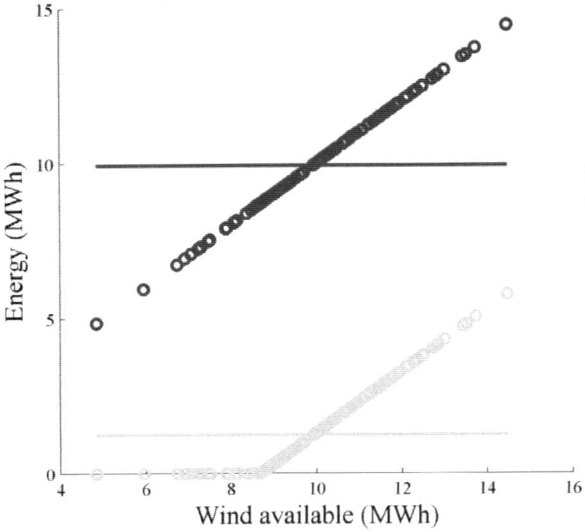

Figure 4. Energy bid in the day-ahead (E^c) and balancing stage (E^*) for both approaches under the normal operation case.

The reserve bids in day-ahead and balancing stage for both strategies are shown in Figure 5.

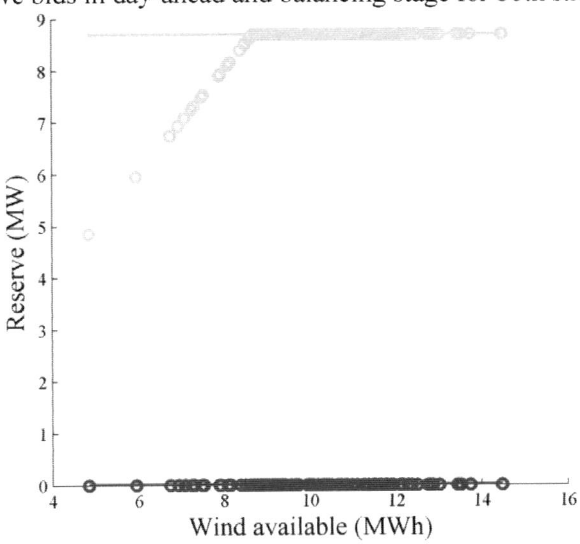

Figure 5. Reserve bid in the day-ahead (P^c) and balancing stage (P^*) for both approaches under the normal operation case.

Moreover, the expected revenue that the WPPs may achieve by participating in energy and reserve market with different behavior of the share between energy and reserve in day-ahead and balancing market is 387 € and 395 €, respectively. In this case, the opportunity to change the energy and reserve share in the balancing market improves the revenue of the WPPs about 2%.

4.2. Special operation – single market participation

In cases of occur different schemes of prices and penalties, the participation in the market behaves differently, as expected. For instance, in cases where the capacity price is higher than the spot price ($\lambda^{cap} \geq \lambda^{sp}$) and the reserve penalty lower than the energy penalty ($\lambda^{bpt,-} \leq \lambda^{*,-}$), both strategies fully offers in the reserve only market. In this case, make total sense to offer only in the reserve market since there is no gain on participating in the energy market.

On the opposite case, when the capacity price is lower than the spot price ($\lambda^{cap} \leq \lambda^{sp}$) and the reserve penalty is higher than the energy penalty ($\lambda^{bpt,-} \geq \lambda^{*,-}$), both strategies assumes the same behavior by participating in the energy only market. One can notice that participating in the energy market will results in higher revenues in the day-ahead market and less expected costs in the balancing stage.

Both special cases implies a logical participation in a single market, however, these cases are unlikely to happen in future electricity markets with competitive integration of wind power generation in the reserve market.

5. Conclusions

With the introduction of new business models where the WPPs can provide energy and reserve bids in the day-ahead market while accounting for the expected cost from the balancing market, new strategic bidding for WPPs is crucial to increase their profit.

This work presents two ways for WPPs to submit their bids in the energy and reserve markets based on the assumption of WPP behaving as a price-taker. One of the approaches considers fixed share of energy and reserve in the stochastic general problem, i.e. the share parameter in both day-ahead and balancing stages remains the same. The other approach sets the share parameter to be free between the day-ahead and balancing stages. Although the strategy with flexible share parameter can increase the revenue of the WPP, this requires a certain level of perfect information of the balancing stage, since the "non-anticipativity" constraint between first and second stage of the problem is not applicable in this case. However, this strategy allows the WPP to change their bids in the market when getting close to the market closure gate, where information of its available production is more reliable.

Notwithstanding, future electricity market may face some changes on this topic, since new behavior and market opportunities for WPPs may influence the market design and mechanisms specially in reserve markets. On the one hand, system operators can require some guarantees from the WPPs, controlling somehow the level of uncertainty in the reserve product and maintaining proper levels of system reliability. On the other hand, market operators must develop mechanisms to ensure a fair participation of all type of market participants in both energy and reserve products and avoiding market power. Pushing decisions close to real-time is of the most interest of WPPs, since it will improve the quality of their decisions and to some extent reduce the lead time effect between the day-ahead decisions and the energy and services delivered.

References

[1] Pinson P, Chevallier C and Kariniotakis G N 2007 Trading wind generation from short-term probabilistic forecasts of wind power *IEEE Trans. Power Syst.* **22** 1148–56

[2] Dent C J, Bialek J W and Hobbs B F 2011 Opportunity cost bidding by wind generators in forward markets: analytical results *IEEE Trans. Power Syst.* **26** 1600–8

[3] Morales J M, Conejo A J and Pérez-ruiz J 2010 Short-term trading for a wind power producer *IEEE Trans. Power Syst.* **25** 554–64

[4] Matevosyan J and Söder L 2006 Minimization of imbalance cost trading wind power on the short-term power market *IEEE Trans. Power Syst.* **21** 1396–404

[5] Botterud A, Zhou Z, Wang J, Bessa R J, Keko H, Sumaili J and Miranda V 2012 Wind Power Trading under Uncertainty in LMP Markets *IEEE Trans. Power Syst.* **27** 894–903

[6] Baringo L and Conejo A J 2013 Strategic offering for a wind power producer *IEEE Trans. Power Syst.* **28** 4645–54
[7] Zugno M, Morales J M, Pinson P, Madsen H and Sets A 2013 Pool strategy of a price-maker wind power producer *IEEE Trans. Power Syst.* **28** 3440–50
[8] Kazempour S J and Zareipour H 2014 Equilibria in an oligopolistic market with wind power production *IEEE Trans. Power Syst.* **29** 686–97
[9] Ela E, Kirby B, Navid N and Smith J C 2012 Effective ancillary services market designs on high wind power penetration systems *IEEE Power and Energy Society General Meeting* (San Diego, California) pp 1–8
[10] Nock D, Krishnan V and McCalley J D 2014 Dispatching intermittent wind resources for ancillary services via wind control and its impact on power system economics *Renew. Energy* **71** 396–400
[11] Zeni L, Rudolph A J, Münster-Swendsen J, Margaris I, Hansen A D and Sørensen P 2013 Virtual inertia for variable speed wind turbines *Wind Energy* **16** 1225–39
[12] Michalke G and Hansen A D 2013 Grid support capabilities of wind turbines *Handbook of Wind Power Systems* ed M P Pardalos, S Rebennack, F M V Pereira, A N Iliadis and V Pappu (Berlin, Heidelberg: Springer Berlin Heidelberg) pp 569–90
[13] Liang J, Grijalva S and Harley R G 2011 Increased wind revenue and system security by trading wind power in energy and regulation reserve markets *IEEE Trans. Sustain. Energy* **2** 340–7
[14] Soares T, Pinson P, Jensen T V. and Morais H 2016 Optimal offering strategies for wind power in energy and primary reserve markets *IEEE Trans. Sustain. Energy* **7** 1–10
[15] Roald L, Misra S, Backhaus S, Chertkov M and Andersson G 2016 Optimal power flow with wind power control and limited expected risk of overloads *19th Power systems computation Conf.* (Genoa) pp 1–7
[16] Krohn S, Morthorst P E and Awerbuch S 2009 The economics of wind energy *EWEA - Eur. Wind Energy Assoc.* 1–155
[17] Skytte K 1999 The regulating power market on the Nordic Power Exchange Nord Pool: an econometric analysis *Energy Econ.* **21** 295–308
[18] Conejo A J, Carrión M and Morales J M 2010 *Decision making under uncertainty in electricity markets* ed F S Hillier, C C Price and S F Austin (Springer US)
[19] Vandezande L, Meeus L, Belmans R, Saguan M and Glachant J-M 2010 Well-functioning balancing markets: A prerequisite for wind power integration *Energy Policy* **38** 3146–54
[20] Energinet.dk 2008 *Regulation C2 : The balancing market and balance settlement*
[21] PLMJ 2014 Renewable Energies - 2013 Legislative changes in the sector *Energy Nat. Resour. - PLMJ* **4** 1–4
[22] Singh S N and Erlich I 2008 Strategies for wind power trading in competitive electricity markets *IEEE Trans. Energy Convers.* **23** 249–56
[23] MIBEL 2009 *Description of the operation of the MIBEL*
[24] Wang Y, Bayem H, Giralt-Devant M, Silva V, Guillaud X and Francois B 2015 Methods for assessing available wind primary power reserve *IEEE Trans. Sustain. Energy* **6** 272–80
[25] Energinet.dk 2015 *Technical regulation 3.2.5 for wind power plants with a power output above 11 kW*
[26] EirGrid 2015 *EirGrid Grid Code, version 6*
[27] Rosenthal R 2008 *GAMS – a user's guide* (Washington, DC: GAMS Development Corporation)
[28] Drud A 1996 *CONOPT: a system for large scale nonlinear optimization*
[29] Bukhsh W A, Zhang C and Pinson P Data for stochastic multiperiod opf problems https://sites.google.com/site/datasmopf/